T0212623

Lecture Notes in Computer Science   9536

Commenced Publication in 1973
Founding and Former Series Editors:
Gerhard Goos, Juris Hartmanis, and Jan van Leeuwen

Editorial Board

More information about this series at http://www.springer.com/series/7411

Prosenjit Bose · Leszek Antoni Gąsieniec
Kay Römer · Roger Wattenhofer (Eds.)

Algorithms for Sensor Systems

11th International Symposium on Algorithms and Experiments
for Wireless Sensor Networks, ALGOSENSORS 2015
Patras, Greece, September 17–18, 2015
Revised Selected Papers

Springer

Editors
Prosenjit Bose
Carleton University
Ottawa, ON
Canada

Leszek Antoni Gąsieniec
University of Liverpool
Liverpool
UK

Kay Römer
Graz University of Technology
Graz
Austria

Roger Wattenhofer
ETH Zurich
Zürich
Switzerland

ISSN 0302-9743   ISSN 1611-3349   (electronic)
Lecture Notes in Computer Science
ISBN 978-3-319-28471-2   ISBN 978-3-319-28472-9   (eBook)
DOI 10.1007/978-3-319-28472-9

Library of Congress Control Number: 2015958836

LNCS Sublibrary: SL5 – Computer Communication Networks and Telecommunications

Printed on acid-free paper

This Springer imprint is published by SpringerNature
The registered company is Springer International Publishing AG Switzerland

Preface

ALGOSENSORS, the International Symposium on Algorithms and Experiments for Wireless Sensor Networks, is an international forum dedicated to the algorithmic aspects of wireless networks, static or mobile. The 11th edition of ALGOSENSORS was held during September 17–18, 2015, in Patras, Greece, within the ALGO annual event.

Originally focused solely on sensor networks, ALGOSENSORS now covers more broadly algorithmic issues arising in all wireless networks of computational entities, including sensor networks, sensor-actuator networks, and systems of autonomous mobile robots. In particular, it focuses on the design and analysis of discrete and distributed algorithms, on models of computation and complexity, on experimental analysis, in the context of wireless networks, sensor networks, and robotic networks and on all foundational and algorithmic aspects of the research in these areas. This year papers were solicited into three tracks: Distributed and Mobile, Experiments and Applications, and Wireless and Geometry.

In response to the call for papers, 30 submissions were received, out of which 16 papers were accepted after a rigorous reviewing process by the (joint) Program Committee, which involved at least three reviewers for each accepted paper. This volume contains the technical papers as well as an invited paper of the keynote talk by Thomas Kesselheim (Max Planck Institute for Informatics).

We would like to thank all Program Committee members, as well as the external reviewers, for their fundamental contribution in selecting the best papers resulting in a strong program. We would also like to warmly thank the ALGO/ESA 2015 organizers for kindly accepting to co-locate ALGOSENSORS with some of the leading events on algorithms in Europe. Furthermore, we would like to thank the local ALGO Organization Committee for their help regarding various administrative tasks, especially the Local Chair, Christos Zaroliagis. Last but not least, we would like thank the Publicity Chair, Klaus-Tycho Foerster, the Web Chair, Laura Peer, and the Steering Committee Chair, Sotiris Nikoletseas, for their help in ensuring a successful ALGOSENSORS 2015.

October 2015

Prosenjit Bose
Leszek Antoni Gąsieniec
Kay Römer
Roger Wattenhofer

Organization

Program Committee

Prosenjit Bose	Carleton University, Canada (Chair Track Wireless and Geometry)
Leszek Antoni Gąsieniec	University of Liverpool, UK (Chair Track Distributed and Mobile)
Kay Römer	TU Graz, Austria (Chair Track Experiments and Applications)
Roger Wattenhofer	ETH Zurich, Switzerland (General Chair)
Carlo Boano	TU Graz, Austria
Nicolas Bonichon	University of Bordeaux, France
Paz Carmi	Ben-Gurion University, Israel
Jérémie Chalopin	CNRS and Aix-Marseille Université, France
Jean-Lou De Carufel	Carleton University, Canada
Stephane Durocher	Manitoba University, Canada
Anna Foerster	University of Bremen, Germany
Martin Gairing	University of Liverpool, UK
Konstantinos Georgiou	University of Waterloo, Canada
Tomasz Jurdzinski	Wroclaw University, Poland
Matias Korman	National Institute of Informatics, Japan
Olaf Landsiedel	Chalmers University of Technology, Sweden
Andreas Loukas	TU Berlin, Germany
George Mertzios	University of Durham, UK
Luca Mottola	Politecnico di Milano, Italy
Merav Parter	MIT, USA
Ljubomir Perkovic	DePaul University, USA
Andreas Reinhardt	TU Clausthal, Germany
Olga Saukh	ETH Zurich, Switzerland
Laura Peer	ETH Zurich, Switzerland (Web Chair)
Klaus-Tycho Foerster	ETH Zurich, Switzerland (Publicity Chair)

Steering Committee

Sotiris Nikoletseas	University of Patras and CTI, Greece (Chair)
Josep Diaz	U.P. Catalunya, Spain
Magnus M. Halldorsson	Reykjavik University, Iceland
Bhaskar Krishnamachari	University of Southern California, USA
P.R. Kumar	Texas A&M University, USA

Jose Rolim   University of Geneva, Switzerland
Paul Spirakis   University of Patras and CTI, Greece
Adam Wolisz   T.U. Berlin, Germany

Additional Reviewers

Agathangelou, Chrysovalandis
Avin, Chen
Czyzowicz, Jurek
Das, Shantanu
Deligkas, Argyrios
Dieudonne, Yoann
Durmus, Yunus
Emek, Yuval
Garncarek, Paweł
Gawrychowski, Pawel
Godard, Emmanuel
Jeż, Łukasz
Klasing, Ralf
Korman, Amos
Korzeniowski, Miroslaw
Kosowski, Adrian

Kranakis, Evangelos
Labourel, Arnaud
Lemiesz, Jakub
Martin, Russell
Michail, Othon
Navarra, Alfredo
Raptopoulos, Christoforos
Roeloffzen, Marcel
Shalom, Mordechai
Tixeuil, Sebastien
Valicov, Petru
van Renssen, André
Wong, Prudence W.H.
Yogev, Eylon
Young, Adam

Online Packing Beyond the Worst Case (Invited Paper)

Thomas Kesselheim

Max-Planck-Institut für Informatik and Saarland University,
Saarbrücken, Germany
thomas.kesselheim@mpi-inf.mpg.de

For a number of online optimization problems, standard worst-case competitive analysis is very pessimistic or even pointless. Sometimes, even a trivial algorithm might be considered optimal because an adversary would be able to trick any algorithm.

An interesting way to avoid these pathological effects is to slightly reduce the power of the adversary by introducing stochastic components. For example, the adversary might still define the instance but not the order in which it is presented to the algorithm. This order is drawn uniformly at random from all possible permutations.

We consider online packing problems and show that this small transition beyond worst-case analysis can have a big impact. We focus on the online independent-set problem in graph classes motivated by wireless networks and on online packing LPs, which among other applications also play a big role in web advertising.

1 Online Independent-Set Problems

In the *online independent-set problem*, a graph is revealed to the algorithm stepwise. In each step, one node is revealed including its edges to previously arrived nodes. This way, many online problems in the domains of scheduling and admission control can be captured. For example, the graph might represent wireless interferences and the task is to select a maximum non-interfering set of transmitters.

A very simple way to model wireless interference is by a disk graph: Each transmitter covers a certain circular area in the plane. The interference constraint requires the areas of no two transmitters to be intersecting. Although NP hard, the independent-set problem in disk graphs can be approximated very well. A very simple greedy algorithm is a constant-factor approximation, and there is even a PTAS [2]. More generally, many graph classes of relevance for practical applications, particularly wireless interference, induce a bounded inductive independence number ρ [4, 6]. The greedy algorithm has an approximation ratio of ρ, thus giving a constant-factor approximation in all these examples.

In a traditional (worst-case) competitive analysis of the online problem, one would have an adversary choosing the instance (i.e., the graph) and the order in which the input is presented. The performance of an algorithm is measured in terms of its competitive ratio $\frac{|\mathrm{ALG}|}{|\mathrm{OPT}|}$, where ALG is the set of vertices chosen by the algorithm and OPT is a maximum independent set in the graph. Unfortunately, even in disk graphs, this approach can only yield devastating results, indicating that no algorithm can be any better than the trivial algorithm, which only selected the first vertex of the input.

In [3], we present a stochastic analysis of this problem. Instead of focusing on a particular stochastic input model, we introduce a generic sampling approach that enables us to devise online algorithms achieving performance guarantees for a variety of input models. In particular, we cover the random-order model, in which the adversary still chooses the graph but cannot determine the order in which the graph is presented to the algorithm. Instead, the order is drawn uniformly at random from all possible permutation after the adversary's choice.

We present an online algorithm for maximum independent set achieving a competitive ratio of $O(1/\rho^2)$ in the random-order model and a number of further stochastic online models. We prove that this result can be extended towards maximum-weight independent set by losing only a factor of $O(1/\log n)$ in the competitive ratio with n denoting the (expected) number of nodes. This upper bound is complemented by a lower bound of $\Omega(\log^2 \log n / \log n)$, showing that our approach achieves nearly the optimal competitive ratio. In addition, we present various extensions of our approach e.g. towards admission control in wireless networks modeled by SINR graphs.

2 Online Packing LPs

In *online packing LPs*, the columns of a packing LP are presented to the algorithm one after the other. The corresponding variables have to be set irrevocably at the arrival of the corresponding column. Again, we assume that the instance is chosen by an adversary but the order in which columns are presented is drawn uniformly at random from all permutations.

In [5], we present a $1 - O(\sqrt{(\log d)/B})$-competitive online algorithm. Here d denotes the *column sparsity*, i.e., the maximum number of resources that occur in a single column, and B denotes the *capacity ratio B*, i.e., the ratio between the capacity of a resource and the maximum demand for this resource. In other words, we achieve a $(1 - \epsilon)$-approximation if the capacity ratio satisfies $B = \Omega(\frac{\log d}{\epsilon^2})$, which is best possible for any (randomized) online algorithms [1].

Our result improves exponentially on previous work with respect to the capacity ratio. In contrast to existing results on packing LP problems, our algorithm does not use dual prices to guide the allocation of resources over time. Instead, the algorithm simply solves, for each request, a scaled version of the partially known primal program and randomly rounds the obtained fractional solution to obtain an integral allocation for this request. We show that this simple algorithmic technique is not restricted to packing LPs with large capacity ratio of order $\Omega(\log d)$, but it also yields close-to-optimal competitive ratios if the capacity ratio is bounded by a constant. In particular, we prove an upper bound on the competitive ratio of $\Omega(d^{-1/(B-1)})$, for any $B \geq 2$.

References

1. Agrawal, S., Wang, Z., Ye, Y.: A dynamic near-optimal algorithm for online linear programming. Oper. Res. **62**(4), 876–890 (2014)
2. Erlebach, T., Jansen, K., Seidel, E.: Polynomial-time approximation schemes for geometric intersection graphs. SIAM J. Comput. **34**(6), 1302–1323 (2005)
3. Göbel, O., Hoefer, M., Kesselheim, T., Schleiden, T., Vöcking, B.: Online independent set beyond the worst-case: secretaries, prophets, and periods. In: Esparza, J., Fraigniaud, P., Husfeldt, T., Koutsoupias, E. (eds.) ICALP 2014, Part II. LNCS, vol. 8573, pp. 508–519 (2014)
4. Hoefer, M., Kesselheim, T., Vöcking, B.: Approximation algorithms for secondary spectrum auctions. ACM Trans. Internet Technol. **14**(2–3), 16:1–16:24 (2014)
5. Kesselheim, T., Radke, K., Tönnis, A., Vöcking, B.: Primal beats dual on online packing LPs in the random-order model. In: Proceedings of 46th Symposium on Theory of Computing (STOC), pp. 303–312 (2014)
6. Ye, Y., Borodin, A.: Elimination graphs. ACM Trans. Algorithms **8**(2), 14 (2012)

Contents

Plane and Planarity Thresholds for Random Geometric Graphs

Ahmad Biniaz$^{(\boxtimes)}$, Evangelos Kranakis, Anil Maheshwari, and Michiel Smid

Carleton University, Ottawa, Canada
ahmad.biniaz@gmail.com

Abstract. A random geometric graph, $G(n, r)$, is formed by choosing n points independently and uniformly at random in a unit square; two points are connected by a straight-line edge if they are at Euclidean distance at most r. For a given constant k, we show that $n^{\frac{-k}{2k-2}}$ is a distance threshold function for $G(n, r)$ to have a connected subgraph on k points. Based on that, we show that $n^{-2/3}$ is a distance threshold function for $G(n, r)$ to be plane, and $n^{-5/8}$ is a distance threshold function for $G(n, r)$ to be planar.

1 Introduction

Wireless networks are usually modeled as disk graphs in the plane. Given a set P of points in the plane and a positive parameter r, the *disk graph* is the geometric graph with vertex set P which has a straight-line edge between two points $p, q \in P$ if and only if $|pq| \leq r$, where $|pq|$ denotes the Euclidean distance between p and q. If $r = 1$, then the disk graph is referred to as *unit disk graph*. A *random geometric graph*, denoted by $G(n, r)$, is a geometric graph formed by choosing n points independently and uniformly at random in a unit square; two points are connected by a straight-line edge if and only if they are at Euclidean distance at most r, where $r = r(n)$ is a function of n and $r \to 0$ as $n \to \infty$.

We say that two line segments in the plane *cross* each other if they have a point in common that is interior to both edges. Two line segments are *non-crossing* if they do not cross. Note that two non-crossing line segments may share an endpoint. A geometric graph is said to be *plane* if its edges do not cross, and *non-plane*, otherwise. A graph is *planar* if and only if it does not contain K_5 (the complete graph on 5 vertices) or $K_{3,3}$ (the complete bipartite graph on six vertices partitioned into two parts each of size 3) as a minor. A *non-planar graph* is a graph which is not planar.

A graph property $\mathcal{P}$ is *increasing* if a graph G satisfies $\mathcal{P}$, then by adding edges to G, the property $\mathcal{P}$ remains valid in G. Similarly, $\mathcal{P}$ is *decreasing* if a graph G satisfies $\mathcal{P}$, then by removing edges from G, the property $\mathcal{P}$ remains valid in G. $\mathcal{P}$ is called a *monotone* property if $\mathcal{P}$ is either increasing or decreasing. Connectivity and "having a clique of size k" are increasing monotone properties,

Research supported by NSERC.

P. Bose et al. (Eds.): ALGOSENSORS 2015, LNCS 9536, pp. 1–12, 2015.
DOI: 10.1007/978-3-319-28472-9_1

while planarity and "being plane" are decreasing monotone properties in $G(n, r)$, where the value of r increases.

By [13] any monotone property of a random geometric graphs has a threshold function. The thresholds in random geometric graphs are expressed by the distance r. In the sequel, the term w.h.p. (with high probability) is to be interpreted to mean that the probability tends to 1 as $n \to \infty$. For an increasing property $\mathcal{P}$, the threshold is a function $t(n)$ such that if $r = o(t(n))$ then w.h.p. $\mathcal{P}$ does not hold in $G(n, r)$, and if $r = \omega(t(n))$ then w.h.p. $\mathcal{P}$ holds in $G(n, r)$. Symmetrically, for a decreasing property $\mathcal{P}$, the threshold is a function $t(n)$ such that if $r = o(t(n))$ then w.h.p. $\mathcal{P}$ holds in $G(n, r)$, and if $r = \omega(t(n))$ then w.h.p. $\mathcal{P}$ does not hold in $G(n, r)$. Note that a threshold function may not be unique. It is well known that $\sqrt{\ln n / n}$ is a connectivity threshold for $G(n, r)$; see [14, 19, 20]. In this paper we investigate thresholds in random geometric graphs for having a connected subgraph of constant size, being plane, and being planar.

1.1   Related Work

Random graphs were first defined and formally studied by Gilbert in [10] and Erdös and Rényi [8]. It seems that the concept of a random geometric graph was first formally suggested by Gilbert in [11] and for that reason is also known as Gilbert's disk model. These classes of graphs are known to have numerous applications as a model for studying communication primitives (broadcasting, routing, etc.) and topology control (connectivity, coverage, etc.) in idealized wireless sensor networks as well as extensive utility in theoretical computer science and many fields of the mathematical sciences.

An instance of Erdös-Rényi graph [8] is obtained by taking n vertices and connecting any two with probability p, independently of all other pairs; the graph derived by this scheme is denoted by $G_{n,p}$. In $G_{n,p}$ the threshold is expressed by the edge existence probability p, while in $G(n, r)$ the threshold is expressed in terms of r. In both random graphs and random geometric graphs, property thresholds are of great interest [4, 7, 9, 13, 18]. Note that edge crossing configurations in $G(n, r)$ have a geometric nature, and as such, have no analogues in the context of the Erdös-Rényi model for random graphs. However, planarity, and having a clique of specific size are of interest in both $G_{n,p}$ and $G(n, r)$.

Bollobás and Thomason [5] showed that any monotone property in random graphs has a threshold function. See also a result of Friedgut and Kalai [9], and a result of Bourgain and Kalai [6]. In the Erdös-Rényi random graph $G_{n,p}$, the connectivity threshold is $p = \log n / n$ and the threshold for having a giant component is $p = 1/n$; see [1]. The planarity threshold for $G_{n,p}$ is $p = 1/n$; see [4, 23].

A general reference on random geometric graphs is [22]. There is extensive literature on various aspects of random geometric graphs of which we mention the related work on coverage by [15, 16] and a review on percolation, connectivity, coverage and colouring by [3]. As in random graphs, any monotone property in geometric random graphs has a threshold function [7, 13, 17, 18].

Random geometric graphs have a connectivity threshold of $\sqrt{\ln n/n}$; see [14, 19, 20]. Gupta and Kumar [14] provided a connectivity threshold for points that are uniformly distributed in a disk. By a result of Penrose [21], in $G(n,r)$, any threshold function for having no isolated vertex (a vertex of degree zero) is also a connectivity threshold function. Panchapakesan and Manjunath [19] showed that $\sqrt{\ln n/n}$ is a threshold for being an isolated vertex in $G(n,r)$. This implies that $\sqrt{\ln n/n}$ is a connectivity threshold for $G(n,r)$. For $k \geq 2$, the details on the k-connectivity threshold in random geometric graphs can be found in [21, 22]. Connectivity of random geometric graphs for points on a line is studied by Godehardt and Jaworski [12]. Appel and Russo [2] considered the connectivity under the L_∞-norm.

1.2   Our Results

In this paper we investigate thresholds for some monotone properties in random geometric graphs. In Sect. 2 we show that for a constant k, the distance threshold for having a connected subgraph on k points is $n^{\frac{-k}{2k-2}}$. We show that the same threshold is valid for the existence of a clique of size k. Based on that, we prove the following thresholds for a random geometric graph to be plane or planar. In Sect. 3, we prove that $n^{-2/3}$ is a distance threshold for a random geometric graph to be plane. In Sect. 4, we prove that $n^{-5/8}$ is a distance threshold for a random geometric graph to be planar.

2   The Threshold for Having a Connected Subgraph on k Points

In this section, we look for the distance threshold for "existence of connected subgraphs of constant size"; this is an increasing property. For a given constant k, we show that $n^{\frac{-k}{2k-2}}$ is the threshold function for the existence of a connected subgraph on k points in $G(n,r)$. Specifically, we show that if $r = o(n^{\frac{-k}{2k-2}})$, then w.h.p. $G(n,r)$ has no connected subgraph on k points, and if $r = \omega(n^{\frac{-k}{2k-2}})$, then w.h.p. $G(n,r)$ has a connected subgraph on k points. We also show that the same threshold function holds for the existence of a clique of size k.

Theorem 1. *Let $k \geq 2$ be an integer constant. Then, $n^{\frac{-k}{2k-2}}$ is a distance threshold function for $G(n,r)$ to have a connected subgraph on k points.*

Proof. Let $P_1, \ldots, P_{\binom{n}{k}}$ be an enumeration of all subsets of k points in $G(n,r)$. Let $DG[P_i]$ be the subgraph of $G(n,r)$ that is induced by P_i. Let X_i be the random variable such that

$$X_i = \begin{cases} 1 & \text{if } DG[P_i] \text{ is connected,} \\ 0 & \text{otherwise.} \end{cases}$$

Let the random variable X count the number of sets P_i for which $DG[P_i]$ is connected. It is clear that

$$X = \sum_{i=1}^{\binom{n}{k}} X_i. \tag{1}$$

Observe that $\mathrm{E}[X_i] = \Pr[X_i = 1]$. Since the random variables X_i have identical distributions, we have

$$\mathrm{E}[X] = \binom{n}{k}\mathrm{E}[X_1]. \tag{2}$$

We obtain an upper bound and a lower bound for $\Pr[X_i = 1]$. First, partition the unit square into squares of side equal to r. Let $\{s_1, \ldots, s_{1/r^2}\}$ be the resulting set of squares. For a square s_t, let S_t be the $kr \times kr$ square which has s_t on its left bottom corner; see Fig. 1(a). S_t contains at most k^2 squares each of side length r (each S_t on the boundary of the unit square contains less than k^2 squares). Let $A_{i,t}$ be the event that all points in P_i are contained in S_t. Observe that if $DG[P_i]$ is connected then P_i lies in S_t for some $t \in \{1, \ldots, 1/r^2\}$. Therefore,

if $DG[P_i]$ is connected, then $(A_{i,1} \vee A_{i,2} \vee \cdots \vee A_{i,1/r^2})$,

and hence we have

$$\Pr[X_i = 1] \leq \sum_{t=1}^{1/r^2} \Pr[A_{i,t}] \leq \sum_{t=1}^{1/r^2} (k^2 r^2)^k = k^{2k} r^{2k-2}. \tag{3}$$

Now, partition the unit square into squares with diagonal length equal to r. Each such square has side length equal to $r/\sqrt{2}$. Let $\{s_1, \ldots, s_{2/r^2}\}$ be the resulting set of squares. Let $B_{i,t}$ be the event that all points of P_i are in s_t. Observe that if all points of P_i are in the same square, then $DG[P_i]$ is a complete graph and hence connected. Therefore,

if $(B_{i,1} \vee B_{i,2} \vee \cdots \vee B_{i,2/r^2})$, then $DG[P_i]$ is connected,

and hence we have

$$\Pr[X_i = 1] \geq \sum_{t=1}^{2/r^2} \Pr[B_{i,t}] = \sum_{t=1}^{2/r^2} \left(\frac{r^2}{2}\right)^k = \frac{1}{2^{k-1}} r^{2k-2}. \tag{4}$$

Since $k \geq 2$ is a constant, Inequalities (3) and (4) and Eq. (2) imply that

$$\mathrm{E}[X_i] = \Theta(r^{2k-2}), \tag{5}$$
$$\mathrm{E}[X] = \Theta(n^k r^{2k-2}). \tag{6}$$

If $n \to \infty$ and $r = o(n^{\frac{-k}{2k-2}})$ we conclude that the following inequalities are valid

$$\Pr[X \geq 1] \leq \mathrm{E}[X] \ (\mathrm{by\,Markov's\,Inequality})$$
$$= \Theta(n^k r^{2k-2}) \ (\mathrm{by}\ (6))$$
$$= o(1). \tag{7}$$

Therefore, w.h.p. $G(n, r)$ has no connected subgraph on k points.

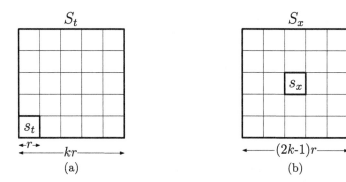

Fig. 1. (a) The square S_t has s_t on its left bottom corner. (b) The square S_x which is centered at s_x.

In the rest of the proof, we assume that $r = w(n^{\frac{-k}{2k-2}})$. In order to show that w.h.p. $G(n, r)$ has at least one connected subgraph on k vertices, we show, using the second moment method [1], that $\Pr[X = 0] \to 0$ as $n \to \infty$. Recall from Chebyshev's inequality that

$$\Pr[X = 0] \leq \frac{\text{Var}(X)}{\text{E}[X]^2}. \tag{8}$$

Therefore, in order to show that $\Pr[X = 0] \to 0$, it suffices to show that

$$\frac{\text{Var}(X)}{\text{E}[X]^2} \to 0. \tag{9}$$

In view of Identity (1) we have

$$\text{Var}(X) = \sum_{1 \leq i,j \leq \binom{n}{k}} \text{Cov}(X_i, X_j), \tag{10}$$

where $\text{Cov}(X_i, X_j) = \text{E}[X_i X_j] - \text{E}[X_i]\text{E}[X_j] \leq \text{E}[X_i X_j]$. If $|P_i \cap P_j| = 0$ then $DG[P_i]$ and $DG[P_j]$ are disjoint. Thus, the random variables X_i and X_j are independent, and hence $\text{Cov}(X_i, X_j) = 0$. It is enough to consider the cases when P_i and P_j are not disjoint. Assume $|P_i \cap P_j| = w$, where $w \in \{1, \ldots, k\}$. Thus, in view of Eq. (10), we have

$$\text{Var}(X) = \sum_{w=1}^{k} \sum_{|P_i \cap P_j| = w} \text{Cov}(X_i, X_j)$$

$$\leq \sum_{w=1}^{k} \sum_{|P_i \cap P_j| = w} \text{E}[X_i X_j]. \tag{11}$$

The computation of $\text{E}[X_i, X_j]$ involves some geometric considerations which are being discussed in detail below. Since X_i and X_j are 0–1 random variables, $X_i X_j$ is a 0–1 random variable and

$$X_i X_j = \begin{cases} 1 & \text{if both } DG[P_i] \text{ and } DG[P_j] \text{ are connected,} \\ 0 & \text{otherwise.} \end{cases}$$

By the definition of the expected value we have

$$\begin{aligned} \mathrm{E}[X_i X_j] &= \Pr[X_j = 1 | X_i = 1] \Pr[X_i = 1] \\ &= \Pr[X_j = 1 | X_i = 1] \mathrm{E}[X_i]. \end{aligned} \tag{12}$$

By (5), $\mathrm{E}[X_i] = \Theta(r^{2k-2})$. It remains to compute $\Pr[X_j = 1 | X_i = 1]$, i.e., the probability that $DG[P_j]$ is connected given that $DG[P_i]$ is connected. Consider the k-tuples P_i and P_j under the condition that $DG[P_i]$ is connected. Let x be a point in $P_i \cap P_j$. Partition the unit square into squares of side length equal to r. Let s_x be the square containing x. Let S_x be the $(2k-1)r \times (2k-1)r$ square centered at s_x. S_x contains at most $(2k-1)^2$ squares each of side length r (if S_x is on the boundary of the unit square then it contains less than $(2k-1)^2$ squares); see Fig. 1(b). The area of S_x is at most $(2kr)^2$, and hence the probability that a specific point of P_j is in S_t is at most $4k^2 r^2$. Since P_i and P_j share w points, in order for $DG[P_j]$ to be connected, the remaining $k - w$ points of P_j must lie in S_x. Thus, the probability that $DG[P_j]$ is connected given that $DG[P_i]$ is connected is at most $(4k^2 r^2)^{k-w} \le c_w r^{2k-2w}$, for some constant $c_w > 0$. Thus, $\Pr[X_j = 1 | X_i = 1] \le c_w r^{2k-2w}$. In view of Eq. (12), we have

$$\mathrm{E}[X_i X_j] \le c_w' \cdot r^{2k-2w} \cdot r^{2k-2} = c_w' r^{4k-2w-2}, \tag{13}$$

for some constant $c_w' > 0$.

Since P_i and P_j are k-tuples which share w points, $|P_i \cup P_j| = 2k - w$. There are $\binom{n}{2k-w}$ ways to choose $2k - w$ points for $P_i \cup P_j$. Since we choose w points for $P_i \cap P_j$, $k - w$ points for P_i alone, and $k - w$ points for P_j alone, there are $\binom{2k-w}{w, k-w, k-w}$ ways to split the $2k - w$ chosen points into P_i and P_j. Based on this and Inequality (13), Inequality (11) turns out to

$$\begin{aligned} \mathrm{Var}(X) &\le \sum_{w=1}^{k} \sum_{|P_i \cap P_j| = w} \mathrm{E}[X_i X_j] \\ &\le \sum_{w=1}^{k} \binom{n}{2k-w} \binom{2k-w}{w, k-w, k-w} c_w' r^{4k-2w-2} \\ &\le \sum_{w=1}^{k} c_w'' n^{2k-w} r^{4k-2w-2}. \end{aligned}$$

for some constants $c_w'' > 0$. Consider (9) and note that by (6), $\mathrm{E}[X]^2 \ge c'' n^{2k} r^{4k-4}$, for some constant $c'' > 0$. Thus,

$$\begin{aligned} \frac{\mathrm{Var}(X)}{\mathrm{E}[X]^2} &\le \sum_{w=1}^{k} \frac{c_w'' n^{2k-w} r^{4k-2w-2}}{c'' n^{2k} r^{4k-4}} = \sum_{w=1}^{k} \frac{c_w''}{c''} \cdot \frac{1}{n^w r^{2w-2}} \\ &= \frac{c_1''}{c''} \cdot \frac{1}{n^1 r^0} + \frac{c_2''}{c''} \cdot \frac{1}{n^2 r^2} + \cdots + \frac{c_k''}{c''} \cdot \frac{1}{n^k r^{2k-2}} \end{aligned} \tag{14}$$

Since $r = \omega(n^{\frac{-k}{2k-2}})$, all terms in (14) tend to zero. This proves the convergence in (9). Thus, $\Pr[X = 0] \to 0$ as $n \to \infty$. This implies that if $r = \omega(n^{\frac{-k}{2k-2}})$, then $G(n, r)$ has a connected subgraph on k vertices with high probability. $\blacksquare$

In the following theorem we show that if $k = O(1)$, then $n^{\frac{-k}{2k-2}}$ is also a threshold for $G(n, r)$ to have a clique of size k; this is an increasing property.

Theorem 2. *Let $k \geq 2$ be an integer constant. Then, $n^{\frac{-k}{2k-2}}$ is a distance threshold function for $G(n, r)$ to have a clique of size k.*

Proof. By Theorem 1, if $r = o(n^{\frac{-k}{2k-2}})$, then w.h.p. $G(n, r)$ has no connected subgraph on k vertices, and hence it has no clique of size k. This proves the first statement. We prove the second statement by adjusting the proof of Theorem 1, which is based on the second moment method. Assume $r = \omega(n^{\frac{-k}{2k-2}})$. Let $P_1, \ldots, P_{\binom{n}{k}}$ be an enumeration of all subsets of k points. Let X_i be equal to 1 if $DG[P_i]$ is a clique, and 0 otherwise. Let $X = \sum X_i$.

Partition the unit square into a set $\{s_1, \ldots, s_{1/r^2}\}$ of squares of side length r. Let S_t be the $2r \times 2r$ square which has s_t on its left bottom corner. If $DG[P_i]$ is a clique then P_i lies in S_t for some $t \in \{1, \ldots, 1/r^2\}$. Therefore,

$$\Pr[X_i = 1] \leq 4^k r^{2k-2}.$$

Now, partition the unit square into a set $\{s_1, \ldots, s_{2/r^2}\}$ of squares with diagonal length r. If all points of P_i fall in the square s_t, then $DG[P_i]$ is a clique. Thus,

$$\Pr[X_i = 1] \geq \frac{1}{2^{k-1}} r^{2k-2}.$$

Since $k \geq 2$ is a constant, we have

$$E[X_i] = \Theta(r^{2k-2}),$$
$$E[X] = \Theta(n^k r^{2k-2}).$$

In view of Chebyshev's inequality we need to show that $\frac{\text{Var}(X)}{E[X]^2}$ tends to 0 as n goes to infinity. We bound $\text{Var}(X)$ from above by Inequality (11). Consider the k-tuples P_i and P_j under the condition that $DG[P_i]$ is a clique. Let $|P_i \cap P_j| = w$, and let x be a point in $P_i \cap P_j$. Partition the unit square into squares of side length r. Let s_x be the square containing x. Let S_x be the $3r \times 3r$ square centered at s_x. In order for $DG[P_j]$ to be a clique, the remaining $k - w$ points of P_j must lie in S_x. Thus,

$$E[X_i X_j] \leq c'_w r^{4k-2w-2},$$

for some constant $c'_w > 0$. By a similar argument as in the proof of Theorem 1, we can show that for some constants $c'', c''_w > 0$ the followings inequalities are valid:

$$\text{Var}(X) \leq \sum_{w=1}^{k} c''_w n^{2k-w} r^{4k-2w-2},$$

$$\frac{\text{Var}(X)}{E[X]^2} \leq \sum_{w=1}^{k} \frac{c''_w}{c''} \cdot \frac{1}{n^w r^{2w-2}}.$$

Since $r = \omega(n^{\frac{-k}{2k-2}})$, the last inequality tends to 0 as n goes to infinity. This completes the proof for the second statement. ∎

As a direct consequence of Theorem 2, we have the following corollary.

Corollary 1. n^{-1} *is a threshold for* $G(n,r)$ *to have an edge, and* $n^{-\frac{3}{4}}$ *is a threshold for* $G(n,r)$ *to have a triangle.*

3   The Threshold for $G(n,r)$ to be Plane

In this section we investigate the threshold for a random geometric graph to be plane; this is a decreasing property. Recall that $G(n,r)$ is plane if no two of its edges cross. As a warm-up exercise we first prove a simple result which is based on the connectivity threshold for random geometric graphs, which is known to be $\sqrt{\ln n/n}$.

Theorem 3. If $r \geq \sqrt{\frac{c\ln n}{n}}$, with $c \geq 36$, then w.h.p. $G(n,r)$ is not plane.

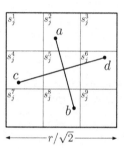

proof In order to prove that w.h.p. $G(n,r)$ is not plane, we show that w.h.p. it has a pair of crossing edges. Partition the unit square into squares each with diagonal length r. Then subdivide each such square into nine sub-squares as depicted in Fig. 2. There are $\frac{18}{r^2}$ sub-squares, each of side length $\frac{r}{3\sqrt{2}}$. The probability that no point lies in a specific sub-square is $(1 - \frac{r^2}{18})^n$. Thus, the probability that there exists an empty sub-square is at most

Fig. 2. An square of diameter r which is partitioned into nine sub-squares.

$$\frac{18}{r^2}\left(1 - \frac{r^2}{18}\right)^n \leq n\left(1 - \frac{c\ln n}{18n}\right)^n \leq n^{1-c/18} \leq \frac{1}{n},$$

when $c \geq 36$. Therefore, with probability at least $1 - \frac{1}{n}$ all sub-squares contain points. By choosing four points a, b, c, and d as depicted in Fig. 2, it is easy to see that the edges (a,b) and (c,d) cross. Thus, w.h.p. $G(n,r)$ has a pair of crossing edges, and hence w.h.p. it is not plane. ∎

In fact, Theorem 3 ensures that w.h.p. there exists a pair of crossing edges in each of the squares. This implies that there are $\Omega\left(\frac{n}{\ln n}\right)$ disjoint pair of crossing edges, while for $G(n,r)$ to be not plane we need to show the existence of at least one pair of crossing edges. Thus, the value of r provided by the connectivity threshold seems rather weak. By a different approach, in the rest of this section we show that $n^{-\frac{2}{3}}$ is the correct threshold.

Lemma 1. *Let* (a,b) *and* (c,d) *be two crossing edges in* $G(n,r)$, *and let* Q *be the convex quadrilateral formed by* a, b, c, *and* d. *Then, two adjacent sides of* Q *are edges of* $G(n,r)$.

Proof. Refer to Fig. 3. At least one of the angles of Q, say $\angle cad$, is bigger than or equal to $\pi/2$. It follows that in the triangle $\triangle cad$ the side cd is the longest, i.e., $|cd| \geq \max\{|ac|, |ad|\}$. Since $|cd| \leq r$, both $|ac|$ and $|ad|$ are at most r. Thus, ac and ad—which are adjacent—are edges of $G(n, r)$. ∎

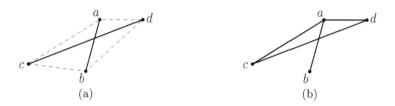

(a)   (b)

Fig. 3. (a) Illustration of Lemma 1. (b) Crossing edges (a, b) and (c, d) form an anchor.

In the proof of Lemma 1, a is connected to b, c, and d. So the distance between a to each of b, c, and d is at most r. Thus, we have the following corollary.

Corollary 2. *The endpoints of every two crossing edges in $G(n, r)$ are at distance at most $2r$ from each other. Moreover, there exists an endpoint which is within distance r from other endpoints.*

Based on the proof of Lemma 1, we define an *anchor* as a set $\{a, b, c, d\}$ of four points in $G(n, r)$ such that three of them form a triangle, say $\triangle cad$, and the fourth vertex, b, is connected to a by an edge which crosses cd; see Fig. 3(b). We call a as the *crown* of the anchor. The crown is within distance r from the other three points. Note that bc and bd may or may not be edges of $G(n, r)$. In view of Lemma 1, two crossing edges in $G(n, r)$ form an anchor. Conversely, every anchor in $G(n, r)$ introduces a pair of crossing edges.

Observation 1. *$G(n, r)$ is plane if and only if it has no anchor.*

Theorem 4. *$n^{-\frac{2}{3}}$ is a threshold for $G(n, r)$ to be plane.*

Proof. In order to show that $G(n, r)$ is plane, by Observation 1, it is enough to show that it has no anchors. Every anchor has four points and it is connected. By Theorem 1, if $r = o(n^{-\frac{2}{3}})$, then w.h.p. $G(n, r)$ has no connected subgraph on 4 points, and hence it has no anchors. This proves the first statement.

We prove the second statement by adjusting the proof of Theorem 1 for $k = 4$. Assume $r = \omega(n^{-\frac{2}{3}})$. Let $P_1, \dots, P_{\binom{n}{4}}$ be an enumeration of all subsets of 4 points. Let X_i be equal to 1 if $DG[P_i]$ contains an anchor, and 0 otherwise. Let $X = \sum X_i$. In view of Chebyshev's inequality we need to show that $\frac{\text{Var}(X)}{\text{E}[X]^2}$ tends to 0 as n goes to infinity.

Partition the unit square into a set $\{s_1, \dots, s_{2/r^2}\}$ of squares with diagonal length r. Then, subdivide each square s_j, into nine sub-squares $s_j^1, \dots, s_j^9$ as depicted in Fig. 2. If each of $s_j^1, s_j^3, s_j^7, s_j^9$ or each of $s_j^2, s_j^4, s_j^6, s_j^8$ contains a point of P_i, then $DG[P_i]$ is a convex clique of size four and hence it contains an anchor. Thus,

$$\Pr[X_i = 1] \geq \frac{r^6}{2^3} \cdot \frac{2}{9^4}.$$

This implies that $E[X_i] = \Omega(r^6)$, and hence $E[X] = \Omega(n^4 r^6)$. Therefore,

$$E[X]^2 \geq c'' n^8 r^{12},$$

for some constant $c'' > 0$. By a similar argument as in the proof of Theorem 1 we bound the variance of X from above by

$$\mathrm{Var}(X) \leq c_1'' n^7 r^{12} + c_2'' n^6 r^{10} + c_3'' n^5 r^8 + c_4'' n^4 r^6.$$

Since $r = \omega(n^{-\frac{2}{3}})$, $\frac{\mathrm{Var}(X)}{E[X]^2}$ tends to 0 as n goes to infinity. That is, w.h.p. $G(n,r)$ has an anchor. By Observation 1, w.h.p. $G(n,r)$ is not plane. ∎

As a direct consequence of the proof of Theorem 4, we have the following:

Corollary 3. *With high probability if a random geometric graph is not plane, then it has a clique of size four.*

Note that every anchor introduces a crossing and each crossing introduces an anchor. Since, every anchor is a connected graph and has four points, by (6) we have the following corollary.

Corollary 4. *The expected number of crossings in $G(n,r)$ is $\Theta(n^4 r^6)$.*

4   The Threshold for $G(n,r)$ to be Planar

In this section we investigate the threshold for the planarity of a random geometric graph; this is a decreasing property. By Kuratowski's theorem, a finite graph is planar if and only if it does not contain a subgraph that is a subdivision of K_5 or of $K_{3,3}$. Note that any plane random geometric graph is planar too; observe that the reverse statement may not be true. Thus, the threshold for planarity seems to be larger than the threshold of being plane. By a similar argument as in the proof of Theorem 3 we can show that if $r \geq \sqrt{c \ln n / n}$, then w.h.p. each square with diagonal length r contains K_5, and hence $G(n,r)$ is not planar.

Theorem 5. $n^{-\frac{5}{8}}$ *is a threshold for $G(n,r)$ to be planar.*

Proof. By Theorem 2, if $r = \omega(n^{-\frac{5}{8}})$, then w.h.p. $G(n,r)$ has a clique of size 5. Thus, w.h.p. $G(n,r)$ contains K_5 and hence it is not planar. This proves the second statement of the theorem.

If $r = o(n^{-\frac{5}{8}})$, then by Theorem 1, w.h.p. $G(n,r)$ has no connected subgraph on 5 points, and hence it has no K_5. Similarly, if $r = o(n^{-\frac{3}{5}})$, then w.h.p. $G(n,r)$ has no connected subgraph on 6 points, and hence it has no $K_{3,3}$.

Since $n^{-\frac{5}{8}} < n^{-\frac{3}{5}}$, it follows that if $r = o(n^{-\frac{5}{8}})$, then w.h.p. $G(n,r)$ has neither K_5 nor $K_{3,3}$ as a subgraph.

Note that, in order to prove that $G(n,r)$ is planar, we have to show that it does not contain any subdivision of either K_5 or $K_{3,3}$. Any subdivision of either K_5 or $K_{3,3}$ contains a connected subgraph on $k \geq 5$ vertices. Since $n^{-5/8} < n^{-k/(2k-2)}$ for all $k \geq 5$, in view of Theorem 1, we conclude that if $r = o(n^{-\frac{5}{8}})$, then w.h.p. $G(n,r)$ has no subdivision of K_5 and $K_{3,3}$, and hence $G(n,r)$ is planar. This proves the first statement of the theorem.   ∎

As a direct consequence of the proof of Theorem 5, we have the following:

Corollary 5. *With high probability if a random geometric graph does not contain a clique of size five, then it is planar.*

5   Conclusion and Further Results

We presented thresholds for random geometric graphs to have a connected subgraph of constant size, to be plane, and to be planar. A natural open problem is to extend Theorem 1 for connected subgraphs of k vertices where k is not necessarily a constant, and for connected subgraphs of k vertices which have diameter δ.

References

1. Alon, N., Spencer, J.H.: The Probabilistic Method, 3rd edn. Wiley, New York (2007)
2. Appel, M.J.B., Russo, R.P.: The connectivity of a graph on uniform points on $[0,1]^d$. Stat. Prob. Lett. **60**(4), 351–357 (2002)
3. Balister, P., Sarkar, A., Bollobás, B.: Percolation, connectivity, coverage and colouring of random geometric graphs. In: Bollobás, B., Kozma, R., Miklós, D. (eds.) Handbook of Large-Scale Random Networks, pp. 117–142. Springer, Heidelberg (2008)
4. Bollobás, B.: Random Graphs. Cambridge University Press, Cambridge (2001)
5. Bollobás, B., Thomason, A.: Threshold functions. Combinatorica **7**(1), 35–38 (1987)
6. Bourgain, J., Kalai, G.: Threshold intervals under group symmetries. Convex Geom. Anal. MSRI Publ. **34**, 59–63 (1998)
7. Bradonjić, M., Perkins, W.: On sharp thresholds in random geometric graphs. In: Approximation, Randomization, and Combinatorial Optimization. Algorithms and Techniques, APPROX/RANDOM, pp. 500–514 (2014)
8. Erdös, P., Rényi, A.: On the evolution of random graphs. Publ. Math. Inst. Hungar. Acad. Sci. **5**, 17–61 (1960)
9. Friedgut, E., Kalai, G.: Every monotone graph property has a sharp threshold. Proc. Am. Math. Soc. **124**(10), 2993–3002 (1996)
10. Gilbert, E.: Random graphs. Ann. Math. Stat. **30**(4), 1141–1144 (1959)
11. Gilbert, E.: Random plane networks. J. Soc. Ind. Appl. Math. **9**(4), 533–543 (1961)
12. Godehardt, E., Jaworski, J.: On the connectivity of a random interval graph. Random Struct. Algorithms **9**(1–2), 137–161 (1996)

13. Goel, A., Rai, S., Krishnamachari, B.: Sharp thresholds for monotone properties in random geometric graphs. In: Proceedings of STOC, pp. 580–586. ACM (2004)
14. Gupta, P., Kumar, P.R.: Critical power for asymptotic connectivity in wireless networks. In: McEneaney, W.M., George Yin, G., Zhang, Q. (eds.) Stochastic Analysis, Control, Optimization and Applications, pp. 547–566. Springer, New York (1998)
15. Hall, P.: On the coverage of k-dimensional space by k-dimensional spheres. Ann. Prob. **13**(3), 991–1002 (1985)
16. Janson, S.: Random coverings in several dimensions. Acta Mathematica **156**(1), 83–118 (1986)
17. Krishnamachari, B., Wicker, S.B., Béjar, R., Pearlman, M.: Critical density thresholds in distributed wireless networks. In: Bhargava, V.K., Vincent Poor, H., Tarokh, V., Yoon, S. (eds.) Communications, Information and Network Security, vol. 712, pp. 279–296. Springer, USA (2002)
18. Mccolm, G.L.: Threshold functions for random graphs on a line segment. Comb. Prob. Comput. **13**, 373–387 (2001)
19. Panchapakesan, P., Manjunath, D.: On the transmission range in dense ad hoc radio networks. In: Proceedings of IEEE Signal Processing Communication (SPCOM) (2001)
20. Penrose, M.D.: The longest edge of the random minimal spanning tree. Ann. Appl. Prob. **7**(2), 340–361 (1997)
21. Penrose, M.D.: On k-connectivity for a geometric random graph. Random Struct. Algorithms **15**(2), 145–164 (1999)
22. Penrose, M.D.: Random geometric graphs, vol. 5. Oxford University Press, Oxford (2003)
23. Spencer, J.H.: Ten Lectures on the Probabilistic Method, vol. 52. SIAM, Philadelphia (1987)

Connectivity of a Dense Mesh of Randomly Oriented Directional Antennas Under a Realistic Fading Model

Amitabha Bagchi[1]([⊠]), Francesco Betti Sorbelli[2], Cristina Maria Pinotti[2], and Vinay Ribeiro[1]

[1] Computer Science and Engineering Department, IIT, Delhi, India
{bagchi,vinay}@cse.iitd.ac.in
[2] Department of Computer Science and Mathematics,
University of Perugia, Perugia, Italy
francesco.betti.sorbelli@gmail.com, cristina.pinotti@unipg.it

Abstract. We study mesh networks formed by nodes equipped with directional antennas in a high node-density setting. To do so we create a random geometric graph with n nodes placed uniformly at random. The antenna at each node chooses a direction of orientation at random and edges are placed between pairs of nodes based on their distance from each other and their directions of orientation according to the gain function of the antennas. To model the directionality of the antennas we consider a realistic gain function where the signal fades away from the direction of orientation. We also consider an idealised function that concentrates the gain uniformly in a sector of angle 2θ centred around the direction of orientation. In this setting we show theoretically that with probability tending to 1 the optimal power required for achieving connectivity is significantly lower than that needed for connectivity in an omnidirectional setting. We capture mathematically the relationship between this optimal power level and the maximum gain of the antenna, showing that as the directionality of the antenna increases the power needed for connectivity decreases. However this optimal power level is also inversely proportional to the probability of connectivity of two randomly placed nodes, which decreases as directionality increases. We validate these results through simulation.

1 Introduction

Directional antennas are used in several applications including satellite communications, terrestrial microwave communications, VHF and UHF terrestrial TV transmission, cellular communication, and rural mesh networks [8]. Antenna directionality focuses transmission power in a particular direction and improves communication range while simultaneously reducing interference with nearby antennas. However, a drawback is that a priori knowledge of the location of the intended radio receiver and, in some cases, the ability to steer the antenna or switch beams in the relevant direction is required to form connections between

© Springer International Publishing Switzerland 2015
P. Bose et al. (Eds.): ALGOSENSORS 2015, LNCS 9536, pp. 13–26, 2015.
DOI: 10.1007/978-3-319-28472-9_2

pairs of nodes. A larger issue is of network-wide connectivity which becomes especially important in applications such as disaster management and military battlefield communications that require the rapid setup of a wireless multihop (mesh) network [7,18]. In order to benefit from the ability of directional antennas to focus power in such settings it is critical to understand how a connected network can be formed using highly directional antenna beams. Hence the critical question is: *Can fixed (non-steerable) directional antennas also be used to successfully build connected wireless mesh networks in which nodes and antenna orientations are chosen randomly?*

While there has been a lot of discussion on the issue of capacity in highly directional antenna-based networks, there are few works in the literature that have addressed the question of connectivity. Notable among them is the work by Li et al. [13] which addresses this question but under an idealized model of directional transmission that assumes a sector shaped area of transmission (with some back lobes) with uniform power transmitted throughout the sector. In this paper we approach directional transmission in much greater generality and provide a result that holds for a large family of gain functions. We demonstrate how these results help us find the optimal power for connectivity in the dense mesh network setting. Using the gain function as our primary mode of describing the directionality of an antenna, we present a theorem that helps us determine the optimal power for connectivity for all gain functions that satisfy some moderate conditions. We further support our theoretical results through simulation using a particular family of gain functions that have been empirically found to accurately describe the power transmission pattern of directional antennas. Since in such models the gain of the antenna (i.e. the power transmitted per unit solid angle) is maximum in the direction of orientation and fades as we move away from this direction we refer to this model as the *directional fading model* or just the *fading model*. We determine the optimal power level needed to achieve connectivity for a mesh network whose antennas follow the fading model. We also revisit the simpler idealized model studied by Li et al. [13], we call it the *ideal directional model* or just the *ideal model*, and show how to set the parameters in this model so that any two antennas have the same probability of being connected in the ideal model as in the fading model. This result is obtained by equating the half-beam of a directional antenna in the fading model with the width of the sector in the ideal model. We observe that under such an equivalence the optimal power level of the fading model is double of the optimal power level of the ideal model. This is of great help for deciding the antenna setting in the fading model when connected directional meshes are designed subject to constraints on the power transmission level.

Another important novelty of our paper is that our main result on connectivity is completely rigorous. A disadvantage of [13] is that they assume that if all the antennas are positioned randomly and their antennas are oriented randomly then the edges between nodes are formed independently. This assumption clearly does not hold, as we will show in detail in Sect. 3. Our mathematical results do not need the independence assumption. Hence we claim to present the first fully rigorous analysis of connectivity in dense mesh networks built with directional antennas.

Our Contribution. We assume that nodes are deployed randomly in a finite circular area, and that each node is equipped with a directional antenna whose orientation is initially fixed randomly and kept fixed thereafter. The major contribution of our paper is that we show that for random directional mesh networks there is an optimal power level which is necessary and sufficient for connectivity to be achieved. We show that the optimal power level for connectivity is equal to $\alpha^* P_T^o$ where $\alpha^* = \frac{1}{\gamma_G G(0)^2}$, $G(0)$ being the maximum gain of an antenna with gain function G and P_T^o being the optimal power level for connectivity of an omnidirectional antenna. The quantity γ_G is a function of the gain pattern (as captured by the gain function G) and is defined as follows: it is the probability that a node u connects with another node v that is placed uniformly at random in a unit disc centred at u, assuming both nodes are equipped with randomly oriented antennas whose radiation pattern is described by gain function G, and that both antennas have a power level that allows them to communicate only up to a unit distance in the direction of maximum gain. Here, we note that $G(0)$ increases as directionality increases (and, in fact, $G(0)$ is a measure of the directionality of an antenna) while γ_G decreases as directionality increases, but we show that the overall effect is such that the power level required for connectivity is much lower than that of omnidirectional antennas, i.e., $\alpha^* \ll 1$.

Organization. In Sect. 2 we review the literature related to our work. Our model of directional mesh networks is presented in Sect. 3. We discuss the conditions for connectivity in Sect. 4. Our connectivity results are validated through simulations in Sect. 5. Finally we present our conclusions in Sect. 6.

2   Related Work

Connectivity in mesh networks using omnidirectional antennas has been studied in depth since the seminal work of Gupta and Kumar [11]. They proved that for m nodes with omnidirectional antennas randomly placed in a disc of unit area, if transmission power for all nodes was set such that each node could communicate with any other node in a circular vicinity of area $(\log m + c(m))/m$, then the network is asymptotically connected with probability 1 if and only if $c(m) \to \infty$ [11]. Our connectivity result, Theorem 1 is an analog of this result for the more complex setting where the antennas are highly directional. Our work on connectivity benefits from the general theorem proved by Bagchi et al. [1].

Connectivity was widely studied within the omnidirectional model in mobile ad hoc networks [14], in thin finite strips [3], under a physical model for interferences [9], and when nodes are active independently with a certain probability [19,21]. Several authors have studied connectivity of mesh networks equipped with steerable directional antennas in contrast to our work which considers non-steerable antennas. Kranakis et al. consider sensors deployed on a unit line and unit square with steerable directional antennas [12]. Given a set of nodes on a plane, each with a directional antenna, modeled as a sector, Caragiannis et al. investigated the problem of orienting the antennas to get a connected network [5]. Carmi et al. model the communication area of a steerable directional antenna as a wedge of infinite area which captures its directionality [6] and show that a sixty degree directional antenna suffices to form a connected network for arbitrarily located nodes. Xu et al. study the problem of connectivity through simulations when each node is equipped with several different directional antennas oriented uniformly in a circular fashion [20]. Yu et al. consider the problem of placement of wireless sensor nodes, with a view to ensuring connectivity and coverage [22]. In our work node placement is random. Bettstetter et al. considered a scenario of nodes deployed over a finite area and equipped with linear and circular antenna arrays used for random beamforming [4], demonstrating that increasing directionality leads to larger connected components. Our theoretical results support their experimental findings on connectivity.

Li et al. study asymptotic connectivity in a similar network scenario as ours [13]. Although they conjecture the same result as Theorem 1 of our paper their analysis suffers from a critical flaw. In the proof of their main theorem they use a theorem of Penrose, Theorem 3 of [16], which states that in a high density setting all the nodes of a random geometric graph are either isolated or part of a connected component almost surely. However, Penrose makes it clear that this result holds only for the case where *each edge is formed independently of all others* which is clearly not true here (see Sect. 3). Additionally they critically need the condition that in the random geometric graph formed by directional antennas in the infinite plane, when the density is supercritical the giant component is unique. For this they cite Theorem 6.3 from Meester and Roy [15], which also applies only if the independence assumption holds. We will see in Sect. 3 that the independence assumption does not hold in our setting.

3   Modeling Directional Mesh Networks

Directional Antennas. The power received by a receiving antenna, P_{R_x}, at distance r from a transmitting antenna that is transmitting at wavelength λ with power P_{T_x} is described by the Friis transmission equation:

$$P_{R_x} = P_{T_x} G_{R_x} G_{T_x} \left(\frac{\lambda}{4\pi r} \right)^2, \tag{1}$$

where G_{R_x} and G_{T_x} are the receiver and transmitters gains and depend on the orientations of the two antennas. For highly directional antennas these gains can be very high since these antennas tend to concentrate their beams in one direction. Gain is formally defined as *the ratio of the power radiated in a given direction per unit solid angle to the average power radiated per unit solid angle,* (c.f., e.g., [2,17]).

Although the gain function depends on both the polar and azimuthal angles in 3 dimensions we will assume for ease of presentation that the gain function $G : [-\pi, \pi] \to \mathbb{R}_+ \cup \{0\}$ is defined over two dimensions, i.e., depends only on the azimuthal angle. We note that our methods are general and can be transposed to 3 dimensions with suitable modifications. We assume that our gain function has the following properties: (Directionality) $G(\psi) = 0$ for $|\psi| \geq \pi/2$.

(Symmetry around angle of orientation) $G(\psi) = G(-\psi)$. (Monotonicity) $G(\psi) > G(\psi')$ whenever $|\psi| < |\psi'|$. The assumption that G takes non-zero values only in $[-\pi/2, \pi/2]$ neglects back-lobe transmission, which is a simplification we make for ease of presentation. From these properties we can additionally deduce that $G(\cdot)$ reaches its maximum value at 0. Also $G(\cdot)$ is not an invertible function, since it is not one-to-one. So we follow the convention, similar to that of inverse trigonometric functions, that $G^{-1}(x)$ is a positive valued function i.e. if $G(\psi) = x$ then we say that $G^{-1}(x) = |\psi|$. Also, by the reciprocity principle it is known that the receiver gain and transmitter gain of an antenna are identical. In this paper we will deal with settings where all antennas are considered identical to each other and so we will consider only one single gain function at a time.

A Realistic Directional Fading Model. In a realistic antenna setting the gain decreases as we move away from the angle of orientation of the antenna. In this paper we will work with a family of gain functions that satisfy this property. We will refer to this model as the *directional fading model* or simply the *fading model*. This family of functions, which has been mentioned in the antenna theory literature as being of particular interest [2,17], is:

$$G_f^n(\psi) = \begin{cases} G_f^n(0)\cos^n(\psi) & 0 \le |\psi| \le \frac{\pi}{2}, \\ 0 & |\psi| \ge \frac{\pi}{2}, \end{cases} \tag{2}$$

where n takes even values and the f in the subscript of G_f is to indicate the "fading" model and differentiate it from the ideal model we will also study (see below). The angle ψ is relative to the angle of orientation of the antenna. Since, by the definition of gain, the integral of gain over the unit sphere should be 4π, we can compute the normalization constant $G_f^n(0)$ for this family. We omit the exact calculation here only noting that in general $2n + 1$ is a reasonable approximation of $G_f^n(0)$ as n grows.

From now on, we simply denote the realistic gain function $G_f^n(\cdot)$ by $G(\cdot)$.

The Ideal Directional Model. As a theoretical counterpoint we introduce a simple idealised directional gain function that captures the idea of a beam of width 2θ centred at the angle of orientation. The gain everywhere is a uniform non-zero value within this beam is and zero everywhere outside. We denote the ideal gain function $G_i^\theta(\cdot)$, using the subscript i for "ideal" to differentiate it from the fading model above. This gain function can be explicitly computed by integrating the uniform gain over the surface of the sphere centred at the antenna and equating this value to 4π. By doing this we find.

$$G_i^\theta(\psi) = \begin{cases} \frac{2}{1-\cos(\theta)} & 0 \le |\psi| \le \theta, \\ 0 & |\psi| > \theta, \end{cases} \tag{3}$$

The Power Parameter α and Radius of Connectivity. In the omnidirectional case under the assumption of uniform unit gain in all directions, Gupta and Kumar showed that in the setting where m nodes are distributed uniformly at random in a unit disc and if each node can communicate with another node at distance

r from it, then, the random graph thus formed is connected with probability tending to 1 as $m \to \infty$ if and only if the radius within which two nodes can communicate is

$$r_o(m) = \sqrt{\frac{\log(m) + c(m)}{m\pi}} \qquad (4)$$

where $c(m) \to \infty$ as $m \to \infty$ [10]. In the following when the number of nodes m is understood, we will often just use r_o to denote this radius.

Restating this in terms of power, using the Friis transmission equation, we can say that if P_R^* is the minimum received power required for the signal to be correctly received, then, since $G_{R_x} = G_{T_x} = 1$, the omnidirectional transmission power required is

$$P_T^o = P_R^* \left(\frac{4\pi r_o}{\lambda}\right)^2. \qquad (5)$$

In this paper we will use this value of P_T^o as a scaling constant for the transmission power used, and r_o as a scaling constant for distances. In particular we will say that the transmission power used by our directional antennas is $P_T^d = \alpha P_T^o$.

We will use α as a parameter to tune the antenna transmission power for the rest of this paper. To find the furthest distance, $r_G(\alpha)$, that an antenna u with gain function $G(\cdot)$ and power parameter α can communicate we have to find the largest x such that the power received by an antenna v which is at distance x from the transmitting antenna u is at least P_R^*, i.e., we have to find x such that

$$\max_{\beta_1,\beta_2 \in [-\pi/2,\pi/2]} P_T^d G(\beta_1) G(\beta_2) \left(\frac{\lambda}{4\pi x}\right)^2 \geq P_R^* \qquad (6)$$

where β_1 is the angle between the ray defining the angle of orientation of the transmitter and the line segment $u \to v$ and β_2 is defined analogously for the receiver (see Fig. 1.) Solving this by putting the values of P_R^* and P_T^d, and observing that $G(\cdot)$ is maximised at $G(0)$ by definition, we get that

$$r_G(\alpha) = \sqrt{\alpha} \cdot G(0) \cdot r_o. \qquad (7)$$

Hence by varying α we can control the distance to which the connections can be made. Note that the maximum distances for the two models can be derived by using the values of $G_f^n(0)$ and $G_i^\theta(0)$.

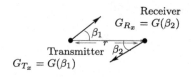

Fig. 1. Connecting transmitter to receiver.

Random Orientations and Connectivity Probability. Unlike in the simple RGG model studied by Gupta and Kumar [10], connectivity between two antennas in the directional setting does not depend only on the distance between them, it also depends on their angles of orientation. We now study the situation where the antennas are located in the 2-d plane and each antenna picks its angle of orientation uniformly at random from $[0, 2\pi]$.

Assuming that the receiver has fixed its angle of orientation (β_2 relative to the line joining receiver to transmitter) we compute the probability of connectivity at distance r by integrating over the range of values of the angle of orientation of the transmitter, β_1, within which the received power is at least P_R^*. This gives us:

$$g_G(r) = \int_{-G^{-1}\left(\frac{r^2}{\alpha r_o^2 \cdot G(0)}\right)}^{G^{-1}\left(\frac{r^2}{\alpha r_o^2 \cdot G(0)}\right)} \frac{1}{2\pi^2} \cdot G^{-1}\left(\frac{r^2}{\alpha r_o^2 \cdot G(\beta_1)}\right) d\beta_1. \tag{8}$$

The above function is non-trivial to compute in the fading model, but in the ideal directional model, under the gain function $G_i^\theta(\cdot)$ it reduces to

$$g_{G_i^\theta}(r) = \begin{cases} \frac{\theta^2}{\pi^2} & 0 < r \le \sqrt{\alpha} \cdot \frac{2}{1-\cos(\theta)} \cdot r_o, \\ 0 & \text{otherwise,} \end{cases} \tag{9}$$

This is simply the probability that the receiver lies in a randomly chosen sector of radius $r_i(\alpha)$ with angle 2θ centred at the transmitter and vice-versa.

We also compute the probability, γ_G, that a node u connects with another node v that is placed uniformly at random in the disc of radius $r_f(\alpha) = \sqrt{\alpha}G(0)r_o$ centred at u in the realistic fading model. This quantity is going to be critical in our study of network connectivity (Sect. 4). Conditioning on the position of u and integrating over the disc we get

$$\gamma_G = \int_{x=0}^{\sqrt{\alpha}r_o G(0)} g_G(x)\frac{2x}{\alpha r_o^2 G(0)^2}dx. \tag{10}$$

An important point to note here is that γ_G does *not* depend on α as long as $\alpha > 0$. This can be seen by changing variables in (10), replacing x with z where $x = \sqrt{\alpha}r_o z$.

For the ideal model we compute the probability, γ_{G_i}, that a node u connects with another node v that is placed uniformly at random in the disc of radius $r_i(\alpha) = \sqrt{\alpha}G_i^\theta(0)r_o$ centred at u. By substituting in Eq. 10 the probability of connectivity at distance x, i.e., $g_{G_i^\theta}(x)$ given by Eq. 9, we get $\gamma_{G_i} = \theta^2/\pi^2$.

A Random Graph Model. We model a mesh network of directional antennas as a random geometric graph, $H = (V, E)$, whose nodes are distributed uniformly at random in a unit disc in $\mathbb{R}^2$. Each node $u \in V$ is equipped with a directional antenna that chooses its angle of orientation ξ_u uniformly at random from $[0, 2\pi]$

independently of all other nodes. The other parameters of the model are a power level α as defined in Sect. 3 and a gain function $G(\cdot)$.

For convenience we will use the following notation to refer to random graphs modeling networks using the directional fading and ideal directional model:

- DF-RGG(m, n, α): a random graph formed as above on m nodes with $G = G_f^n(\cdot)$ and power parameter α, briefly DF-RGG when the parameter values are understood.
- DI-RGG(m, θ, α): a random graph formed as above on m nodes with $G = G_i^\theta(\cdot)$ and power parameter α, briefly DI-RGG.

The Edge-Independence Assumption does not Hold. To show this let us consider the simpler ideal model. Assume there are three nodes x, y and z which are placed such that their pairwise distances are all equal to some $r > 0$, i.e. they are placed at the vertices of an equilateral triangle of side length r. Consider a value of θ that is smaller than 30 degrees and an α large enough to ensure that each pair can communicate if the antenna orientations are correct. For a given pair of nodes, say x, y, the probability that they are connected is θ^2/π^2. But clearly the probability of all three pairs being connected is 0 which is less than θ^6/π^6 which is what it would have been if the probabilities of the edges being formed were independent. Hence, we find that the independence assumption does not hold and so the theory developed under this assumption cannot be used in this case as has been done by Li et al. [13]. We will now show how this problem can be handled.

4   Connectivity

In this section we show that highly directional antennas achieve network connectivity at a much lower power level than omnidirectional antennas. This is a somewhat counterintuitive result that we feel has major implications for the design of mesh networks.

A Connectivity Theorem for Directional Random Mesh Networks. We now present our main theorem on connectivity. The key factor in this theorem is the probability of connectivity γ_G associated with an antenna with gain function G. As we showed in Sect. 3, this probability is independent of the transmission power and hence is a property of the antenna model alone and depends only on the gain function G. Our main theorem is:

Theorem 1. *Suppose we are given a set V of m nodes distributed uniformly at random in a unit disc B of $\mathbb{R}^2$ and each node is equipped with an antenna with gain function G that is (a) non-zero in $[-\pi, \pi]$, (b) symmetric around the angle of orientation and (c) monotonically decreasing away from the angle of orientation. Assume that each antenna has transmission power that allows it to transmit to a distance of $r > 0$ in its direction of maximum gain. Denote by*

γ_G *the probability that two nodes that lie within distance* r *of each other are connected.*

We construct a random graph model D-RGG(m, G, r) by placing edges between each pair of points that can communicate with each other, and for this we have that $P(\text{D-RGG}(m, G, r)$ is connected$) \to 1$ as $m \to \infty$ if and only if

$$\pi r(m)^2 \gamma_G = (\log m + c(m))/m, \tag{11}$$

where $\lim_{m \to \infty} c(m) = \infty$ *as* $m \to \infty$.

Due to shortage of space we omit the proof of this theorem. We note that the optimal radius suggested by Theorem 1 is simply the optimal radius for omnidirectional antennas given by Gupta and Kumar [11] scaled by a factor of $1/\sqrt{\gamma_G}$. This implies that the radius of connectivity is *larger* than that for omnidirectional antennas, since $\gamma_G < 1$, and appears to run counter to our claim that random directional mesh networks require lower power. However, as we have already seen the directionality of an antenna means that it can achieve a much larger transmission range, at least in the direction of orientation, and so we will find that the power required is much lower than that required for omnidirectional antennas.

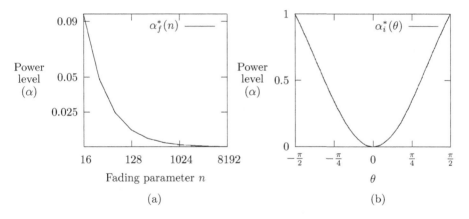

Fig. 2. Optimal power level vs model parameters for (b) the ideal model (parameter θ) and (a) the fading model (parameter n).

Optimal Power for Connectivity. From Theorem 1 we deduce that the optimal radius r_d of connectivity of the directional model with gain function G is given by $r_d = r_o/\sqrt{\gamma_G}$. The power level α that reaches the maximum distance $\sqrt{\alpha G(0)}r_o$ equal to r_d will be called the *optimal power level* α^* and is given in the fading and ideal model by, respectively:

$$\alpha_f^*(n) = \frac{1}{\gamma_{G_f} G_f^n(0)^2} \quad (12) \qquad \alpha_i^*(\theta) = \left(\frac{\pi(1 - \cos(\theta))}{2\theta}\right)^2 \quad (13)$$

In Fig. 2, after computing $\gamma_f(n)$ numerically for $n = 16, 32, 64, \ldots, 8192$, we plot $\alpha_f^*(n)$ versus n and $\alpha_i^*(\theta)$ versus θ. It is worth noting that α^* depends on n in the fading model and on θ in the ideal model. Since the gain in the direction of orientation is a measure of how "directional" the antenna beam is, i.e., how concentrated the signal is in the direction of orientation, the inverse relationship of the optimal power level to $G(0)^2$ implies that the power level required for connectivity decreases as the directionality of the antenna increases.

Table 1. The parameter n, associated angle $\theta(n)$ and corresponding optimal power levels and connectivity probabilities.

n	$\theta(n)$ (degrees)	γ_{G_i}	$\alpha_i^*(n)$	γ_{G_f}	$\alpha_f^*(n)$
16	16.74	0.008652	0.0519	0.009640	0.0952468
32	11.88	0.004357	0.0263	0.004896	0.0483396
64	8.41	0.002186	0.0132	0.0024673	0.024355
128	5.95	0.001095	0.0066	0.0012385	0.0122246
256	4.21	0.000548	0.0033	0.00620	0.00612419
512	2.98	0.000274	0.0016	0.00031	0.00306508
1024	2.10	0.000137	8.34e-04	0.00015	0.00153329
2048	1.49	6.86e-05	4.17e-04	7.76e-05	0.00076683
4096	1.05	3.42e-05	2.08e-04	3.88e-05	0.000383462
8192	0.74	1.71e-05	1.04e-04	1.94e-05	0.000191743

Comparing the Ideal Model and the Fading Model. It is not a priori clear how to determine which of the two models, ideal or fading, is more power efficient. In order to compare them, we propose to study the *half-power beamwidth* (or, simply, the *halfbeam*) for antennas with realistic gain function [17].

For an antenna of parameter n, the halfbeam is defined as the angle 2χ between the two directions in which the gain $G_f^n(\chi)$ is one half the maximum value, that is, χ such that $G_f^n(\chi) = \frac{1}{2}G_f^n(0)\cos^n(0)$. Solving the above equation, we obtain that the halfbeam of an antenna of parameter n is the angle $2\chi = 2\cos^{-1}\left(\sqrt[n]{1/2}\right)$. Thus, we associate the fading model whose gain function has parameter n to the ideal model of parameter $\theta(n) = \cos^{-1}\left(\sqrt[n]{1/2}\right)$.

With this correspondence, we report the optimal power levels in Table 1: we compute $\alpha_i^*(\theta(n))$ by recalling $\gamma_{G_i} = \theta^2/\pi^2$ and using Eq. 3. After computing γ_G by numerical integration (see Eq. 10), we derive $\alpha_f^*(n)$ using Eq. 2. Note that the values of $\alpha_i^*(n)$ in Table 1 zoom into Fig. 2b since $\theta(n)$ lies in $[0.74, 16.74]$.

In Table 1 we report the connectivity probabilities of the fading model with different values of the parameter n and those the associated DI-RGG, i.e. the ideal model with parameter $\theta(n)$. As we see, they almost coincide thus validating the engineering intuition that guided us in making this association. This connectivity probability is for a pair of points but when we come to network-wide

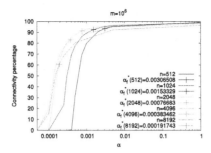

Fig. 3. The percentage of connectivity versus α in the fading model.

connectivity the models differ: the optimal power level for DI-RGG is about half of that for the corresponding version of DF-RGG. This is because each DI-RGG antenna covers at a smaller area (i.e., halfbeam) than the one considered in DF-RGG but with a better (uniform) gain value. This shows that for network connectivity the halfbeam assumption is overly optimistic and gives us lower power levels than required. Nevertheless for all values of n the optimal power for the fading model is double that of the ideal model, and we can state as a rule of thumb that $\alpha_f^*(n) = 2\alpha_i^*(\theta(n))$. This is an important input for the design of a connected directional mesh in which the directional antennas transmit at power level at most α.

5   Simulation Results for the Fading and Ideal Models

In this section we experimentally test our results on connectivity in directional meshes. We built our own simulator and we ran the experiments on a 2.2 GHz Intel i3 processor with 4 GB of main memory. We implemented the algorithm in C++. We followed the communication model for the DF-RGGs and DI-RGGs described in Sect. 3. Our main metric in this study is what we call the *percentage of connectivity* or *connectivity percentage*, which is defined as the percentage of nodes in the largest connected component. First we validate our main result on the optimal power level for the fading model. Figure 3 shows the percentage of connectivity versus power level α for several values of the fading parameter n. For each value of n, the optimal power level $\alpha_f^*(n)$ is highlighted with a small cross. As one can see, whenever $n \leq 4096$, the optimal power level derived in Eq. 12 is very accurate. Indeed, at $\alpha_f^*(n)$, the percentage of connectivity reaches the maximum value and after that, it remains stable. In other words, extra power would not significantly improve the connectivity. For $n = 8192$, $\alpha_f^*(8192)$ is less accurate since the percentage of connectivity increases for $\alpha > \alpha_f^*(8192)$. This eventually shows that the connectivity probability is slightly overestimated in such extreme value of n. The remaining experiments test the percentage of connectivity in DF-RGG and DI-RGG at the optimal power level α^*, reported in Table 1. Figure 4 shows that the percentage of connectivity achieved in directional mesh is high and comparable to that of omnidirectional mesh, although

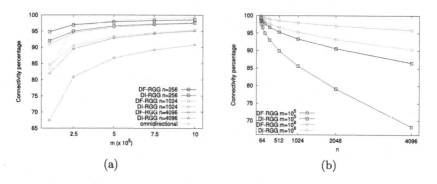

Fig. 4. The percentage of connectivity when: (a) m varies, (b) n varies.

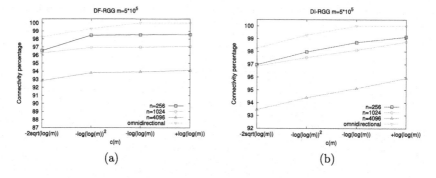

Fig. 5. The percentage of connectivity vs $c(m)$: (a) in DF-RGG (b) in DI-RGG.

the power used by directional models is well below P_T^o which is conventionally set to 1 in our experiments. It also appears that for a more directional model to achieve a high connectivity percentage, we need a higher density than we need for a less directional model. Nonetheless, it is interesting to point out that when m is small, moderate directionality may achieve higher connectivity than omnidirectional networks, i.e., reaching further nodes within a (sufficiently wide) sector is more effective for achieving connectivity than reaching nodes that do not lie as far but are located all around the antenna. We then verified whether the power level derived by Eqs. 12 and 13 is necessary for achieving connectivity. For this purpose, we varied the connectivity radius in Eq. 4 below the optimal threshold using $c(m) = \{-\log\log(m), -\log^2\log(m), -2\sqrt{\log(m)}\}$. Changing $c(m)$, the radius reduces from r_o to r, and the directional optimal power level is scaled by factor $F = (\frac{r}{r_o})^2$. The scale coefficients F used in Fig. 5 for the three values of $c(m)$ are $\{0.64984, 0.41382, 0.37443\}$. We take $m = 5 \cdot 10^5$ here. We note in Fig. 5 that the more we decrease the power level, the greater the loss in connectivity. The trends of the connectivity curves are the same for all values of n, sharpening for higher values of n.

6   Conclusions

In this paper we have argued that connected mesh networks can be built using directional antennas and that such mesh networks can operate with much lower power than mesh networks built with isotropic omnidirectional antennas. We have also demonstrated how a simple idealised gain function can be used to approach mesh network design where the antennas have a more realistic and complex gain function.

References

1. Bagchi, A., Pinotti, C.M., Galhotra, S., Mangla, T.: Optimal radius for connectivity in duty-cycled wireless sensor networks. ACM Trans. Sens. Netw. **11**(2), Article no. 36, 1–37 (2015)
2. Balanis, C.A.: Antenna Theory: Analysis and Design. Wiley, New York (2012)
3. Balister, P., Bollobas, B., Sarkar, A., Kumar, S.: Reliable density estimates for coverage and connectivity in thin strips of finite length. In: Proceedings of 13th Annual ACM International Conference on Mobile Computing and Networking (Mobicom 2007), pp. 75–86. ACM (2007)
4. Bettstetter, C., Hartmann, C., Moser, C.: How does randomized beamforming improve the connectivity of ad hoc networks? In: IEEE International Conference on Communications (ICC 2005), vol. 5, pp. 3380–3385. IEEE (2005)
5. Caragiannis, I., Kaklamanis, C., Kranakis, E., Krizanc, D., Wiese, A.: Communication in wireless networks with directional antennas. In: Proceedings of 20th Annual Symposium Parallelism in Algorithms and Architectures (SPAA 2008), pp. 344–351. ACM (2008)
6. Carmi, P., Katz, M.J., Lotker, Z., Rosén, A.: Connectivity guarantees for wireless networks with directional antennas. Comput. Geom. **44**(9), 477–485 (2011)
7. Chawla, A., Yadav, V., Dev Sharma, V., Bajaj, J., Nanda, E., Ribeiro, V., Saran, H.: RODEO: robust and rapidly deployable TDM mesh with QoS differentiation. In: Proceedings of 4th International Conference Communication Systems and Networks (COMSNETS 2012), pp. 1–6. IEEE (2012)
8. Chebrolu, K., Raman, B.: FRACTEL: a fresh perspective on (rural) mesh networks. In: Proceedings of 2007 Workshop on Networked Systems for Developing Regions, p. 8. ACM (2007)
9. Dousse, O., Baccelli, F., Thiran, P.: Impact of interferences on connectivity in ad hoc networks. IEEE/ACM Trans. Netw. **13**(2), 425–436 (2005)
10. Gupta, P., Kumar, P.R.: Critical power for asymptotic connectivity. In: Proceedings of 37th IEEE Conference on Decision and Control, pp. 1106–1110. IEEE (1998)
11. Gupta, P., Kumar, P.R.: Critical power for asymptotic connectivity in wireless networks. In: McEneaney, W.M., George Yin, G., Zhang, Q. (eds.) Stochastic Analysis, Control, Optimization and Applications. A Volume in Honor of W.H. Fleming, pp. 547–566. Springer, New York (1999)
12. Kranakis, E., Krizanc, D., Williams, E.: Directional versus omnidirectional antennas for energy consumption and k-connectivity of networks of sensors. In: Higashino, T. (ed.) OPODIS 2004. LNCS, vol. 3544, pp. 357–368. Springer, Heidelberg (2005)
13. Li, P., Zhang, C., Fang, Y.: Asymptotic connectivity in wireless ad hoc networks using directional antennas. IEEE/ACM Trans. Netw. **17**(4), 1106–1117 (2009)

14. Madsen, T.K., Fitzek, F.H., Prasad, R., Schulte, G.: Connectivity probability of wireless ad hoc networks: definition, evaluation, comparison. Wirel. Pers. Commun. **35**(1–2), 135–151 (2005)

15. Meester, R., Roy, R.: Continuum Percolation. Number 119 in Cambridge Tracts in Mathematics. Cambridge University dss, Cambridge (1996)

16. Penrose, M.D.: On a continuum percolation model. Adv. Appl. Probab. **23**(3), 546–556 (1991)

17. Silver, S.: Microwave Antenna Theory and Design. McGraw-Hill, New York (1949)

18. Souryal, M.R., Wapf, A., Moayeri, N.: Rapidly-deployable mesh network testbed. In: Proceedings of Global Telecommunications Conference (GLOBECOM 2009), pp. 1–6. IEEE (2009)

19. Wan, P.-J., Yi, C.-W.: Asymptotic critical transmission ranges for connectivity in wireless ad hoc networks with bernoulli nodes. In: Proceedings of IEEE Wireless Communications and Networking Conference (WCNC 2005), vol. 4, pp. 2219–2224. IEEE (2005)

20. Xu, H., Dai, H.-N., Zhao, Q.: On the connectivity of wireless networks with multiple directional antennas. In: 18th IEEE International Conference on Networks (ICON 2012), pp. 155–160. IEEE (2012)

21. Yi, C.-W., Wan, P.-J., Li, M., Frieder, O.: Asymptotic distribution of the number of isolated nodes in wireless ad hoc networks with bernoulli nodes. IEEE Trans. Commun. **54**(3), 510–517 (2006)

22. Yu, Z., Teng, J., Bai, X., Xuan, D., Jia, W.: Connected coverage in wireless networks with directional antennas. ACM Trans. Sens. Netw. **10**(3), 51 (2014)

Maintaining Intruder Detection Capability in a Rectangular Domain with Sensors

Evangelos Kranakis[1], Danny Krizanc[2], Flaminia L. Luccio[3]([⊠]), and Brett Smith[2]

[1] School of Computer Science, Carleton University, Ottawa, ON, Canada
[2] Department of Mathematics and Computer Science, Wesleyan University, Middletown, CT, USA
[3] DAIS, Università Ca' Foscari Venezia, Venice, Italy
luccio@unive.it

Abstract. In order to detect intruders that attempt to pass through a rectangular domain, sensors are placed at nodes of a regular spaced grid laid out over the rectangle. An intruder that steps within the sensing range of a sensor will be detected. It is desired that we prevent potential attacks in either one dimension or two dimensions. A one-dimensional attack succeeds when an intruder enters from the top (North) side and exits out the bottom (South) side of the domain without being detected. Preventing attacks in two dimensions requires that we simultaneously prevent the intruder from either entering North and exiting South or entering East (left side) and exiting West (right side) undetected.

Initially, all of the sensors are working properly and the domain is fully protected, i.e., attacks will be detected, in both dimensions (assuming the grid points are such that neighboring sensors have overlapping sensing ranges and include all four boundaries of the domain). Over time, the sensors may fail and we are left with a subset of working sensors. Under these conditions we wish to (1) determine if one or two-dimensional attack detection still persists and (2) if not, restore protection by adding the least number of sensors required to ensure detection in either one or two dimensions.

Ideally, the set of currently working sensors would provide some amount of fault-tolerance. In particular, it would be advantageous if for a given k, the set of sensors maintains protection (in one or two dimensions) even if up to k of the sensors fail. This leads to the problems of (1) deciding if a subset of the sensors provides protection with up to k faults and (2) if not, finding the minimum number of grid points to add sensors to in order to achieve k fault-tolerance.

In this paper, we provide algorithms for deciding if a set sensors provides k-fault tolerant protection against attacks in both one and two dimensions, for optimally restoring k-fault tolerant protection in one dimension and for restoring protection in two dimensions (optimally for $k = 0$ and approximately otherwise).

Research supported in part by NSERC grant, and by MIUR project Security Horizons. Work partially done while the first two authors were visiting Ca' Foscari University.

P. Bose et al. (Eds.): ALGOSENSORS 2015, LNCS 9536, pp. 27–40, 2015.
DOI: 10.1007/978-3-319-28472-9_3

Keywords: Rectangular grid · Intruder detection · Sensors · Fault tolerance

1 Introduction

In order to detect intruders that attempt to pass through a rectangular domain, sensors are placed at nodes of a regular spaced grid laid out over the rectangle. An intruder that steps within the sensing range of a sensor will be detected. It is desired that we prevent potential attacks in either one dimension or two dimensions. A one-dimensional attack succeeds when an intruder enters from the top (North) side and exits out the bottom (South) side of the domain without being detected. Preventing attacks in two dimensions requires that we simultaneously prevent the intruder from either entering North and exiting South or entering East (left side) and exiting West (right side) undetected.

Initially, all of the sensors are working properly and the domain is fully protected, i.e., attacks will be detected, in both dimensions (assuming the grid points are such that neighboring sensors have overlapping sensing ranges and include all four boundaries of the domain). Over time, the sensors may fail and we are left with a subset of working sensors. Under these conditions we wish to (1) determine if one- or two-dimensional attack detection still persists and (2) if not, restore protection by adding the least number of sensors required to ensure detection in either one or two dimensions.

Ideally, the set of currently working sensors would provide some amount of fault-tolerance. In particular, it would be advantageous if for a given k, the set of sensors maintains protection (in one or two dimensions) even if up to k of the sensors fail. This leads to the problems of (1) deciding if a subset of the sensors provides protection with up to k faults and (2) if not, finding the minimum number of grid points to add sensors to in order to achieve k fault-tolerance.

In this paper, we provide algorithms for deciding if a set sensors provides k-fault tolerant protection against attacks in both one and two dimensions, for optimally restoring k-fault tolerant protection in one dimension and for restoring protection in two dimensions (optimally for $k = 0$ and approximately otherwise). The rest of this introduction provides more precise definitions for our problems, a description of our results and a discussion of related work. The following section presents our results in detail and we finish with a discussion of extensions and open problems.

1.1 Preliminaries and Notation

In order to present our results we first provide some definitions. As many of our results depend upon results concerning shortest paths and network flow in directed graphs, we will, at times, represent an undirected graph, say H, with a directed graph, denoted $H_{\leftrightarrow}$, by replacing each edge with two edges of opposite orientation. Any directed path in $H_{\leftrightarrow}$ will then correspond to an undirected path in H and vice versa. Moreover, two directed paths are vertex-disjoint in $H_{\leftrightarrow}$ if and only if the corresponding paths in H are vertex-disjoint.

Also, we will need a binary weight function on the vertices of our graphs to indicate whether or not a vertex can be occupied by an adversary without detection by a sensor. As we will see, it is convenient to add 'dummy' vertices to our grid, G, which will not be accessible by an adversary. These vertices will be given weight zero, as will any vertex on which we place a sensor. Any grid vertex which does not contain a sensor will have weight one.

Many of the cut/flow results we would like to use depend on edge weights to determine the length of a path. As we are interested in the presence of sensors on a path, for any plane graph, H, with weighted vertices, we give edge weights to $H_{\leftrightarrow}$ determined by the terminal vertex of each edge. Then the weight of a directed path in $H_{\leftrightarrow}$ will be the sum of the edge weights plus the weight (in H) of the initial vertex on the path. This provides an equivalent weight on undirected paths in H and a distance function in both graphs.

Fix $m, n > 0$ and consider an $m \times n$-grid, G, embedded in the plane. Label the four vertex sets corresponding to the four sides of the grid, $North, South, East$ and $West$, so that $North$ and $South$ have size m; $East$ and $West$ have size n. Let $C = (c_0, ..., c_{2n+2m-5})$ be the cycle peripheral to the unbounded face of G so that

$$North = \{c_0, ..., c_{m-1}\},$$
$$East = \{c_{m-1}, ..., c_{m+n-2}\},$$
$$South = \{c_{m+n-2}, ..., c_{2m+n-3}\},$$
$$West = \{c_{2m+n-3}, ..., c_{2m+2n-5}, c_0\}.$$

A $(North, South)$-**path** in $G_{\leftrightarrow}$ is a directed path whose initial vertex is in $North$ and whose terminal vertex is in $South$; similarly, an $(East, West)$-**path** in $G_{\leftrightarrow}$ is a directed path whose initial vertex is in $East$ and whose terminal vertex is in $West$. We refer to undirected paths in G as $(North, South)$ and $(East, West)$-paths as well.

Definition 1 ((North, South)- and (East, West)-Blocking Set). *A $(North, South)$- blocking set, $B \subseteq V(G)$, is a set of vertices such that there is no $(North, South)$-path in $G - B$. Similarly, an $(East, West)$-blocking set, $B \subseteq V(G)$, is a set of vertices such that there is no $(East, West)$-path in $G - B$.*

Placing sensors on such a blocking set will detect an adversary's movement along any path between the identified pair of sides in the grid.

We generalize these blocking sets to allow for a number, k, of faults amongst sensors.

Definition 2 (k-Blocking Set). *For any fixed $k \geq 0$, a $(North, South)$-k-blocking set, $B \subseteq V(G)$, is a set of vertices such that for any $F \subseteq B$ with $|F| \leq k$, there is no $(North, South)$-path in $G - (B \backslash F)$. Similarly, an $(East, West)$-k-blocking set, $B \subseteq V(G)$, is a set of vertices such that for any $F \subseteq B$ with $|F| \leq k$, there is no $(East, West)$-path in $G - (B \backslash F)$. For any fixed $k \geq 0$, a k-blocking set, $B \subseteq V(G)$, is a set of vertices so that for any $F \subseteq B$ with $|F| \leq k$, there is neither a $(North, South)$-path nor an $(East, West)$-path in $G - (B \backslash F)$.*

Placing sensors on a $(North, South)$-k-blocking set (respectively, $(East, West)$-k-blocking set) will detect movement along any path between the identified pair of sides in the grid, allowing for up to k faulty sensors. Moreover, a k-blocking set detects both $(North, South)$ and $(East, West)$ movement at the same time, allowing for up to k faulty sensors.

We now define our problems.

Definition 3 (One-Dimensional k-Protection Decision Problem). *Given an $m \times n$-grid, G, a subset $B \subseteq V(G)$, and a non-negative integer k, does B form a $(North, South)$-k-blocking set? The case $k = 0$ will be referred to simply as the one-dimensional protection decision problem.*

Definition 4 (Two-Dimensional k-Protection Decision Problem). *Given an $m \times n$-grid, G, a subset $B \subseteq V(G)$, and a non-negative integer k, does B form a k-blocking set? The case $k = 0$ will be referred to simply as the two-dimensional protection decision problem.*

Definition 5 (One-Dimensional k-Protection Placement Problem). *Given an $m \times n$-grid, G, a subset $B \subseteq V(G)$, and a non-negative integer k, find a set $B' \subseteq V(G) \backslash B$ of minimum size such that $B \cup B'$ forms a $(North, South)$-k-blocking set. The case $k = 0$ will be referred to simply as the one-dimensional protection placement problem.*

Definition 6 (Two-Dimensional k-Protection Placement Problem). *Given an $m \times n$-grid, G, a subset $B \subseteq V(G)$, and a non-negative integer k, find a set $B' \subseteq V(G) \backslash B$ of minimum size such that $B \cup B'$ forms a k-blocking set. The case $k = 0$ will be referred to simply as the two-dimensional protection placement problem.*

1.2 Our Results

We show the following:

1. There exist $O(mn)$ time algorithms for solving the one- and two-dimensional k-protection decision problems. See Theorems 2 and 5. (Note: the one dimensional case follows from a result in [13]. We present our own version of the proof for completeness.)
2. There exists a $O(kmn \log(mn))$ time algorithm for solving the one-dimensional k-protection placement problem. See Theorem 3.
3. There exists a $O(m^2 n^2)$ time algorithm for solving the two-dimensional protection placement problem. See Theorem 6.
4. There exists a $O(kmn \log(mn))$ time 2-approximation algorithm for solving the two-dimensional k-protection placement problem. See Theorem 7.

In all of the above we assume $k < \min\{m, n\}$ as we shall see that the problems can not be solved otherwise. Further we discuss extensions of these results for solving more general versions of these problems including protecting:

1. domains containing impassable regions,
2. non-rectangular domains,
3. and against more general attacks than just North-South or East-West.

1.3 Related Work

As far as we know, we are the first to study these problems as formalized above. A closely related problem concerning the placement of sensors to accomplish coverage of a given region has been studied extensively in the literature. Generally, one assumes sensors can sense a limited region defined by their sensing radius. To monitor a larger region against potential threats every point of the region must be within the sensing range of at least one of the sensors. This has been studied in several papers, and includes research on *area coverage* whereby one ensures monitoring of an entire region [11,14], and on *perimeter* or *barrier coverage* whereby a region is monitored via its perimeter thus sensing intrusions or exits from the interior [1,13]. The fault tolerance of such placements has also been studied [4,13]. For the case where the sensors may be moved, the complexity of minimizing the sensor displacement has also been studied in some detail. For example, for sensors placed on a line [6] shows that there is an $O(n^2)$ algorithm for minimizing the max displacement of a sensor while the problem becomes NP-complete if there are two separate (non-overlapping) barriers (cf. also [5] for arbitrary sensor ranges). Similar research is known if one is interested in sum of sensor displacements [7], or the number of sensors moved [15]. Further, [9] considers the complexity of several natural generalizations of the barrier coverage problem with sensors of arbitrary ranges, including when the initial positions of sensors are arbitrary points in the two-dimensional plane, as well as multiple barriers that are parallel or perpendicular to each other. Perhaps the most closely related work to ours is that of [16] where the authors look at how to best randomly distribute additional sensors in order to maintain barrier coverage under the potential for faults.

2 Main Results

Our main results are based upon establishing a characterization of minimal k-blocking sets in grids. To do this, it is easier to work in the more general setting of graphs embedded in the plane and use some ideas derived from the graph theory literature. We begin by establishing some definitions and important lemmas.

2.1 Connectedness and Surface Graphs

Definition 7. *Let H be a simple, 2-connected graph embedded in the plane. The* **surface graph** *of H, $\widehat{H}$, is obtained from H as follows. In each face, f, of H, add a new node, v_f, and edges from v_f to each vertex of H which is peripheral to f. For each $X \subseteq V(H)$, the* **bounded surface set** *of X, called $\overline{X}$, is equal to $X \cup \{v_f \in V(\widehat{H}) : f$ is a bounded face of $H\} \subseteq V(\widehat{H})$.*

That is, we obtain a subset of vertices of $\widehat{H}$ from X by including each vertex which corresponds to a bounded face of H. The subgraph of $\widehat{H}$ induced by $\overline{X}$ is denoted by $\widehat{H}[\overline{X}]$.

Notice that $\widehat{H}$ is a maximal plane graph. We observe the following connectedness property for any maximal plane graph.

Lemma 1. *Let H be a maximal plane graph, and let $X \subseteq V(H)$. If f is a face of $H - X$, then the set,*

$$F := \{u \in V(H) : u \text{ is in the face } f\},$$

is a connected set of vertices in H.

Proof. Let $u, v \in F$. Then there is some simple polygonal (u, v)-curve, say γ, contained in the face, f. Moreover, we may assume that γ does not meet any vertices from H other than its endpoints, and γ's intersection with each edge consists of an isolated point. Consider the finite multi-sequence of edges from H which intersect γ, beginning with the edge closest to u on γ, $(e_1, e_2, ..., e_n)$. For each $i \in \{1, ..., n\}$, let v_i be an endpoint of e_i contained in F (choosing arbitrarily if both endpoints of e_i lie in F). We want to show that $(u, v_1, ..., v_n, v)$ is a (u, v)-walk in H.

Suppose $1 \leq i \leq n - 1$. If we restrict γ to the curve between its identified intersection with e_i and e_{i+1}, then the interior of the resulting curve does not intersect any edges or vertices of H. Thus, its interior is contained in a single face of H. Since H is a maximal plane graph, there are three vertices incident with this face, and they form a clique in H. In particular, either $v_i = v_{i+1}$ or v_i is adjacent to v_{i+1}. Similar arguments show that u is adjacent to v_1 and v is adjacent to v_n. Therefore, $(u, v_1, ..., v_n, v)$ is a (u, v)-walk in H consisting entirely of vertices in F, so F is connected in H. ∎

Our characterization of k-blocking sets in one or two dimensions will depend upon the existence of a set of paths in G with certain properties. The next lemma will be useful in establishing this correspondence.

Let H be a simple, 2-connected graph embedded in the plane and let $C = (c_0, ..., c_{t-1})$ be the cycle in H which is peripheral to the unbounded face. Following Robertson and Seymour's treatment of the DISJOINT CONNECTING SUBGRAPHS problem in [17], for any 4-tuple, (i, j, i', j'), such that

$$0 \leq i < j \leq i' \leq j' \leq t - 1$$

we say that $\{c_i, c_{i'}\}$ **crosses** $\{c_j, c_{j'}\}$ in H. Notice that we allow the degenerate cases, where $i = i'$ or $j = j'$. For convenience, we may say that $\{a, b\}$ crosses $\{c, d\}$ in H without referring to indices. In this case, given an (a, b)-path, P, in H, it is a straightforward consequence of the Jordan Curve Theorem that any (c, d)-path must contain a vertex in P.

Lemma 2. *Let H be a 2-connected plane graph, let $C = (c_0, ..., c_{t-1})$ be the cycle in H which is peripheral to the unbounded face, and let $X \subseteq V(H)$. For $c_i, c_{i'} \in V(H) \backslash X$ such that $0 \leq i \leq i' \leq t - 1$, there is a $(c_i, c_{i'})$-path in $H - X$ if and only if for every j, j' such that $i \leq j \leq i' \leq j' \leq t - 1$, there is no $(c_j, c_{j'})$-path in $\widehat{H}[\overline{X}]$.*

Proof. The forward direction is a consequence of the Jordan Curve Theorem. For the backward direction, we prove the contrapositive. Suppose there is no

$(c_i, c_{i'})$-path in $H - X$. Define j so that c_{j-1} is the last vertex on the path, $(c_i, c_{i+1}..., c_{i'-1})$, which is in the same component of $H - X$ as c_i, and define j' so that $c_{j'+1}$ is the first vertex on the path, $(c_{i'+1}, c_{i'+2}..., c_i)$, which is in the same component of $H - X$ as c_i (here, our indices are modulo t). By definition, both $c_j, c_{j'} \in X$. We want to show that c_j and $c_{j'}$ lie in the same face of $\widehat{H} - \overline{X}$.

If not, there is some pair, $c_k, c_{k'} \in V(C) \cap (V(H) \backslash X)$, with

$$c_k \in \{c_{j+1}, c_{j+2}, ..., c_{j'-1}\} \text{ and } c_{k'} \in \{c_{j'+1}, c_{j'+2}, ..., c_{j-1}\},$$

such that there is a $(c_k, c_{k'})$-path in $H - X$ and $\{c_k, c_{k'}\}$ crosses $\{c_j, c_{j'}\}$ in H. We consider three cases.

Case 1: Suppose $k \in \{j+1, ..., i'-1\}$. If $k' \in \{j'+1, ..., i\}$, then $\{c_k, c_{k'}\}$ crosses $\{c_{j'+1}, c_i\}$. Therefore c_k is in the same component of $H - X$ as c_i, contradicting the maximality of $j - 1$. Otherwise, $k' \in \{i + 1, ..., j - 1\}$, and $\{c_k, c_{k'}\}$ crosses $\{c_{j-1}, c_i\}$. Again, c_k is in the same component of $H - X$ as c_i, contradicting the maximality of $j - 1$.

Case 2: Suppose $k \in \{i' + 1, ..., j'-1\}$. If $k' \in \{j'+1, ..., i\}$, then $\{c_k, c_{k'}\}$ crosses $\{c_{j'+1}, c_i\}$. Therefore c_k is in the same component of $H - X$ as c_i, contradicting the minimality of $j'+1$. Otherwise, $k' \in \{i, ..., k\}$, and $\{c_k, c_{k'}\}$ crosses $\{c_{j+1}, c_i\}$. Again, c_k is in the same component of $H - X$ as c_i, contradicting the minimality of $j' + 1$.

Case 3: Suppose $k = i'$. Then $\{c_k, c_{k'}\}$ crosses $\{c_{j-1}, c_{j'+1}\}$, so c_i is in the same component of $H - X$ as $c_{i'}$. This contradicts our assumption that there is no $(c_i, c_{i'})$-path in $H - X$.

Therefore, c_j and $c_{j'}$ lie in the same face, say f, of $\widehat{H} - \overline{X}$, where $\widehat{H}$ is a maximal plane graph. By lemma 1, the collection of vertices from $\overline{X}$ which lie in f is connected in $\widehat{H}$. In particular, there is a $(c_j, c_{j'})$-path in $\widehat{H}[\overline{X}]$. This completes the proof. ∎

We are now prepared to prove our main results.

2.2 One-Dimensional Blocking

For an $m \times n$-grid, G, define the plane graph, G', obtained from G by adding two new vertices, e and w, to the unbounded face so that e is adjacent to each vertex in $East$ and w is adjacent to each vertex in $West$. For each $B \subseteq V(G)$, we can think of $\widehat{G}[\overline{B}]$ as a subgraph of $\widehat{G'}[\overline{B}]$ because every bounded face of G is a bounded face of G'. Moreover, there are $k + 1$ vertex-disjoint $(East, West)$-path in $\widehat{G}[\overline{B}]$ just in case there are $k + 1$ internally vertex-disjoint (e, w)-path in $\widehat{G'}[B \cup \{e, w\}]$. Define $B' = B \cup \{e, w\}$.

Theorem 1. *Let $m, n > 0$, let G be the $m \times n$-grid and let B be a subset of vertices from G. For each $k \geq 0$, B is a $(North, South)$-k-blocking set if and only if $\widehat{G}[\overline{B}]$ contains $k + 1$ vertex-disjoint $(East, West)$-paths.*

Proof. Suppose B is a $(North, South)$-k-blocking set in G. For the sake of contradiction, suppose $\widehat{G'}[\overline{B'}]$ contains at most k vertex-disjoint (e, w)-paths. By Menger's Theorem, there is some $X \subseteq \overline{B}$ separating e and w in $\widehat{G'}[\overline{B'}]$ with $|X| \leq k$. We claim there is some $F \subseteq B$ separating e and w in $\widehat{G'}[\overline{B'}]$ with $|F| = |X|$.

We prove the claim by induction on the number of vertices in $X \backslash B$. Let X be a minimum size set of vertices separating e and w in $\widehat{G'}[\overline{B'}]$. If $|X \backslash B| = 0$, then $X \subseteq B$ and we are done. Otherwise, there is some face, f, of G' such that $v_f \in X \backslash B$. By the minimality of X, there is an (e, w)-path P in $\widehat{G'}[\overline{B'}] - (X \backslash \{v_f\})$. Moreover, it must be the case that $v_f \in V(P)$, and the two neighbors of v_f in P are not adjacent in $\widehat{G'}$.

Clearly, neither e nor w is peripheral to f since each such face is bounded by a 3-cycle. Therefore, f is a bounded face in G; let $C_f = (a, b, c, d)$ be the cycle peripheral to f. Without loss of generality, we may assume that a and c are the two neighbors of v_f in P and $b, d \notin B \backslash X$. Let $X' = X \backslash \{v_f\} \cup \{a\}$. Notice that v_f has degree 1 in $\widehat{G'}[\overline{B'}] - X'$. Therefore, no (e, w)-path in $\widehat{G'}[\overline{B'}] - X'$ uses v_f. By the choice of X, there is no (e, w)-path in $\widehat{G'}[\overline{B'}] - (X \backslash \{v_f\})$ which does not use v_f. Hence, there is no (e, w)-path in $\widehat{G'}[\overline{B'}] - X'$. Finally, $|X| = |X'|$ and $|X \backslash B| - 1 = |X' \backslash B|$. This completes the induction.

We have shown the existence of some $F \subseteq B$ separating e and w in $\widehat{G'}[\overline{B'}]$ with $|F| \leq k$. That is, there is no (e, w)-path in $\widehat{G'}[\overline{B' \backslash F}]$. But, since B is a $(North, South)$-k-blocking set, there is no $(North, South)$-path in $G' - (B' \backslash F)$. This contradicts Lemma 2.

For the backward direction, suppose $\widehat{G'}[\overline{B'}]$ contains $k + 1$ vertex disjoint (e, w)-paths. Then for any $F \subseteq B$ with $|F| \leq k$, there is an (e, w)-path in $\widehat{G'}[\overline{B' \backslash F}]$. By Lemma 2, there is no $(North, South)$-path in $G' - (B' \backslash F)$. Therefore, B is a $(North, South)$-k-blocking set. ∎

Theorem 2. *Let $m, n > 0$, let G be the $m \times n$-grid and let B be a subset of vertices from G. For each $k \geq 0$, one can decide whether B is a $(North, South)$-k-blocking set in $O(mn)$ time.*

Proof. Theorem 1 implies that B is a $(North, South)$-k-blocking set if and only if there are $k + 1$ internally vertex-disjoint (e, w)-paths in $\widehat{G'}[\overline{B'}]$. Letting ℓ equal the number of vertices in $\widehat{G'}[\overline{B'}]$, we can compute the vertex connectivity between two vertices in a planar graph in $O(\ell)$ time [2, 10]. Moreover, Euler's formula tells us that

$$\begin{aligned} \ell &\leq |V(\widehat{G'}) \backslash \{v_{f_\infty}\}| \\ &= |V(G')| + |F(G')| = |E(G')| + 2 \\ &= n(m-1) + m(n-1) + 2n + 2 \\ &= 2mn + n - m + 2. \end{aligned}$$

Here, $F(G')$ is the set of faces in G' and f_∞ is the unbounded face of G'. ∎

Define a weight function, σ, on $V(\widehat{G})$ so that

$$\sigma(v) = \begin{cases} 1, & v \in V(G) \\ 0, & v \in V(\widehat{G}) \backslash V(G). \end{cases}$$

From σ, we obtain a weight function, $\sigma_{\leftrightarrow}$, on the vertices of $\widehat{G}_{\leftrightarrow}$ and a distance function δ on the edges of $\widehat{G}_{\leftrightarrow}$, as described in the preliminaries. These functions are easily extended to $\widehat{G}'$ and $\widehat{G}'_{\leftrightarrow}$ by giving each new vertex weight zero.

Now suppose we are given some initial set of sensors in the grid. We are interested in placing additional sensors to obtain a $(North, South)$-k-blocking set. Moreover, we would like to minimize the number of sensors used to obtain this result. Let $B_0 \subseteq V(G)$ be a set of sensors initially placed on the grid. We define a new weight function, σ_0, by altering σ so that $\sigma_0(v) = 0$ for each $v \in B_0$. The functions $\sigma_{0\leftrightarrow}$ and δ_0 are defined naturally from these new weights.

Theorem 3. *Let $m, n > 0$, let G be the $m \times n$-grid and let $B_0 \subseteq V(G)$ be given weight zero. For each $k \geq 0$, there is an $O(kmn \log(mn))$ algorithm to find a minimum size set, $B_1 \subseteq V(G)\backslash B_0$, such that $B_0 \cup B_1$ is a $(North, South)$-k-blocking set.*

Proof. By Theorem 1, $B \subseteq V(G)$ is a $(North, South)$-k-blocking set if and only if there exist $k + 1$ internally-disjoint (e, w)-paths in $\widehat{G}'_{\leftrightarrow}$. Given the weight function, $\sigma_{0\leftrightarrow}$, on the edges of $\widehat{G}'_{\leftrightarrow}$, we can use Suurballe's algorithm [19] to find $k + 1$ internally vertex-disjoint (e, w)-paths of minimum total length. Since $\sigma(e), \sigma(w) = 0$, the total length of these $k+1$ paths will be equal to the number of vertices used which are in G and do not have a sensor placed on them. Let B be the set of vertices in these $k + 1$ internally vertex-disjoint paths, and let B_1 be the set of vertices in B which have weight 1. Then $\overline{B_0 \cup B_1} \supseteq B$ and $\widehat{G}[\overline{B_0 \cup B_1}]$ contains $k+1$ vertex-disjoint $(East, West)$-paths. Therefore, $B_0 \cup B_1$ is a $(North, South)$-k-blocking set. By construction, no set smaller than B_1 has this property.

Since $\widehat{G}_{\leftrightarrow}$ is a plane graph, we can use Borradaile and Klein's shortest directed path algorithm from [2] in the implementation of Suurballe's algorithm. Borradaile and Klein's algorithm runs in $O(mn \log(mn))$ time, and Suurballe requires $k + 1$ iterations. ∎

2.3   Two-Dimensional Blocking

It is straightforward to extend Theorems 1 and 2 to two dimensions. First we characterize the two-dimensional solution in terms of disjoint paths.

Theorem 4. *Let $m, n > 0$, let G be the $m \times n$-grid and let B be a subset of vertices from G. For each $k \geq 0$, B is k-blocking set if and only if $\widehat{G}[\overline{B}]$ contains $k+1$ vertex-disjoint $(North, South)$-paths and $k+1$ vertex-disjoint $(East, West)$-paths.*

From this, the solution to the decision version follows:

Theorem 5. *Let $m, n > 0$, let G be the $m \times n$-grid and let B be a subset of vertices from G. For each $k \geq 0$, one can decide whether B is a k-blocking set in $O(mn)$ time.*

Using the distance function, δ, defined on $\widehat{G}$, we now describe a property of a minimum weight 0-blocking set.

Lemma 3. *Let $m, n > 0$, let G be the $m \times n$-grid and let $B \subseteq V(G)$ be given weight zero. If B is a minimum weight 0-blocking set, then $\widehat{G}[\overline{B}]$ contains a tree, T, such that $B \subseteq V(T)$, and there exist special vertices $u, v \in V(T)$ (possibly $u = v$) such that T is the union of five shortest paths, a (u, v)-path, a $(North, \{u\})$-path, a $(South, \{v\})$-path, and either a $(East, \{u\})$-path and a $(West, \{v\})$-path or an $(East, \{v\})$-path and a $(West, \{u\})$-path.*

Proof. If B is a 0-blocking set, $\widehat{G}[\overline{B}]$ contains a $(North, South)$-path, say P, and an $(East, West)$-path, say Q. The endpoints of P cross the endpoints of Q; therefore, $S = P \cup Q$ is a connected graph. Moreover, S is a 0-blocking set, so by minimality, $B \subseteq V(S)$. Let P_1 be the subpath of P beginning with the initial vertex in $North$ and ending with the first vertex in $V(Q)$, say u. Let P_2 be the subpath of P beginning with the last vertex of P in $V(Q)$, say v, and ending with the terminal vertex in $South$. Either u occurs before v in Q, u occurs after v in Q or $u = v$.

In the first case, let Q_1 be the subpath of Q beginning with the initial vertex in $East$ and ending with u and let Q_2 be the subpath of Q beginning with v and ending with the terminal vertex in $West$. Let R be the subpath of P beginning with u and ending with v. Then $S' = P_1 \cup P_2 \cup Q_1 \cup Q_2 \cup R$ is a subgraph of S and is also a 0-blocking set. By minimality, $V(S') \cap B = V(S) \cap B$. Moreover, their definition ensures that these subpaths are internally disjoint. The minimality of B implies that P_1 is a shortest $(North, \{u\})$-path, P_2 is a shortest $(\{v\}, South)$-path, Q_1 is a shortest $(East, \{u\})$-path, Q_2 is a shortest $(\{v\}, West)$-path and R is a shortest (u, v)-path. The second and third cases follow similarly. ∎

We use the above characterization to describe an algorithm for finding special vertices and a minimum 0-blocking set.

Algorithm: $\mathcal{A}_1$, Minimum sensor 0-blocking in G.

Input: Fixed integers, $m, n > 0$, the plane graph $\widehat{G}$ obtained from the $m \times n$-grid, G, and a set of vertices $B_0 \subseteq V(G)$.

Initialization: Order the vertices of $\widehat{G}$, $\{u_1, ..., u_s\}$. Here $s = |V(\widehat{G})|$. Define the distance function, δ_0, on $\widehat{G}$.

1. For $i = 1, ..., s$:
   (a) Run the single-source shortest path algorithm from [10] on $\widehat{G}$ with source, u_i, obtaining a weighted distance, $\delta_0(u_i, v)$, for each $v \in V(\widehat{G})$.

(b) Set

$$r_N(u_i) = \min\{\delta_0(u_i, v) : v \in North\},$$
$$r_S(u_i) = \min\{\delta_0(u_i, v) : v \in South\},$$
$$r_E(u_i) = \min\{\delta_0(u_i, v) : v \in East\},$$
$$r_W(u_i) = \min\{\delta_0(u_i, v) : v \in West\}.$$

2. For $i = 1, ..., s$ and $j = i, ..., s$:
   (a) Set

$$R_{NE}(i, j) = r_N(u_i) + r_E(u_i) + r_S(u_j) + r_W(u_j) + \delta_0(u_i, u_j),$$
$$R_{NW}(i, j) = r_N(u_i) + r_W(u_i) + r_S(u_j) + r_E(u_j) + \delta_0(u_i, u_j),$$
$$R_{SE}(i, j) = r_S(u_i) + r_E(u_i) + r_N(u_j) + r_W(u_j) + \delta_0(u_i, u_j),$$
$$R_{SW}(i, j) = r_S(u_i) + r_W(u_i) + r_N(u_j) + r_E(u_j) + \delta_0(u_i, u_j).$$

   (b) Set

$$R(i, j) = \min\{R_{NE}(i, j), R_{NW}(i, j), R_{SE}(i, j), R_{SW}(i, j)\}$$
$$- 2\sigma_0(u_i) - 2\sigma_0(u_j)$$

   and set $D(i, j) = (X, Y) \in \{North, South\} \times \{East, West\}$ such that $R(i, j) = R_{XY}(i, j)$.
3. Set $\rho(G) = \min\{R(i, j) : 1 \leq i \leq j \leq s\}$.
4. Set $(\alpha, \beta) = \min\{(i, j) : R(i, j) = \rho(G)\}$, given the lexicographic ordering on tuples.
5. Run the single-source shortest path algorithm from [10] on $\widehat{G}$ to find shortest paths for $(u_\alpha, X), (u_\alpha, Y), (u_\beta, X^c), (u_\beta, Y^c)$ and (u_α, u_β), where $D(\alpha, \beta) = (X, Y)$. Here, if $X = North$, then $X^c = South$; if $Y = West$, $Y^c = East$, etc. Stop.

Output: T, the graph obtained from the union of the five shortest paths found in step 5. $\rho(G)$, which gives the number of vertices in $(T \cap V(G))\backslash(B_0)$.

Theorem 6. *Let $m, n > 0$, let G be the the $m \times n$-grid and let $B_0 \subseteq V(G)$ contain sensors. There is an $O(m^2n^2)$ time algorithm for finding a minimum size set that extends B_0 to a 0-blocking set.*

Proof. By Theorem 1, the set, $V(T)$, output by algorithm, $\mathcal{A}_1$ is a 0-blocking set. By Lemma 3, no 0-blocking set has smaller size. Step 1 of $\mathcal{A}_1$ consists of $s = 2mn + n - m + 2$ iterations of the shortest-path algorithm in [10], which runs in $O(mn)$ time. Thus, step 1 runs in $O(m^2n^2)$ time. Step 2 is iterated $\binom{s}{2} + s$ times, running in $O(m^2n^2)$ time. Step 5 is completed in $O(mn)$ time. Therefore, algorithm $\mathcal{A}_1$ runs in $O(m^2n^2)$ time. ∎

While a characterization similar to that Lemma 3 for k-blocking sets $(k > 0)$ is easily derived, unfortunately it does not immediately lead to a polynomial time algorithm for finding the optimal placement.

Instead we describe an efficient 2-approximation algorithm for this case. The graph used is $\widehat{G}$ with weights on the vertices are as above.

1. Using Suurballe's algorithm (with the optimization by Borradaile and Klein) find $k + 1$ disjoint paths of minimum total weight connecting $East$ to $West$ (adding a start node with weight 0 attached to all of the nodes in $East$ and a finish node with weight 0 attached to all of the nodes in $West$). Let those paths be $EW_0, \ldots, EW_k$ with total cost ew.
2. Using Suurballe's algorithm find $k + 1$ disjoint paths of minimum total weight connecting $North$ to $South$. Let those paths be $NS_0, \ldots, NS_k$ with cost ns.
3. Return the combination of paths $EW_0, \ldots, EW_k$ and $NS_0, \ldots, NS_k$ with total weight at most $ew + ns$.

Theorem 7. *Let $m, n > 0$ and let G be the $m \times n$-grid and let $B_0 \subseteq V(G)$ contain sensors. There is an $O(kmn \log(mn))$ algorithm for finding a set of vertices that extends B_0 to a k-blocking set and that is within a factor of 2 of optimal in size.*

Proof. By Theorem 4 and by construction the set output by the algorithm above is a k-blocking set and it clearly runs in $O(kmn \log(mn))$ time.

Let the value of the optimal solution by OPT. Observe that $OPT \geq ns$. This follows from the fact that the optimal solution must contain $k + 1$ disjoint paths from $North$ to $South$ and therefore must have cost at least ns (which is optimal). Similarly, we have $OPT \geq ew$. If follows that the value of our solution is at most $ew + ns \leq 2 \cdot OPT$. ∎

3   Extensions and Open Problems

Lemmas 1 and 2 are written in such generality as to allow us to easily extend our results to other domains and problems. For example, the original (planar) domain need not be rectangular and may contain "holes" representing impassable regions. The attacks detected need not be North-South or East-West paths but an intruder may enter at any contiguous portion of the border of the region and exit any other (disjoint) contiguous region. Multiple such disjoint attacks may be tested for simultaneously generalizing the results of Theorems 5, 6 and 7.

Two major open problems come to mind. The first is extending the result of Theorem 6 to $k > 0$ faults. While it is easy to generalize the characterization given in Lemma 3 for $k = 0$ faults to the case $k > 0$, it is not obvious that this results in a polynomial time algorithm. To make it effective, it seems that one must solve the minimum sum t vertex disjoints paths problem. In particular, a polynomial time solution to that problem would be sufficient to solve the two-dimensional k-protection placement problem in time $O((mn)^{2(k+1)^2})$ using an algorithm analogous to $\mathcal{A}_1$. It is known that for variable t this problem is NP-complete [12]. On the other hand, for fixed t the problem of deciding if the paths exist is in P [18]. While some progress has been made on this question [3], it remains open.

The second problem involves the case of movable sensors. Instead of replacing faulty sensors with new sensors, what if one was allowed to move non-faulty sensors to new points in the grid. The question of moving the least number of sensors the least total distance or the minimum maximum distance may be of interest. Related problems concerning coverage appear to be NP-hard [8]. An experimental study of a greedy strategy for this problem appears in [20].

References

1. Balister, P., Bollobas, B., Sarkar, A., Kumar, S.: Reliable density estimates for coverage and connectivity in thin strips of finite length. In: Proceedings of MobiCom 2007, pp. 75–86. ACM (2007)
2. Borradaile, G., Klein, P.: An $O(n \log n)$ algorithm for maximum st-flow in a directed planar graph. J. ACM **56**(2), 30 (2009). Art. 9
3. Borradaile, G., Nayyeri, A., Zafarani, F.: Towards single face shortest vertex-disjoint paths in undirected planar graphs. In: Bansal, N., Finocchi, I. (eds.) ESA 2015. LNCS, pp. 227–238. Springer, Heidelberg (2015)
4. Chen, A., Lai, T.H., Xuan, D.: Measuring and guaranteeing quality of barrier-coverage in wireless sensor networks. In: Proceedings of the 9th ACM International Symposium on Mobile Ad Hoc Networking and Computing, pp. 421–430. ACM (2008)
5. Chen, D.Z., Gu, Y., Li, J., Wang, H.: Algorithms on minimizing the maximum sensor movement for barrier coverage of a linear domain. In: Fomin, F.V., Kaski, P. (eds.) SWAT 2012. LNCS, vol. 7357, pp. 177–188. Springer, Heidelberg (2012)
6. Czyzowicz, J., Kranakis, E., Krizanc, D., Lambadaris, I., Narayanan, L., Opatrny, J., Stacho, L., Urrutia, J., Yazdani, M.: On minimizing the maximum sensor movement for barrier coverage of a line segment. In: Ruiz, P.M., Garcia-Luna-Aceves, J.J. (eds.) ADHOC-NOW 2009. LNCS, vol. 5793, pp. 194–212. Springer, Heidelberg (2009)
7. Czyzowicz, J., Kranakis, E., Krizanc, D., Lambadaris, I., Narayanan, L., Opatrny, J., Stacho, L., Urrutia, J., Yazdani, M.: On minimizing the sum of sensor movements for barrier coverage of a line segment. In: Wu, K., Nikolaidis, I. (eds.) ADHOC-NOW 2010. LNCS, vol. 6288, pp. 29–42. Springer, Heidelberg (2010)
8. Dobrev, S.: Personal communication
9. Dobrev, S., Durocher, S., Eftekhari, M., Georgiou, K., Kranakis, E., Krizanc, D., Narayanan, L., Opatrny, J., Shende, S., Urrutia, J.: Complexity of barrier coverage with relocatable sensors in the plane. In: Spirakis, P.G., Serna, M. (eds.) CIAC 2013. LNCS, vol. 7878, pp. 170–182. Springer, Heidelberg (2013)
10. Henzinger, M.R., Klein, P., Rao, S., Subramanian, S.: Faster shortest-path algorithms for planar graphs. J. Comput. Syst. Sci. **55**(1, part 1), 3–23 (1997). 26th Annual ACM Symposium on the Theory of Computing (STOC 1994) (Montreal, PQ, 1994)
11. Huang, C.F., Tseng, Y.C.: The coverage problem in a wireless sensor network. In: Proceedings of the 2nd ACM International Conference on Wireless Sensor Networks and Applications, WSNA 2003, pp. 115–121. ACM (2003)
12. Karp, R.: On the complexity of combinatorial problems. Networks **5**, 45–68 (1975)
13. Kumar, S., Lai, T.H., Arora, A.: Barrier coverage with wireless sensors. In: Proceedings of MobiCom 2005, pp. 284–298. ACM (2005)
14. Meguerdichian, S., Koushanfar, F., Potkonjak, M., Srivastava, M.B.: Coverage problems in wireless ad-hoc sensor networks. In: Proceedings of INFOCOM, vol. 3, pp. 1380–1387 (2001)

15. Mehrandish, M., Narayanan, L., Opatrny, J.: Minimizing the number of sensors moved on line barriers. In: Proceedings of IEEE WCNC 2011, pp. 1464–1469 (2011)
16. Park, T., Shi, H.: Extending the lifetime of barrier coverage by adding sensors to a bottleneck region. In: 12th IEEE Consumer Communications and Networking Conference (CCNC). IEEE (2015)
17. Robertson, N., Seymour, P.D.: Graph minors. VI. Disjoint paths across a disc. J. Comb. Theory Ser. B **41**(1), 115–138 (1986)
18. Robertson, N., Seymour, P.D.: Graph minors. XIII. The disjoint paths problem. J. Comb. Theory Ser. B **63**, 65–110 (1995)
19. Suurballe, J.W.: Disjoint paths in a network. Networks **4**, 125–145 (1974)
20. Xie, H., Li, M., Wang, W., Wang, C., Li, X., Zhang, Y.: Minimal patching barrier healing strategy for barrier coverage in hybrid wsns. In: International Symposium on Personal, Indoor, and Mobile Radio Communication (PIMRC), pp. 1558–1563. IEEE (2014)

The Weakest Oracle for Symmetric Consensus in Population Protocols

Joffroy Beauquier[1], Peva Blanchard[2(✉)], Janna Burman[1], and Shay Kutten[3]

[1] LRI, Paris-South 11 University, Orsay, France
{joffroy.beauquier,janna.burman}@lri.fr
[2] LPD, EPFL, Lausanne, Switzerland
peva.blanchard@epfl.ch
[3] Technion, Haifa, Israel
kutten@ie.technion.ac.il

Abstract. We consider the *symmetric consensus* problem, a version of consensus adapted to *population protocols*, a model for large scale networks of resource-limited mobile sensors. After proving that consensus is impossible in the considered model, we look for *oracles* to circumvent this impossibility. An oracle is an external (to the system) module providing some information allowing to solve the problem. We define a class of oracles adapted to population protocols, and we prove that an oracle in this class, namely *DejaVu*, allows to obtain a solution. Finally, and this is the major contribution of the paper, we prove that *DejaVu* is the *weakest* oracle for solving the problem.

Keywords: Networks of mobile sensors · Population protocols · Consensus · Oracles · Weakest oracle

1 Introduction

Consensus is a classical decision problem in distributed computing. In this problem, each process is given initially a value and has to take eventually an irreversible decision (*termination* condition). Processes must decide on a common value depending on the input values, according to *agreement* and *validity (nontriviality)* conditions [3,17,25].

Consensus-related problems are relevant to mobile sensor networks in many different contexts like, for example, flocking (see, e.g., [11]), swarm formation control (see, e.g., [26]), distributed sensor fusion (see, e.g., [22]) and attitude alignment (see, e.g., [16]). See also [21,23,24] for surveys and references on consensus-related problems in mobile wireless networks.

A fundamental result by Fisher, Lynch and Paterson [14] states that in the classical asynchronous message passing model, no deterministic algorithm for consensus exists, even in the case of a unique possible crash (halting) failure.

This work has been partially supported by the Israeli-French Maimonide and the INS2I PEPS JCJC research projects.

© Springer International Publishing Switzerland 2015
P. Bose et al. (Eds.): ALGOSENSORS 2015, LNCS 9536, pp. 41–56, 2015.
DOI: 10.1007/978-3-319-28472-9_4

It is not surprising that the same impossibility holds in the model of population protocols [2]. This model is fundamentally asynchronous, which is also one of the main reasons for the result in [14]. However, some inherent characteristics of population protocols make consensus even more difficult. The *agents*, i.e. the mobile processes in population protocols, are *uniform*, i.e. indistinguishable and executing the same code. They have a constant memory size and thus cannot neither obtain nor store labels or any other information depending on the network size. Agents communicate by asynchronous pair-wise interactions. No broadcast communication is available. Due to all these limitations, the agents are unable to detect which other agents are present but not interacting, even if no crash failure is possible. Hence, in population protocols, even with the assumption of absence of failures, consensus is impossible (Sect. 3).

Like in the message passing model, it seems interesting to study what is missing for solving consensus in population protocols. We adopt the point of view of Chandra and Toueg [10] for defining the possible missing information under the form of oracles, i.e., specific behaviours. Recall that an oracle can be thought as a collection of modules able to provide each process with some information, hopefully useful to solve a given problem. The *failure detectors* [10] are oracles that usually provide each process with failure-related information. In our case, such information seems meaningless since consensus is impossible to solve even without any failure. Moreover, the failure detectors of [10] cannot be used in our case, because they provide lists of process identifiers (estimated to have crashed). As already mentioned, identifiers are absent in population protocols (due to the constant size memory requirement).

Several identity-free oracles exist, though, in the literature. The failure detector introduced in [12] outputs a boolean value at every process, and solves the $(n-1)$-set agreement problem in n-process message passing system. However, this boolean value is mainly used to indicate wether or not the process is "alone", i.e., the other processes have all crashed. Thus, this failure detector does not fit in our case since we do not consider crash failures. Another type of oracles proposed in [19,20] (and used, e.g., in [5,6]) to deal with anonymity, provides information on the number of crashed processes (bounded by $f < n$), and, for the same reason of constant agent state space, cannot be used in the framework of population protocols. A so called "heartbeat" failure detector proposed in [1] requires to maintain unbounded counters at every process, and thus, again, is not suitable in our case. Some other failure detectors used to solve consensus and adapted to anonymous systems, like $A\Omega$ [6], $A\mathcal{L}$ and $A\Sigma'$ [7,8], provide, roughly speaking, information about the number of correct processes, thus breaking the requirement of constant memory. In addition, in message-passing system, these oracles are used in combination with other powerful model assumptions and capabilities (e.g., "terminating" broadcast, unbounded process memory, etc.) which are unavailable in our case.

Thus, defining oracles in the context of population protocols appears as a real challenge. Several attempts have already been made, like the "eventual leader detector" of [13] (generalized in [4]). This oracle is useful to solve self-stabilizing leader election but, intuitively, is not reliable enough for solving the terminating consensus problem. Another interesting oracle proposed in [18] provides a "cover-time service" in the sense that a state hopping from agents to agents can know when it has covered the whole population. Based on this service, the authors propose an abstract oracle, namely an "absence detector", which is able to notify a specific agent about the absence or presence of some states in the population. This abstraction is adapted to study the computational power[1] of population protocols augmented with a cover-time service. Yet, this formulation is not helpful enough to assess the weakest oracle for a given task.

Due to all the above-mentioned reasons, we introduce a new type of oracles. The constraints we had in mind, when designing these oracles, are basically to make them implementable with minimum external assumptions on the system. Each oracle in the proposed class is distributed: it consists of a collection of local modules mapped onto the agents. Each agent's local module provides information only related to the past schedule, and is independent of the protocol being run (somewhat similarly to the classical failure detectors [10]). To communicate this information, the agent's local module outputs a boolean value (as the failure detector in [12]).

Moreover, the proposed oracles are unreliable in the sense that the local modules are not required to provide this information at the precise time when it appears, but only *eventually*, *at least once* and *at least in one agent*. That is, the local modules may be very slow for some agents, and even completely dysfunctional (i.e., providing no information) for some others.

Finally, each oracle in our class is *anonymity-compliant*, in the sense that the information provided by the local modules does not depend on how these modules are mapped onto the agents. Roughly speaking, a permutation of the agents does not affect the possible output of these oracles (see Sect. 2.4 for precise definitions).

Besides this new class of oracles and the appropriately adapted computational model (Sect. 2), the paper presents three results. The first result (Sect. 3) states that consensus is impossible without an oracle. The second contribution (Sect. 4) is the presentation of an oracle in the class, called *DejaVu*, allowing a solution. These two results are relatively easy. The third result is intricate and is the main contribution of the paper (Sect. 5). It states that the proposed oracle is the *weakest* (see Sect. 2.5) for solving a *symmetric* version of consensus, a version adapted to population protocols (see Sect. 2.6). Intuitively, *DejaVu* being the weakest oracle means that it provides the minimum required information for solving the problem (among all the oracles in the proposed oracle class).

2   Model and Definitions

2.1   Population Protocol

Here, we use the definitions of [2] with some slight modifications. A network is represented by a directed graph $G = (V, \mathcal{E})$ with n vertices and no

---

[1] The class of functions computable by a terminating protocol.

multi-edges nor self-loops. Each vertex represents a finite-state sensing device called an *agent*, and an edge $(u, v) \in \mathcal{E}$ indicates the possibility of a communication (meeting/interaction) between two distinct nodes u and v in which u plays the role of the *initiator* and v of the *responder*. The orientation of an edge corresponds to this asymmetry in roles. We often write G instead of V to refer to the vertices of G. In this paper, for the sake of simplicity, we consider only *bidirectional complete* graphs: for any two vertices u, v there is an edge from u to v, and an edge from v to u. An edge e *involves* an agent u if u is an endpoint of e. Two edges are *independent* if they involve no common agent. Otherwise, they are *dependent*. To deal with permutation of agents, we also introduce the group $\mathfrak{S}G$ of permutations of the vertices of G.

A *population protocol* $\mathcal{A}(\mathcal{Q}, \mathcal{I}, Init, Input, \delta)$ consists of a finite *state space* $\mathcal{Q}$, a set $\mathcal{I}$ of *initial values*, a set $Input$ of *input values*, an *initialization map* $Init : \mathcal{I} \to \mathcal{Q}$, and a transition function $\delta : (\mathcal{Q} \times Input)^2 \to \mathcal{Q}^2$ that maps any tuple (q_1, v_1, q_2, v_2) to an element $\delta(q_1, v_1, q_2, v_2)$ in $\mathcal{Q}^2$. A *(transition) rule* of the protocol is a tuple $(q_1, v_1, q_2, v_2, q_1', q_2')$ where $(q_1', q_2') = \delta(q_1, v_1, q_2, v_2)$ and is denoted by $(q_1, v_1)(q_2, v_2) \to (q_1', q_2')$. We refer to $(q_1, v_1)(q_2, v_2)$ (resp. (q_1', q_2')) as the left (resp. right) part of the rule. Note that we only consider *deterministic* population protocols (the right part is uniquely determined by the left part of the rule).

Intuitively, the input values in $Input$ are provided to the agents by some external device like an oracle, continuously along the protocol execution. Besides, the initial values will correspond to the initial values to the consensus instance.

At the beginning of an execution, every agent is assigned an initial value from the set $\mathcal{I}$, formalized by the initialization map. In [2] a population protocol is defined to compute a function of such initial values. In our case, the initial values represent the initial values in the consensus problem. Note that consensus, *a priori*, cannot be represented as a function of input values only (its output depends on the schedule too).

2.2   Schedules and Histories

The characterization of the weakest oracle for consensus relies on a tight analysis of the causal order [15]. In contrast to the usual presentation of schedules as words of events, the following definition embeds an explicit formulation of the causal structure of the events. Formally, a *schedule* S is a (possibly infinite) sequence $E_1|E_2|E_3|\ldots$ where the *Cartier-Foata condition* [9] holds: *(i)* every E_i is a subset of independent edges, *(ii)* for every i, every edge in E_{i+1} depends on some edge in E_i. The empty schedule is denoted by ϵ. When we write $E_1|E_2|\ldots$, it is implicitly assumed that the sequence satisfies the Cartier-Foata condition; otherwise we simply write $E_1, E_2, \ldots$. The *support of a schedule* S, denoted by $supp(S)$, is the subset of agents in G involved in the schedule S. Given a permutation $\alpha \in \mathfrak{S}G$, we write $\alpha S = \alpha E_1|\alpha E_2|\ldots$ where $\alpha E_i = \{\alpha(e), e \in E_i\}$.

A schedule $S = E_1|E_2|\ldots$ is a *factor* of a schedule $S' = E_1'|E_2'|\ldots$ if there exists $i \geq 1$ such that $E_1 \subseteq E_i'$, $E_2 \subseteq E_{i+1}'$, and so on. We also say that S'

contains S. If $i = 1$ above, and if S is finite, then S is a *prefix* of S', and S' is an *extension* of S.

An *event in* S is a pair $p = (i, e)$ such that $e \in E_i$. We denote by $\mathbb{P}(S)$ the set of events in S, which is naturally endowed with a partial order $\leadsto$ defined as the reflexive transitive closure of: $(i, e) \leadsto (i + 1, e')$ if and only if e, e' are dependent. Intuitively, the relation $\leadsto$ encodes the causal order of events during S. We often write e without mentioning the index i to refer to the event (i, e) when it is clear from the context.

The *causal past of* $p = (k, e)$ *in* S is the set $Past(p, S) = \{p' \in \mathbb{P}(S), p' \leadsto p\}$. We can equivalently write $Past(p, S) = F_1 | \ldots | F_k$, where $F_i = \{f, (i, f) \leadsto (k, e)\}$; note that necessarily $F_k = \{e\}$. Since these two presentations are equivalent, we will use the same notation to refer to them. Also, when S is clear from the context, we simply write $Past(p)$.

A finite schedule K is a *past cone* if there exists an event p in K such that all the events in K are in the causal past of p in K. If p involves an agent x, K is said to be a *past cone at* x.

Given a finite schedule S, an event $p = (i, (x, y))$, and an agent $z \neq x$, we say that x *meets indirectly* z at p in S if z is involved in some event in the causal past $past(p, S)$.

An infinite schedule S is *weakly fair* if and only if, for any pair of agents (x, y), there are infinitely many factors K of S such that x meets indirectly y at some event in K. Intuitively, this means that for any pair (x, y) the agent y causally influences the agent x infinitely often. Unless stated otherwise, every infinite schedule in this paper is assumed to be weakly fair.

A *history* $H = (S, h)$ with values in the set $Input$ is a schedule S together with a function h that associates with every event $p = (i, (x, y))$, a pair (v_x, v_y) of values from $Input$. The value v_x (resp. v_y) is the *history value*[2] at x *(resp. at y) in event p.* We also say that the history H *outputs the value* v_x (resp. v_y) at x (resp. y) during p. The schedule S is the *underlying schedule* of the history. The *support of* H is the support of its underlying schedule, $supp(H) = supp(S)$. Given a permutation $\alpha \in \mathfrak{S}G$, we write $\alpha H = (\alpha S, h')$ where $h'(i, \alpha(e)) = h(i, e)$. Finally, given a factor L of the schedule S, we denote by $H|_L = (L, h'')$ the *restriction of H to L*, where h'' is the restriction of h to the events occurring in L.

2.3   Executions

Consider a (deterministic) population protocol $\mathcal{A}$ with state space $\mathcal{Q}$, input values set $Input$, and initial values set $\mathcal{I}$. A *configuration* γ is a function that associates with every agent x a state $\gamma(x)$ in $\mathcal{Q}$. For every assignment κ of initial values from $\mathcal{I}$ to the agents of the graph, we define the corresponding *initial configuration* $\gamma_\kappa = Init \circ \kappa$, where $Init$ is the initialization map of the protocol.

Let γ, γ' be configurations, E a subset of independent edges, and h a function labeling each edge in E with a pair of values from $Input$. We write $\gamma \xrightarrow{E, h} \gamma'$ when, for every edge $(x, y) \in E$ such that $h(x, y) = (v_x, v_y)$,

---

[2] These values will later be provided by an oracle.

$(\gamma(x), v_x)(\gamma(y), v_y) \rightarrow (\gamma'(x), \gamma'(y))$ is a rule of the protocol. In other words, γ' is the configuration that results from γ by applying the labeled events (E, h). Generally speaking, h represents the values provided by the environment as input to the protocol (the information provided by an oracle, in our case) during the corresponding transitions. We use these values to model the oracle output (Sect. 2.4). Note also that, since the edges are independent, the above definition is consistent.

An *execution* is a sequence $\gamma_0 \xrightarrow{E_0, h_0} \gamma_1 \xrightarrow{E_1, h_1} \ldots$ such that the sequence $E_0, E_1, \ldots$ satisfy the Cartier-Foata condition (i.e. it forms a schedule). The sequences $S = E_0 | E_1 \ldots$ and $h_0, h_1, \ldots$ naturally yield a history $H = (S, h)$ where $h(i, e) = h_i(e)$. Since we deal with deterministic population protocols, an execution is entirely determined by a history H and an initial configuration γ_0. Hence we can denote an execution by $H[\gamma_0]$. The history H is referred to as the *input history* of the execution $H[\gamma_0]$.

Consider an output map $Out : Q \rightarrow R$. The *output history* of the execution $H[\gamma_0]$ is the history $H' = (S, h')$ (with the same schedule as H) where h' associates with every event $e_i = (x, y) \in E_i$ the pair $(Out(\gamma_i(x)), Out(\gamma_i(y)))$. Intuitively, the output history is the history produced by the protocol during the execution.

2.4   Oracles

In the following, we define oracles having in mind the classical failure detectors of [10]. Informally, we think of an oracle as a collection of local modules, each of them being attached to an agent. These modules may be different and mapped onto the agents arbitrarily. Each module looks for specific predefined patterns of meetings in the causal past of an agent, and notifies the agent accordingly. A module never notifies wrongly an agent, but the notification can be arbitrarily delayed. Some modules may even never deliver their notifications.

More formally, an oracle is a function that associates with each graph G, each permutation $\sigma \in \mathfrak{S}G$ and each weakly fair schedule S on G, a set $O(G, \sigma, S)$ of *legal histories* that take values in $\{0, 1\}$ and all have S as their underlying schedule. These are the legal histories when the local modules are mapped to the agents according to σ.

More precisely, each oracle O is specified by a family of (possibly empty) sets, $Cones(O, G, \sigma, x)$, of finite schedules, for every complete graph G, every permutation σ of the vertices and every agent x in G. The set $Cones(O, G, \sigma, x)$ represents the patterns that will be looked for in the causal past of x. These sets satisfy: *i. (cone)* every schedule K in $Cones(O, G, \sigma, x)$ is a past cone at x; *ii. (anonymity-compliance)* for every permutation $\alpha \in \mathfrak{S}G$ of the agents, $K \in Cones(O, G, \sigma, x)$ if and only if $\alpha K \in Cones(O, G, \alpha\sigma, \alpha(x))$.

The anonymity-compliance condition ensures that the set of cones detectable at agent x in the original permutation σ ($K \in Cones(O, G, \sigma, x)$) is exactly the set of cones detectable at agent $\alpha(x)$ in the new permutation $\alpha\sigma$, up to a relabeling of the agents ($\alpha K \in Cones(O, G, \alpha\sigma, \alpha(x))$).

Then, a history H is a legal history of G with schedule S given the permutation σ (i.e. $H \in O(G, \sigma, S)$) if and only if $i.$ *(safety)* if H outputs 1 at x in some event p in S, then the causal past $Past(p, S)$ contains some schedule from $Cones(O, G, \sigma, x)$; $ii.$ *(liveness)* if the set of agents x, whose causal past contains at some point a schedule from $Cones(O, G, \sigma, x)$, is not empty, then the history H eventually outputs 1 at least one of these agents in some event during S.

Intuitively, the safety property ensures that if O outputs 1 at x, then the corresponding prefix actually contains a schedule from $Cones(O, G, \sigma, x)$. The liveness property ensures that *at least one* agent is eventually notified about this fact.

Note that the set $Cones(O, G, \sigma, x)$ may be empty, which means that it is possible, *a priori*, for O to permanently output 0 at x. In particular, if all the defined sets $Cones(\dots)$ are empty, then the corresponding oracle always outputs 0 at every agent; which basically amounts to having no oracle at all, since the agents get no useful information.

2.5   Comparison Between Oracles

We say that a protocol $\mathcal{A}$ *uses an oracle* O when the only considered executions of $\mathcal{A}$ are those whose input histories are legal histories of O.

Intuitively, an oracle O_1 is weaker than an oracle O_2 if there exists a population protocol that simulates a history of O_1 using O_2. Formally, an oracle O_1 is *weaker* than an oracle O_2 if there exists a population protocol $\mathcal{A}$ and an output map such that, for every execution $H[\gamma_0]$, H being a legal history of O_2, the corresponding output history H' is a legal history of O_1.

Stating that $H = (S, h)$ and $H' = (S, h')$ are legal histories of O_2 and O_1, respectively, means that there exist permutations σ and τ such that $H \in O_2(G, \sigma, S)$ and $H' \in O_1(G, \tau, S)$. However, the definition does not force σ and τ to be equal. Intuitively, it means that the history computed by the emulation provided by $\mathcal{A}$ is a legal history of O_1 up to a permutation of the local modules.

Given a family $\mathcal{F}$ of oracles, a *weakest oracle in* $\mathcal{F}$ is an oracle O that is weaker than every oracle in $\mathcal{F}$. Note that there is no evidence *a priori* that a weakest oracle exists. Note also that all the weakest oracles, if they exist, are equivalent.

2.6   Symmetric Consensus

Consider a population protocol $\mathcal{A}$ with initial values $\mathcal{I}$. We assume that the agents have an instruction *decide* which causes them to decide irreversibly on some value in $\mathcal{I}$.

The population protocol $\mathcal{A}$ (possibly using an oracle) is said to solve the *consensus* problem if, for each complete graph G, for each initial configuration γ, for any legal execution $H[\gamma]$, it satisfies: $i.$ *(termination)* every agent eventually decides in the execution $H[\gamma]$; $ii.$ *(agreement)* two agents cannot decide on different values; $iii.$ *(validity)* if all the agents have the same initial value v, then an agent can only decide on v.

The protocol $\mathcal{A}$ is said to solve the *symmetric consensus* problem if it solves the consensus problem and, in addition, for each complete graph G, it satisfies an additional condition: *iv. (symmetry)* for any legal execution $H[\gamma]$, for any permutation $\alpha \in \mathfrak{S}G$ of the vertices, any agent decides on the same value in the execution $H[\gamma]$ and in $H[\gamma\alpha]$.

Intuitively, in the symmetric consensus, the decision value in an execution does not depend on the distribution of the initial values between the agents. Note that the condition of symmetry is quite natural for population protocols. In the seminal paper by Angluin et al. [2], the same condition applies to the predicates that are computable.

3   Impossibility of Consensus Without Oracle

We first show that the consensus problem is impossible without an oracle. In particular, the symmetric consensus problem is also impossible. The proof relies on the well-known partitioning argument.

Before proceeding, we introduce a useful notation. Given two schedules $S = E_1|E_2|\ldots$ and $S' = E_1'|E_2'|\ldots$ such that each $E_i \cup E_i'$ is a subset of independent edges, we denote by $S \cup S'$ the schedule $E_1 \cup E_1'|E_2 \cup E_2'|\ldots$.

Proposition 1. *Under weak fairness, there is no population protocol that solves the consensus problem over complete graphs.*

Proof. Assume that there exists a population protocol $\mathcal{A}$ that solves consensus over all complete graphs. Pick a complete graph G of $2 \cdot n$ agents (vertices), and select two complete subgraphs G_0, G_1 of n agents each. Let γ be the initial configuration of $\mathcal{A}$ corresponding to the agents in G_0 (resp. G_1) having the initial value 0 (resp. 1). Let S_v be a weakly fair schedule over G_v. By the validity condition of the consensus problem, in the execution $S_v[\gamma]$, all agents in G_v decide on the value v. Let S_v' be a finite prefix of S_v such that all the agents in G_v decide (on v) in the finite execution $S_v'[\gamma]$. Let S'' be any weakly fair extension of the schedule $S_0' \cup S_1'$. This last schedule is well-defined since the two graphs are disjoint. Then, in the execution $S''[\gamma]$, the agents in G_0 decide on 0, and the agents in G_1 decide on 1; whence a contradiction with the agreement condition. $\square$

4   Symmetric Consensus with *DejaVu*

Proposition 1 motivates the use of oracles. We define a particular oracle, called *DejaVu*. Intuitively, oracle *DejaVu* outputs 1 at some agent x only when x has indirectly met (see definition in Sect. 2.2) every other agent at least once. Formally, a schedule K belongs to $Cones(DejaVu, G, \sigma, x)$ if and only if K is a past cone at x and $supp(K) = G$. The legal histories of *DejaVu* are then defined according to the oracle rules.

The purpose of this section is to show that *DejaVu* is sufficient to solve symmetric consensus. A simple protocol using *DejaVu* is presented under the form of pseudo-code (Algorithm 1), which is equivalent to the representation using transition rules.

We denote by $\mathcal{I}$ the set of initial values in the consensus problem. Every agent x has the following variables: val_x holding an estimate of the consensus value (set to the initial value of x at initialization), a boolean flag $decided_x$ (initially **false**), and a read-only boolean variable $done_x^{DV}$ holding the input value provided by the local module of $DejaVu$ at agent x. We assume that the set $\mathcal{I}$ is totally ordered. When two agents x and y meet, they both select the minimum of val_x and val_y as a new estimate of the consensus value. An agent x decides on its estimate when either its $DejaVu$'s local module outputs **true** ($done_x^{DV} = $ **true**), or agent x meets an agent y that has already decided ($decided_y = $ **true**); agent x then sets its flag $decided_x$ to **true**.

Algorithm 1. Symmetric consensus with $DejaVu$

---

1 $done_x^{DV}$: output of the oracle $DejaVu$ at x;
2 **Initialization:**;
3      $val_x \leftarrow$ a value in $\mathcal{I}$;
4      $decided_x \leftarrow$ **false**;
5 **On a meeting event** (x, y) **of the agents** x **and** y:;
6      $val_x \leftarrow \min(val_x, val_y)$;
7   **if** $\neg decided_x \wedge (done_x^{DV} \vee decided_y)$ **then**
8   decide on val_x;
9          $decided_x \leftarrow$ **true**;

Lemma 1 (Termination and Validity). *Let $H \in DejaVu(G, \sigma, S)$ be a legal history and γ be an initial configuration. Then, in the execution $H[\gamma]$, every agent eventually decides on some initial value present in γ.*

Proof. Since an agent x can only decide on its estimate val_x, and since every update of val_x assigns a value of some agent, x can only decide on a value present in γ. The liveness property of the oracle $DejaVu$ and weak fairness imply that the oracle eventually outputs **true** at some agent x, which thus decides. Then, thanks to weak fairness, every agent will eventually indirectly meet x, and decide too (if it has not decided already). □

Lemma 2. *Let $H = (S, h)$ be any history with values in $\{0, 1\}$, and γ be an initial configuration. Consider the causal past $Past(p)$ for some event p in S, and let x be an agent involved in p. Then, at the end of the finite execution $H|_{Past(p)}[\gamma]$, the value of val_x at x is equal to the minimum of the initial values of the agents in the support of the causal past of p.*

Proof. For any event p in S, for any agent z involved in p, we denote by $val(p, z)$ the value of val_z right after p. We denote by $val(\perp, z)$ the initial value of the agent z.

Let x, y be the agents involved in the event p. Let p_x (resp. p_y) be the immediate predecessor[3] of p in $Past(p)$ that involves the agent x (resp. the agent y). If such an immediate predecessor does not exist (i.e. p is the first event involving x (resp. y)), then we set $p_x = \bot$ (resp. $p_y = \bot$). By line 6 in Algorithm 1, $val(p, x) = \min(val(p_x, x), val(p_y, y))$. By iterating, we get $val(p, x) = \min\{val(\bot, z), z \in supp(Past(p))\}$. □

Lemma 3. *Consider* Algorithm 1 *using DejaVu. Let* $H \in DejaVu(G, \sigma, S)$ *be a legal history of DejaVu, and* γ *an initial configuration. In the execution* $H[\gamma]$ *of* Algorithm 1, *if some agent* x' *decides in some event* p', *then* $supp(Past(p')) = G$.

Proof. When x' decides, it is either because of the meeting with an agent which has already decided, or because the oracle has output 1 at x' (Algorithm 1, line 7). Hence, there is an event p (in S) involving some agent x such that $p \rightsquigarrow p'$ and the oracle has output 1 at x during p (note that p and p' may be the same event).

By the safety property of *DejaVu*, $Past(p)$ contains some schedule from $Cones(DejaVu, G, \sigma, x)$. Hence, $supp(Past(p')) = supp(Past(p)) = G$, by the definition of *DejaVu*. □

Proposition 2. Algorithm 1 *using DejaVu solves the symmetric consensus.*

Proof. The termination and validity conditions are satisfied thanks to Lemma 1. The agreement and symmetry conditions are satisfied thanks to Lemmas 2 and 3. □

5   Weakest Oracle for Symmetric Consensus

In this section, we prove an intricate property: any oracle O allowing to solve symmetric consensus can be used to implement *DejaVu*. With the result of Sect. 4, this proves that *DejaVu* is the weakest oracle to solve symmetric consensus.

The following lemma states that if an oracle O allows to solve consensus then, for any (weakly fair) schedule, the corresponding "zero history", i.e., the history which always outputs 0 at every agent, cannot be a legal history. This shows, in particular, that any legal history of O eventually outputs 1 at least one agent, which in turn implies that the set $Cones(O, G, \sigma, x)$ is not empty for at least one agent x.

The proof relies on a partitionning argument similar to the one used in the impossibility proof (Proposition 1). The argument though is not exactly the same because, *a priori*, the zero history may be a legal history only on a single specific graph. The partition argument usually consists in building an execution on a twice larger graph. To do so, we have to ensure that the history built on the larger graph is still legal.

---

[3] Immediate means that, if p' involves x and $p_x \rightsquigarrow p' \rightsquigarrow p$, then $p' = p_x$ or $p' = p$.

Lemma 4. *Let $\mathcal{A}$ be a population protocol that solves the consensus[4] problem using an oracle O. For every graph G, and every permutation $\sigma \in \mathfrak{S}G$, for any weakly fair schedule S, the history H with schedule S which outputs the value 0 in every event is not a legal history, i.e., $H \notin O(G, \sigma, S)$. In particular, there exists an agent x such that $Cones(O, G, \sigma, x) \neq \emptyset$.*

Proof. We assume that there exist some graph G, and a legal history H of O on G which permanently outputs 0 everywhere, and we prove a contradiction. Let γ_0 (resp. γ_1) be the initial configuration where all the agents have initial value 0 (resp. 1). In $H[\gamma_v]$, every agent eventually decides on v. Let L be a prefix of S such that by the end of both executions $H[\gamma_0]$ and $H[\gamma_1]$, all the processes have decided.

Consider a graph G' containing two copies G_0 and G_1 of G (with disjoint sets of vertices). Let L_0 (resp. L_1) be the analog of the schedule L applied to G_0 (resp. G_1). Let S' be any weakly fair extension of $L_0 \cup L_1$ (see Sect. 3 for this notation), and $H' \in O(G', \sigma, S')$ be any legal history which outputs the value 0 during $L_0 \cup L_1$. This is possible since the oracle output can be arbitrarily delayed.

Consider now the initial configuration g' on G' in which all agents in G_0 (resp. G_1) have the initial value 0 (resp. 1). Then, by construction, in $H'[g']$, all agents in G_0 (resp. G_1) decide on 0 (resp. 1); whence a contradiction. This proves the first claim.

Moreover, assume that for some graph G, for some σ, for every agent x, $Cones(O, G, \sigma, x) = \emptyset$. Then, by anonymity-compliance, for every agent x, for every permutation τ, $Cones(O, G, \tau, x) = \emptyset$. Thus, the only legal history of O is the one always outputting 0 at every agent; which contradicts the first claim. This proves the second claim. □

The following crucial lemma shows that if O allows to solve symmetric consensus then the sets $Cones(O, \ldots)$ defining O are the subsets of those defining $DejaVu$. In other words, the lemma states that the support of each past cone in $Cones(O, \ldots)$ is the entire graph G. Informally, this means that, if O is strong enough to solve symmetric consensus, then O cannot notify an agent before this agent has indirectly met every other.

The proof relies on the fact that if an agent x can be notified before having indirectly met every other agent, then we can design a legal history H of O so that agent x decides without ever knowing about the initial value of some other agent y. In other words, the decision value of x is left unchanged if we flip agent y's initial value. Because of the validity condition of consensus, there exist two initial configurations g_0, g_1 which only differ by the initial value at one specific agent a, e.g., $g_0(a) = 1 - g_1(a) = 0$, and such that agent x decides on 0 (resp. 1) in the execution $H[g_0]$ (resp. $H[g_1]$). Obviously, a is not y. Let's assume that the initial value of y in g_0 (and g_1) is 1. By considering the configuration g obtained by swapping the initial values of a and y in the configuration g_0, we see that, on one hand, agent x has to decide on 0 by the symmetry condition of symmetric

---

[4] Not necessarily symmetric, in this lemma.

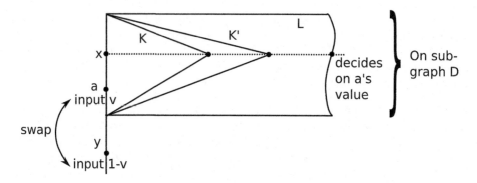

Fig. 1. Illustration to the proof of Lemma 5: x decides on the value v of a, but swapping the values of a and y makes x decide on the value $1 - v$, what contradicts the definition of symmetric consensus.

consensus, and, on the other hand, agent x has to decide on 1, since it cannot distinguish between the configurations g and g_1. This yields a contradiction. The main difficulty of the proof is to build a legal history of the (unknown) oracle O so that x decides without (indirectly) meeting some agent.

Lemma 5. *Let $\mathcal{A}$ be a population protocol that solves the symmetric consensus problem over all complete graphs using an oracle O. Then, for every complete graph G, every permutation σ, and every agent x in G, $Cones(O, G, \sigma, x) \subseteq Cones(DejaVu, G, \sigma, x)$.*

Proof. In this proof, for sake of clarity, we use the same notation for the initial value of an agent, and the corresponding initial state. Figure 1 illustrates the core idea of the proof.

Assume that there is some schedule $K \in Cones(O, G, \sigma, x)$ that is not in $Cones(DejaVu, G, \sigma, x)$, i.e., K is a past cone at x whose support $D = supp(K)$ is a strict subgraph of G.

By Lemma 4, for some agent w (not necessarily distinct from x), the set $Cones(O, D, \sigma, w) \neq \emptyset$. Let $\alpha \in \mathfrak{S}G$ be the permutation that swaps x and w, and $\beta = \alpha\sigma$. Then, by the anonymity-compliance property of the cone sets (Sect. 2.4), $Cones(O, D, \beta, x) \neq \emptyset$. Thus, there is some $K' \in Cones(O, D, \beta, x)$.

Let S be any weakly fair extension of K on D containing the schedule K' as well. We build a history H with schedule S as follows: the history always outputs 0 everywhere except at x, for which it permanently outputs 1 only after the occurrences of K and K' in S. Since $K' \in Cones(O, D, \beta, x)$, we have $H \in O(D, \beta, S)$, i.e. H is a legal history of O on D.

For any initial configuration γ on D, we have an execution $H[\gamma]$ of $\mathcal{A}$ in which every agent in D decides. By the validity property of the consensus, if all the agents have the same initial value 0 (resp. 1), then all agents decide on 0 (resp. 1). Hence, there exist two initial configurations γ_0 and γ_1 on D such that, for some agent a in D, $\gamma_0(a) = 0$, $\gamma_1(a) = 1$ and for every $z \in D - \{a\}$,

$\gamma_0(z) = \gamma_1(z)$, and the agents decide on the value 0 (resp. 1) in the execution $H[\gamma_0]$ (resp. $H[\gamma_1]$).

In particular, x decides on 0 in $H[\gamma_0]$ after some event p_0 in S, and decides on 1 in $H[\gamma_1]$ after some event p_1 in S. Let L be the a prefix of S that contains both $Past(p_0, S)$ and $Past(p_1, S)$. By the end of the finite execution $H|_L[\gamma_0]$ (resp. $H|_L[\gamma_1]$) x decides on the value 0 (resp. 1).

We can extend $H|_L$ to get a weakly fair legal history H' of O on the graph G as follows. Consider any weakly fair extension S' of L on G. In L, the history H' outputs the same values as $H|_L$; and in the complement of L, the history H' outputs 0 everywhere except at x, where it outputs 1. Since $K \in Cones(O, G, \sigma, x)$, we have $H' \in O(G, \sigma, x)$, i.e., H' is a legal history of O on G.

For $v \in \{0, 1\}$, let g_v be the initial configuration on G such that g_v is equal to γ_v on D, and 1 elsewhere. In $H'[g_0]$, agent x decides by the end of L. The support of the causal past of the event preceding its decision, is included in D. Hence, since g_0 and γ_0 are equal on D, x decides on 0 in $H'[g_0]$. For similar reasons, x decides on 1 in $H'[g_1]$. Now pick an agent y in $G - D$, and let g be the initial configuration obtained from g_0 by permuting the values of a and y. In other words, $g(a) = g_0(y) = 1$, $g(y) = g_0(a) = \gamma_0(a) = 0$, and, for every $b \in G - \{a, y\}$, $g(b) = g_0(b)$. The restriction of g to D is equal to γ_1. Hence, in $H'[g]$, the agent x decides on the value 1. On the other hand, since the protocol solves the symmetric consensus, x decides on the value 0; whence a contradiction.   □

A consequence of the previous lemma is that, if an oracle O is strong enough to solve symmetric consensus, then every legal history of O is a legal history of $DejaVu$. Then, roughly speaking, by taking the protocol that simply outputs the same information provided to it by O, we show that $DejaVu$ is weaker than O (according to the definitions in Sect. 2.5). The idea is formalized in the following theorem.

Theorem 1 (Weakest Oracle). *The DejaVu oracle is the weakest oracle for solving symmetric consensus in population protocols.*

Proof. Consider an oracle O such that some protocol solves symmetric consensus using O. By Lemma 5, we have $Cones(O, G, \sigma, x) \subseteq Cones(DejaVu, G, \sigma, x)$ for every triple (G, σ, x). We claim that every legal history of O is a legal history of $DejaVu$.

Indeed, let S be some schedule, and $H \in O(G, \sigma, S)$. We first show that H satisfies the safety property (see Sect. 2.4) of $DejaVu$. If H outputs 1 at x during some event p, then, by the safety property of O, the causal past of p in S contains some schedule $K \in Cones(O, G, \sigma, x)$. Since $K \in Cones(DejaVu, G, \sigma, x)$ by Lemma 5, H also satisfies the safety property of $DejaVu$.

Second, we show that H satisfies the liveness property (see Sect. 2.4) of $DejaVu$. Precisely, we have to show that O eventually outputs 1 at *at least one* of the agents whose causal pasts contain some cones defining $DejaVu$. But since the underlying schedule S is weakly fair, every agent x eventually has some cone from $Cones(DejaVu, G, \sigma, x)$ in its causal past. Therefore, we only have to show that O eventually output 1 at *at least one* agent. But this is a consequence of Lemma 4. Therefore, we have proved that every legal history of O is a legal history of $DejaVu$.

We define a protocol $\mathcal{A}$ with a state space $\{0,1\}$ and an input space $\{0,1\}$ by the following rules

$$(q_1, v_1)(q_2, v_2) \rightarrow (v_1, v_2)$$

There is a unique initial state 0. Intuitively, during a transition, each agent simply copies its input (here provided by the oracle O) into its state. The output map $Out : \{0,1\} \rightarrow \{0,1\}$ is defined as the identity map.

Let $H[\gamma_0] = \gamma_0 \xrightarrow{E_0, h_0} \gamma_1 \xrightarrow{E_1, h_1} \ldots$ be an execution of $\mathcal{A}$ with $H = (S, h)$ a legal history of O. Consider $H' = (S, h')$ be the output history of this execution (see Sect. 2.3). Then

$$\forall (x, y) \in E_0, \; h'(0, (x, y)) = (Out(\gamma_0(x)), Out(\gamma_0(y)))$$
$$= (0, 0)$$
$$\forall i \geq 1, \forall (x, y) \in E_i, \; h'(i, (x, y)) = (Out(\gamma_i(x)), Out(\gamma_i(y)))$$
$$= (h(p_{i,x}), h(p_{i,y}))$$

where $p_{i,x}$ (resp. $p_{i,y}$) is the latest event (different from $(i, (x, y))$) involving agent x (resp. agent y) in the causal past of the event $(i, (x, y))$.

We now prove that H' is a legal history of $DejaVu$. First, H' satisfies the safety property of $DejaVu$. Indeed, if H' outputs 1 at x during some event p, then this implies that the history H outputs 1 at x during some event p_{old} in the causal past of p. Since H is a legal history of $DejaVu$, this implies that the causal past of p_{old} (and thus of p) has a support equal to G. In other words, by event p, agent x has indirectly met with every other agent.

Second, H' satisfies the liveness property of $DejaVu$. Indeed, we have shown above that H eventually outputs 1 at least one agent, say, x during some event p. Therefore, H' outputs 1 at x during the next event involving x (which eventually occurs since the schedule is weakly fair).

Thus, H' is a legal history of $DejaVu$, and we have proved that $DejaVu$ is the weakest oracle for solving symmetric consensus (see Sect. 2.5).   □

6   Conclusion and Perspectives

Designing oracles and searching for the weakest among them in the model of population protocols is especially hard and challenging for several reasons. Anonymity is the first reason. Although oracles have already been studied in anonymous networks, this was done mostly assuming a point-to-point communication model. There, a process can distinguish between two messages arriving from two different communication links. In population protocols there are no links. A second reason is that, for population protocols, the size of the network is unknown, because the available memory for an agent is uniformly bounded. Being unaware of the total number of agents is an important restriction that leads to oracles completely different from those already existing in the literature. Moreover, we want to highlight that in this work we introduce oracles whose output depends only on the past interactions and that are protocol independent, in contrast with previously proposed oracles for population protocols. Thus, exhibiting oracles and the weakest between them, in this context, is especially challenging, because no known technique can be used.

We note that all the results of the paper can be extended to every family of graphs where the partitioning argument is valid (e.g., bounded degree graphs or trees). Moreover, all the results are easily extended also to other fairness conditions, e.g., to the classical, for population protocols, local and global fairness [2,13].

Many difficult problems remain open though. For instance, we did not introduce crash failures, because, even without such failures, consensus is impossible for population protocols. Yet, is it possible to define a variant of *DejaVu* to solve consensus with crash failures? Would this variant still be the weakest? The case of Byzantine failures seems even more problematic. On the other hand, we have focused on the symmetric version of consensus that better suits the model of population protocols. Still, one may want to search for other oracles allowing to solve non symmetric consensus, and look for the existence of a weakest oracle.

References

1. Aguilera, M.K., Chen, W., Toueg, S.: Using the heartbeat failure detector for quiescent reliable communication and consensus in partitionable networks. Theor. Comput. Sci. **220**(1), 3–30 (1999)
2. Angluin, D., Aspnes, J., Diamadi, Z., Fischer, M.J., Peralta, R.: Computation in networks of passively mobile finite-state sensors. Distrib. Comput. **18**(4), 235–253 (2006)
3. Attiya, H., Welch, J.: Distributed Computing. McGraw-Hill, Hightstown (1998)
4. Beauquier, J., Blanchard, P., Burman, J.: Self-stabilizing leader election in population protocols over arbitrary communication graphs. In: Baldoni, R., Nisse, N., van Steen, M. (eds.) OPODIS 2013. LNCS, vol. 8304, pp. 38–52. Springer, Heidelberg (2013)
5. Bonnet, F., Raynal, M.: The price of anonymity: optimal consensus despite asynchrony, crash and anonymity. In: Keidar, I. (ed.) DISC 2009. LNCS, vol. 5805, pp. 341–355. Springer, Heidelberg (2009)
6. Bonnet, F., Raynal, M.: Anonymous asynchronous systems: the case of failure detectors. In: Lynch, N.A., Shvartsman, A.A. (eds.) DISC 2010. LNCS, vol. 6343, pp. 206–220. Springer, Heidelberg (2010)
7. Bouzid, Z., Travers, C.: Anonymity, failures, detectors and consensus. Technical report (2012)
8. Bouzid, Z., Travers, C.: Brief announcement: anonymity, failures, detectors and consensus. In: Aguilera, M.K. (ed.) DISC 2012. LNCS, vol. 7611, pp. 427–428. Springer, Heidelberg (2012)
9. Cartier, P., Foata, D.: Problèmes combinatoire de commutation et réarrangements. Lect. Notes Math. **85** (1969)
10. Chandra, T.D., Toueg, S.: Unreliable failure detectors for reliable distributed systems. J. ACM **43**(2), 225–267 (1996)

11. Cortés, J., Martínez, S., Karatas, T., Bullo, F.: Coverage control for mobile sensing networks. IEEE Trans. Robot. Autom. **20**(2), 243–255 (2004)
12. Delporte-Gallet, C., Fauconnier, H., Guerraoui, R., Tielmann, A.: The weakest failure detector for message passing set-agreement. In: Taubenfeld, G. (ed.) DISC 2008. LNCS, vol. 5218, pp. 109–120. Springer, Heidelberg (2008)
13. Fischer, M., Jiang, H.: Self-stabilizing leader election in networks of finite-state anonymous agents. In: Shvartsman, M.M.A.A. (ed.) OPODIS 2006. LNCS, vol. 4305, pp. 395–409. Springer, Heidelberg (2006)
14. Fischer, M.H., Lynch, N.A., Paterson, M.S.: Impossibility of consensus with one faulty process. J. ACM **32**(2), 374–382 (1985)
15. Lamport, L.: Time, clocks, and the ordering of events in a distributed system. Commun. ACM **21**(7), 558–565 (1978)
16. Lawton, J.R., Beard, R.W.: Synchronized multiple spacecraft rotations. Automatica **38**(8), 1359–1364 (2002)
17. Lynch, N.: Distributed Algorithms. Morgan Kaufmann, San Francisco (1996)
18. Michail, O., Spirakis, P.G.: Terminating population protocols via some minimal global knowledge assumptions. J. Parallel Distrib. Comput. **81–82**, 1–10 (2015)
19. Mostéfaoui, A., Rajsbaum, S., Raynal, M., Travers, C.: The combined power of conditions and information on failures to solve asynchronous set agreement. SIAM J. Comput. **38**(4), 1574–1601 (2008)
20. Mostéfaoui, A., Rajsbaum, S., Raynal, M., Travers, C.: On the computability power and the robustness of set agreement-oriented failure detector classes. Distrib. Comput. **21**(3), 201–222 (2008)
21. Olfati-Saber, R., Fax, J.A., Murray, R.M.: Reply to "comments on "consensus and cooperation in networked multi-agent systems"". Proc. IEEE **98**(7), 1354–1355 (2010)
22. Olfati-Saber, R., Shamma, J.S.: Consensus filters for sensor networks and distributed sensor fusion. In: 44th IEEE Conference Decision and Control and 2005 European Control Conference (CDC-ECC 2005), pp. 6698–6703, December 2005
23. Oshman, R.:. Distributed computation in wireless and dynamic networks. Ph.D. thesis, MIT, Department of Electrical Engineering and Computer Science (2012)
24. Ren, W., Beard, R.W., Atkins, E.M.: A survey of consensus problems in multi-agent coordination. In: American Control Conference, pp. 1859–1864 (2005)
25. Tel, G.: Introduction to Distributed Algorithms, 2nd edn. Cambridge University Press, Cambridge (2000)
26. Xi, W., Tan, X., Baras, J.S.: A stochastic algorithm for self-organization of autonomous swarms. In: Proceedings of 44th IEEE Conference Decision and Control and 2005 European Control Conference (CDC-ECC 2005), pp. 765–770 (2005)

Exact and Approximation Algorithms for Data Mule Scheduling in a Sensor Network

Gui Citovsky[✉], Jie Gao, Joseph S.B. Mitchell, and Jiemin Zeng

Stony Brook University, Stony Brook, NY, USA
gui.citovsky@stonybrook.edu

Abstract. We consider the fundamental problem of scheduling data mules for managing a wireless sensor network. A data mule tours around a sensor network and can help with network maintenance such as data collection and battery recharging/replacement. We assume that each sensor has a fixed data generation rate and a capacity (upper bound on storage size). If the data mule arrives after the storage capacity is met, additional data generated is lost. In this paper we formulate several fundamental problems for the best schedule of single or multiple data mules and provide algorithms with provable performance. First, we consider using a single data mule to collect data from sensors, and we aim to maximize the data collection rate. We then generalize this model to consider k data mules. Additionally, we study the problem of minimizing the number of data mules such that it is possible for them to collect all data, without loss. For the above problems, when we assume that the capacities of all sensors are the same, we provide exact algorithms for special cases and constant-factor approximation algorithms for more general cases. We also show that several of these problems are NP-hard. When we allow sensor capacities to differ, we have a constant-factor approximation for each of the aforementioned problems when the ratio of the maximum capacity to the minimum capacity is constant.

1 Introduction

A number of sensor network designs integrate both static sensor nodes and more powerful mobile nodes, called *data mules*, that serve and help to manage the sensor nodes [25,26,33,34]. The motivation for such designs are twofold. First, there are fundamental limitations with the flat topology of static sensors using short range wireless communication. It is known that such a topology does not scale – the network throughput will diminish if the number of sensors goes to infinity [23], while allowing node mobility will help [22]. Second, a number of fundamental network operations can benefit substantially from mobile nodes.

G. Citovsky and J. Mitchell are partially supported by a grant from the US-Israel Binational Science Foundation (project 2010074) and the National Science Foundation (CCF-1018388, CCF-1540890). J. Gao and J. Zeng are partially supported by grants from AFOSR (FA9550-14-1-0193) and NSF (DMS-1418255, DMS-1221339, CNS-1217823).

© Springer International Publishing Switzerland 2015
P. Bose et al. (Eds.): ALGOSENSORS 2015, LNCS 9536, pp. 57–70, 2015.
DOI: 10.1007/978-3-319-28472-9_5

We consider two example scenarios: sensor data collection and battery recharging. In both cases, data mules that tour around the sensors periodically can be used to maintain the normal functionality of the sensors. In addition, data collection by sensors using multi-hop routing to a fixed base station often suffers from the bottleneck issue near the base station, both in terms of communication and energy usage. Using short range wireless communication with a mobile base station can fundamentally remove such dependency and avoid the single point of failure [37].

Despite the potential benefits of introducing data mules with static sensors, a lot of new challenges emerge at the interface of coordinating the data mules with sensors. One of the most prominent challenges is the scheduling of data mule mobility to serve the sensors in a timely and energy efficient manner. This has been an active research topic for the past few years. However, as surveyed later, most prior work is evaluated by simulations or experiments [3]; algorithms with provable guarantees are scarce. In this paper we make contributions in this direction. We formulate data mule scheduling problems with natural objective functions and provide exact and approximation algorithms.

Our Problem. Suppose there are n sensors and a data mule traveling at a constant speed s to collect data from these sensors. A sensor i generates data at a fixed rate of r_i and has a storage ("bucket") capacity of c_i where $c_i \geq r_i$. When a data mule visits a sensor, all current data stored in the sensor is collected onto the mule. We assume that the mule has unbounded storage capacity. We also assume that data collection at each sensor happens instantaneously, i.e., we ignore the time of data transmission, which is typically much smaller than the time taken by the mule to move between the sensors. If the amount of data generated at a sensor goes beyond its capacity (i.e., its bucket is full), additional data generated is lost. Thus, a natural objective is to schedule data mules to efficiently collect the continuously generated data.

We assume that the data collection and the data mule movement continues indefinitely in time. Therefore, we are mainly concerned about the long-term data gathering efficiency by periodic schedules.

The same problem arises in the case of battery recharging and energy management. In that case, each sensor i uses its battery with capacity c_i at a rate of r_i. When the battery at a sensor is depleted the sensor becomes ineffective. Thus, one would like to minimize the total amount of time of ineffectiveness, over all sensors. We formulate the following three problems.

- **Single Mule Scheduling:** Find a route for a single data mule to collect data from the sensors that maximizes the data collection rate (the average amount of data collected per time unit).
- **k-Mule Scheduling:** Given a budget of k data mules, find routes for them to maximize the rate of data collected from the sensors.
- **No Data Loss Scheduling:** Find the minimum number of data mules, and their schedules, such that all data from all sensors is collected (there is no data loss).

Our Results. We report hardness results, exact algorithms for a few special cases, and approximation algorithms for all three problems. Our algorithmic results are summarized in Table 1. When we assume that the capacities of all sensors are the same, we provide results for the different cases where the sensors lie in different metrics. For the case where the capacities of the sensors are different, we provide general results.

Without loss of generality, we assume that the minimum data rate is 1 and the mule velocity is 1. In fact, we can further assume that all sensors have a data rate of 1; if a sensor has data rate $r_i > 1$, we can replicate this sensor with r_i copies, each with unit data rate and capacity c_i/r_i. Thus, in the following discussion we focus on the case of all sensors having unit data rates and possibly different capacities. When we consider the case where all sensors have the same capacity, we simplify notation and let the capacity of all sensors be c.

We give the first algorithms for such data mule scheduling problems with provable guarantees. In addition, we provide upper and lower bounds on the optimal solution for both problems, and we evaluate the performance using simulations, for a variety of sensor distributions and densities.

Table 1. Our approximation algorithm results for different settings. Note that $m \leq \log(\frac{c_{max}}{c_{min}})$ where c_{max} is the largest capacity and c_{min} is the smallest capacity. For the results in the first four rows, we assume that the sensor capacities are all the same. ε is any positive constant.

With sensors	Single mule	k-mule	No data loss
On a line	Exact	$\frac{1}{3}$	Exact
On a tree	Exact pseudo-polynomial	$\frac{1}{3}(1 - 1/e^{\frac{1}{2+\varepsilon}})$	12
General metric space	$1/6 - \varepsilon$		
Euclidean space	$1/3 - \varepsilon$	$\frac{1}{3}(1 - 1/e^{1-\varepsilon})$	
With different capacities	$O(\frac{1}{m})$		$O(m)$

2   Related Work

Vehicle Routing Problems. The problems we study belong to the general family of vehicle routing problems (VRPs) and traveling salesman problems (TSPs) with constraints [9,29,32,39]. But our problem is the first one considering periodically regenerated rewards/prizes and thus is the first of this type.

Related TSP variations stem from the Prize-Collecting Traveling Salesman Problem (PCTSP) [10,13] which was originally defined by Balas [11] as the problem, given a set of cities with associated prizes and a prize quota to reach, find a path/tour on a subset of the cities such that the quota is met, while minimizing the total distance plus penalties for the cities skipped. (Some recent formulations of this problem do not include penalties for skipped cities.) Archer et al. [4] provided a $(2 - \varepsilon)$-approximation algorithm for this formulation of PCTSP where $\varepsilon \approx 0.007024$.

The Orienteering Problem [21,36] assigns a prize to each city and, given a constraint on the length of the path, aims to maximize the total prize collected. For the rooted version in a general metric space, Blum et al. [14] had proven that the problem is APX-hard and provided a 4-approximation algorithm which was improved to a 3-approximation by Bansal et al. [12] and finally to a $(2+\varepsilon)$-approximation by Chekuri et al. [15]. For the rooted version in $\Re^2$, Arkin et al. [6] give a 2-approximation, which was improved to a $(1+\varepsilon)$-approximation (PTAS) [16] for fixed dimension Euclidean space. For additional information we refer to the review papers [20,36].

Similar to our problems, the Profitable Tour Problem [8] balances the two competing objectives of maximizing total prize collected and minimizing tour length. In some problems, the profit collected is dependent on the latency [17].

Our problems are also very similar to many multi-vehicle routing problems [5,18,19,28]. Arkin et al. [7] give constant-factor approximation algorithms for some types of multiple vehicle routing problems including a 3-approximation for the problem of finding a minimum number of tours shorter than a given bound that cover a given graph. Nagarajan and Ravi [31] provide a 2-approximation for tree metrics and a bicriteria approximation algorithm for general metrics. Khani and Salavatipour [27] present a 2.5-approximation algorithm for the problem of finding, for a given graph and bound λ, the minimum number of trees, each of weight at most λ, to cover the graph (improving on a bound given in [7]).

Data Mule Scheduling. Increasingly, there has been interest in using mobile data mules to collect data in sensor networks. A common question that has arisen is how to schedule multiple mules effectively and efficiently. Many heuristics have been proposed to schedule multiple mules with various constraints and objective functions (e.g., evenly distributing loads [25], scheduling short path lengths [30,38], and minimizing energy [2]). Somasundara et al. [35] address a very similar problem to ours, but with different methods; we obtain provable polynomial-time algorithms, while they employ (worst case exponential-time) integer linear programming and explore heuristics.

3   Single Mule Scheduling

Given a single mobile data mule with unit velocity, n sensors with uniform capacity and unit data rate, the goal is to route the mule in effort to maximize its data gathering rate. We explore this problem with sensors on a line, on a tree, and in space.

3.1   Exact Algorithms on a Line or a Tree

Line Case. We first look at the case when the sensors are on a line. We assume that the input data is integral; specifically, the sensors p_i are located at integer coordinates and the capacities c_i for all i are integers. With this assumption, the optimal schedule can be shown to be periodic.

Lemma 1. *The optimal schedule that minimizes data loss is periodic, assuming integral input data.*

Proof. If the sensors are located at integral positions, the distances between any two of them are integers as well. Thus, all states of the problem can be encoded by the position of the mule and the current amount of data at each sensor i. All of these values are integers. Thus, the total number of possible states is finite; after a state reappears we realize that the robot must follow the same schedule, making the schedule periodic. □

Theorem 1. *Let there be n sensors, $p_1, p_2, \ldots, p_n$ on a line. Assume that the capacities and rates of all sensors are the same: $c_i = c$ and $r_i = 1$, for $1 \leq i \leq n$. Then there exists an optimal path that minimizes data loss with the following properties: (1) its leftmost and rightmost points are at sensors, (2) it is a path making U-turns only at the leftmost and rightmost sensors.*

Despite the simple and clean statement, the proof is in fact fairly technical. To provide intuition for Theorem 1, note that paths that have U-turns not at the outermost points are making a tradeoff of collecting more data from middle sensors at the cost of having more overflow at the outer sensors. If this tradeoff is worth it, then we can show it is also worth it to forgo collecting data from some of the outer sensors. The main technical challenge is to figure out and compare the data rate between the two choices. For the full proof, we refer to a more detailed version of this paper posted on arXiv.

The immediate consequence from Theorem 1 is that one can find the optimal schedule in $O(n^2)$ time, enumerating all possible pairs of extreme points.

It is important to note that it is sometimes necessary for the mule in the optimal solution to gather data more than once from a given sensor in a period. In Fig. 1, sensors are split into six groups, where each group has either k or $2k$ sensors. Within each group, each sensor has the same x-coordinate. In the optimal solution, the data mule traverses the entire interval back and forth, picking up data whenever it reaches a sensor. This solution has data gathering rate $\frac{10.5k}{2} = 5.25k$. In comparison, the best solution that gathers data from a sensor at most once per period has rate $4k$.

Tree Case. We extend our results to a tree topology, with the sensors placed on a tree network embedded in the plane. Then, we show that the structure of an optimal schedule for the mule is to follow (repeatedly) a simple cycle

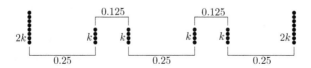

Fig. 1. The optimal solution repeats sensors

(a doubling of a subtree). Again we assume that all sensors have the same capacity c and the same rate, 1, of data accumulation. We also assume that the input is integral, i.e., c is an integer and the distance between any two sensors on the tree network is an integer.

Theorem 2. *Let there be n sensors, $p_1, p_2, \ldots, p_n$ on a tree G. For all p_i, $1 \leq i \leq n$, let $c_i = c$ and $r_i = 1$, i.e. let the capacity and rates of all sensors be the same. There exists an optimal path that minimizes data loss with the following properties: (1) it only changes direction at sensors, (2) it is a cycle obtained by doubling a subtree.*

For the full proof, we refer to a more detailed version of this paper posted on arXiv. A consequence of Thereom 2 is that we can compute an optimal mule route (we can identify an optimal subtree of G) in time that is pseudo-polynomial, using a dynamic programming algorithm.

It is unlikely that there is a strongly polynomial time algorithm for an exact solution, since we show that the problem is weakly NP-hard.

3.2 Hardness

We show that single mule scheduling on a tree is weakly NP-hard. Further, we show that the data gathering problem for a single mule and sensors in Euclidean (or any metric) space is NP-hard.

Theorem 3. *The data gathering problem scheduling a single mule among uniform capacity sensors on a tree is (weakly) NP-hard.*

Proof. Our reduction is from PARTITION (or SUBSET-SUM): given a set $S = \{x_1, \ldots, x_n\}$ of n integers, does there exist a subset, $S' \subset S$, such that $\sum_{x_i \in S'} x_i = M/2$, where $M = \sum_i x_i$? Given an instance of PARTITION, we construct a tree as follows: There is a node v connected to a node u by an edge of length $M/2$. Incident on v are n additional edges, of lengths x_i; the edge of length x_i leads to a node where there are exactly x_i sensors placed. Also, at node u there are M^2 sensors placed. (If one disallows multiple ($x > 1$) sensors to be at a single node w of the tree, we can add x very short (length $\Theta(1/n)$) edges incident to w, each leading to a leaf with a single sensor.) Consider the problem of computing a maximum data-rate tour in this tree, assuming each sensor has capacity $2M$. Then, in order to decide if it is possible to achieve data collection rate of $M^2 + M/2$ we need to decide if it is possible to find a subtree that includes node u (where the large number, M^2, of sensors lie) and a subset of nodes having x_i sensors each, with the sum of these x_i's totalling exactly $M/2$. (If the sum is any less than $M/2$, we fail to collect enough data during the cycle of length $2M$ that is allowed before data overflow; if the sum is any more than $M/2$, we lose data to overflow at u, which cannot be compensated for by additional data collected at the x_i nodes, since M^2 is so large compared to x_i.)   □

Theorem 4. *The data gathering problem scheduling a single mule among uniform capacity sensors in the Euclidean (or any metric) space is NP-hard.*

Proof. We reduce from the Hamiltonian cycle problem in a grid graph where n points are on an integer grid and an edge exists between two points if and only if they are unit distance apart. If we place a sensor at each point with capacity n, it follows that there exists a Hamiltonian cycle in this graph if and only if there exists a data gathering solution with no data loss.   □

NP-hardness for this problem also holds for any general metric space.

3.3   Approximation Algorithm

Theorem 5. *For uniform capacity sensors in fixed dimension Euclidean space, there exists a $(1/3 - \varepsilon)$-approximation for maximizing the data gathering rate of a single mule. For general metric spaces, a $(1/6 - \varepsilon)$-approximation exists.*

Proof. In order to achieve this, we approximate the maximum number of distinct sensors a mule can cover in time $c/2$, the amount of time for sensors to fill from empty to half capacity (it can be shown that one half capacity is the optimal choice). The result will be a path, to which we assign one mule to traverse back and forth. The data gathering rate of this solution is equal to the number of distinct sensors covered as a mule on a schedule with period t will collect exactly t units of data from each sensor. We denote R to be the maximum number of distinct sensors that can be covered by a path of length $c/2$. Note that R can be approximated to within a factor of $1+\varepsilon$ in fixed dimension Euclidean space using the PTAS for orienteering [16]. In general metric spaces, R can be approximated to within a factor of $2 + \varepsilon$ [15]. Let R^* be the data gathering rate of the optimal solution. We now show that $R^* \leq 3R$.

Consider the interval of time $c/2$ in the optimal solution that has the highest data gathering rate. This is an upper bound on R^*. In this time period, we know that the number of distinct sensors visited is at most R. We also know that during this time period at most $\frac{3}{2}c$ units of data can be downloaded from any visited sensor (at most c units of data immediately downloaded and at most $c/2$ units of data downloaded after $c/2$ units of time have passed). Therefore, the total amount of data collected in the optimal solution during this period of time is at most $\frac{3}{2}cR$. Averaging the data collected over the time interval $c/2$, the data gathering rate of the optimal solution is at most $3R$.   □

4 k-Mule Scheduling

Given a budget of k data mules, we now consider the problem of maximizing the total data gathering rate of these mules. We assume the n sensors have uniform capacity, unit data rate, and unit velocity. It is important to note that even with sensors on a line, the optimal solution may not assign mules to private tours; sensors may need to be visited by multiple mules. Consider an input with two mules and sensors uniformly spaced $c/4$ apart from one another. Any time a mule makes a U-turn, it will gather only $c/2$ data from the next sensor it visits. In order to maximize the frequency of full buckets collected, we want to minimize

Fig. 2. With two data mules and sensors uniformly spaced $c/4$ apart, many sensors will be visited by both mules in the optimal solution.

the frequency of U-turns made. This can be done by maintaining separation of length c between the mules and having the mules zig-zag across (nearly) the entire line (see Fig. 2). Interestingly, this example also shows that mules can travel arbitrarily far distances.

4.1   Sensors on a Line

Theorem 6. *Given a budget of k data mules, for uniform capacity sensors on a line, there exists a 1/3-approximation for maximizing the data gathering rate.*

Proof. Similar to the case when $k = 1$, we find the maximum amount of distinct sensors that k mules can cover in time $c/2$ (it can be shown that half capacity is the optimal choice). The result will correspond to a set of disjoint intervals; we assign one mule to each interval. The duration of a cycle for each mule is the length of time a sensor fills up to capacity so no sensor is allowed to overflow. Therefore, the data gathering rate of this solution, call it R, is equal to the number of sensors covered. Note that R, the maximum amount of distinct sensors that can be covered by k disjoint intervals of length at most $c/2$, can be computed exactly in polynomial time using dynamic programming. Let R^* be the data gathering rate of the optimal solution. It follows from the same argument given for the $k = 1$ case (Theorem 5) that $R^* \leq 3R$.   □

4.2   Sensors in a General Metric Space

Theorem 7. *Given a budget of k data mules, for uniform capacity sensors in a general metric space, there exists a $\frac{1}{3}(1 - \frac{1}{e^\beta})$-approximation with $\beta = \frac{1}{2+\varepsilon}$ for maximizing the data gathering rate. In fixed dimension Euclidean space there exists a $\frac{1}{3}(1 - \frac{1}{e^\beta})$-approximation with $\beta = 1 - \varepsilon$.*

Proof. The proof is similar to the proof of Theorem 5. In order to approximate the maximum amount of distinct sensors that k mules can cover in $c/2$ time, we compute an orienteering path with a travel distance budget of $c/2$ on the uncovered sensors. We repeat this operation for a total of k times. In the Maximum Coverage problem, one is given a universe of elements, a collection of subsets, and an integer k. The objective is to maximize the number of elements covered using k subsets. It has been shown by Hochbaum et al. [24] that greedily choosing the set with the largest number of uncovered elements k times yields a $(1 - \frac{1}{e})$-approximation. Interestingly, Hochbaum et al. also show that using a β-approximation for covering the maximum amount of uncovered elements in each of the k rounds yields a $(1 - \frac{1}{e^\beta})$-approximation. Computing orienteering k times on only the remaining uncovered sensors, we achieve a $\frac{1}{(2+\varepsilon)}$-approximation each round and therefore a $(1 - \frac{1}{e^\beta})$-approximation for $\beta = \frac{1}{2+\varepsilon}$ for covering the maximum amount of sensors with k mules. Using similar arguments as the case where $k = 1$ (Theorem 5), it is now easy to see that having mules traverse the k orienteering paths back and forth yields a $\frac{1}{3}(1 - \frac{1}{e^\beta})$-approximation. $\square$

5   No Data Loss Scheduling

In situations in which it is not possible for a fixed number of data mules to collect all data in the network, it is natural to increase the number of data mules and let them collectively finish the data collection task. In the no data loss scheduling problem, we seek to minimize the number of mules in order to avoid data loss. Throughout this section we assume that all sensors have unit data rate, unit velocity, and uniform capacity.

5.1   Exact Algorithm on a Line

When sensors all lie along a line, we show that the problem can be solved in polynomial time. As before, we can assume that the sensors lie at integer coordinates so that, by the same argument as in Lemma 1, the mules in an optimal solution follow periodic schedules.

Lemma 2. *For the minimum cardinality data mule problem with no data loss, if the sensors have uniform capacity and lie on a line, there is an optimal schedule in which all mules follow periodic cycles, zigzaging within disjoint intervals, each with length at most $c/2$.*

   We refer to a more detailed version of this paper posted on arXiv for the full proof. By the above structural lemma, we can use a simple greedy algorithm to minimize the number of data mules necessary to collect data, without loss, for sensors on a line: Starting at the leftmost sensor, schedule a mule to zigzag within an interval of length $c/2$ whose left endpoint is the leftmost sensor, and then continue to the right, adding further intervals of length $c/2$ until all sensors are covered. This is an $O(n)$ algorithm for n (sorted) sensors.

5.2   Hardness

Even when the sensors lie on a tree, the problem of minimum cardinality data mule scheduling with no data loss is already NP-hard.

Theorem 8. *For uniform capacity sensors on a tree, the problem of minimum cardinality data mule scheduling with no data loss is (strongly) NP-hard.*

Proof. We reduce from 3-PARTITION. Given a multiset, $S = \{x_1, x_2, \cdots, x_{3n}\}$, of $3n$ integers with total sum M, 3-PARTITION asks whether there is a partition of S into n subsets, $S_1, \ldots, S_n$, such that each subset sums to exactly $B = M/n$. It is known that we can assume that the integers x_i satisfy $B/4 < x_i < B/2$, so that each subset S_i must consist of a triple of elements ($|S_i| = 3$). We create an instance of a star having a hub (center node) incident on $3n$ edges ("spokes") to $3n$ sensors, with edge lengths equal to x_i. Each sensor has capacity $2M/n$. Thus if there is a partitioning of S into triples of integers that each sum to M/n, then one (unit-speed) mule can traverse each corresponding 3-spoke subtree in time exactly $2M/n$, resulting in no data loss using n mules. On the other hand, any solution using exactly n data mules and having no data loss determines a valid partition of S into triples S_i. Thus, the 3-PARTITION instance has a solution if and only if n data mules suffice.   □

For the general case, with sensors at points of a metric space or in a Euclidean space, the problem of determining the minimum number of data mules necessary to collect all data is also NP-hard. This can be seen by using a similar reduction from Hamiltonian cycle, as in Theorem 4.

5.3   Approximation Algorithm

In the following we describe an algorithm achieving a constant-factor approximation.

It is tempting to think that an optimal solution will allocate each mule to cover an exclusive set of sensors S', that are not covered by other mules. We denote such a set of tours as a *private tour set* on S'. However, the following example shows that this is no longer the case when sensors lie in the plane. Consider $n > 2$ sensors placed on a circle, uniformly spaced with adjacent sensors at (Euclidean) distance exactly $c - 1/n$. The convex hull of these sensors is a regular n-gon of perimeter $n(c - 1/n) = nc - 1$. The optimal solution would use $n - 1$ mules, with each mule touring periodically at constant speed (1) along the boundary of this n-gon, with time/distance separation of exactly c between consecutive mules. This ensures that each sensor is visited exactly when its storage (bucket) becomes full. However, any solution using private tours will have to use n mules, since no mule can use a private tour to cover two or more sensors (since it would have length at least $2c - 2/n > c$).

While it may be that no private tour set is optimal, we now argue that the optimal schedule using only private tours is provably close to optimal (in terms of minimizing the number of data mules). Denote by k^* the minimum number of cycles, each of length at most c, to cover all nodes, which is denoted as a *light cycle cover*. And denote by m^* the minimum number of data mules required to collect all data.

Lemma 3. $m^* \leq k^* \leq 2m^*$.

Proof. First, note that using k^* mules, each traversing a (private) light cycle, results in all data being collected; thus, $m^* \leq k^*$.

Now consider an optimal schedule of m^* data mules. Mule i moves along a schedule C_i. Consider any particular time t. Each sensor j is visited by at least one mule. We assign it to the mule that visits it first, i.e., at the earliest time after t. We know this time is at most c, since no data is lost at sensor j. Thus, consider mule i, at current position p_i (at time t) and all the sensors along C_i that are assigned to it. They all lie on a path (along C_i) of length at most c. Let s_i be the sensor furthest away from p_i, measuring distance along C_i. Let γ_i be the corresponding path along C_i, from p_i to s_i. Let b_i be the midpoint of this path. Place a clone of mule i at point s_i, and create two private cycles for mule i and his clone: one cycle goes from p_i to b_i along γ_i, then returns to p_i directly (along a shortest path or a straight segment), the other goes from b_i to s_i along γ_i, then returns to b_i directly. Mule i traverses the first cycle; his clone traverses the second cycle. Do this for all mules.

We have doubled (via cloning) the number of mules, but now each mule/clone has a private cycle, of sensors assigned only to it, and these cycles are each of total length at most c. Thus, this is a valid solution to the light cycle cover problem. Thus, the minimum number of light cycles, k^* is no greater than $2m^*$.   □

By the above lemma, an α-approximation for the minimum light cycle cover gives a 2α-approximation for the minimum number of data mules. Arkin $et\ al.$ [7] gave a 6-approximation algorithm for the minimum light cycle cover problem; thus, we have a 12-approximation for minimum data mule scheduling. This is summarized in the following theorem.

Theorem 9. *For uniform capacity sensors within a general metric space or in the Euclidean plane, computing the minimum number of data mules to collect all data is NP-hard. There is a polynomial-time 12-approximation algorithm for sensors in a general metric space.*

6   Different Capacities

We now consider both the k-mule scheduling problem and the no data loss scheduling problem on n sensors with potentially different sensor storage capacities. Each sensor has unit data rate. The result for the k-mule scheduling problem obviously holds for the single mule problem (i.e. when $k = 1$).

6.1 k-Mule Scheduling

Lemma 4. *With m groups of sensors, each group having the same storage capacity, optimally solving each group independently and taking the solution with the highest data gathering rate yields a $O(1/m)$-approximation to the k-mule scheduling problem.*

Proof. Let $r(\cdot)$ be the data gathering rate of a solution. Let OPT_i be the optimal solution to group i and let OPT be the schedule with highest data rate. $r(OPT) \leq \sum_{i=1}^{m} r(OPT_i) \leq m \cdot \max_i\{r(OPT_i)\}$. The first inequality is from the following observation. Consider the optimal schedule OPT and modify it such that we only visit the nodes in group i. This is obviously a solution for collecting data from group i and thus has data rate no greater than $r(OPT_i)$.   □

Let c_{max} and c_{min} be the storage capacities of the largest and smallest sensors respectively. We round the storage capacity of each sensor down to its nearest power of two. Doing so, we create m groups of sensors where m is at most $\log(\frac{c_{max}}{c_{min}})$. Note that m may be significantly smaller than $\log(\frac{c_{max}}{c_{min}})$. In the rounding down process, the storage capacity of each sensor is at most halved, thus the optimal solution on the new sensors has data gathering rate of at least $1/2$ of the same solution before rounding. We approximate the optimal solution to each of the groups within a constant factor and choose the one with highest data gathering rate. By Lemma 4, we have the following.

Theorem 10. *By rounding down the sensor capacities into $m \leq \log(\frac{c_{max}}{c_{min}})$ groups, the group with highest data gathering rate has rate at least $O(1/m) \cdot r(OPT)$ where OPT is the optimal solution to the k-mule scheduling problem.*

6.2   No Data Loss Scheduling

Theorem 11. *By rounding down the sensor capacities into $m \leq \log(\frac{c_{max}}{c_{min}})$ groups and solving each group independently, at most $O(m) \cdot |OPT|$ mules are used in total, where $|OPT|$ is the minimum number of mules needed to avoid data loss.*

Proof. Using the same rounding technique as the previous section, we again obtain m groups of sensors with $m \leq \log(\frac{c_{max}}{c_{min}})$. In the rounding down process, the capacity of any sensor is at most halved. Thus, the optimal solution on the rounded down sensors requires at most two times the number of mules as the optimal solution to the original set of sensors. Let $|OPT_i|$ be the minimum number of mules needed for no data loss to occur in group i and let $|OPT|$ be the number of mules in the optimal solution. Since $|OPT| \geq |OPT_i|$ for $1 \leq i \leq m$, we have that $m \cdot |OPT| \geq \sum_{i=1}^{m} |OPT_i|$. Approximating $|OPT_i|$ within a constant factor for all i, we use $O(m) \cdot |OPT|$ mules.   □

7   Conclusion

Our exact and approximation algorithms for single mule scheduling, k-mule scheduling, and no data loss scheduling represent the state of the art results on mule scheduling problems and greatly deepens our understanding of vehicular routing with constraints. For future work, we would like to see if the approximation ratios can be improved, especially with sensors in Euclidean space.

References

1. National traveling salesman problems. http://www.math.uwaterloo.ca/tsp/world/countries.html. Accessed 15 Dec 2014
2. Almi'ani, K., Viglas, A., Libman, L.: Tour and path planning methods for efficient data gathering using mobile elements. Int. J. Ad Hoc Ubiquitous Comput. (2014)
3. Anastasi, G., Conti, M., Di Francesco, M.: Data collection in sensor networks with data mules: an integrated simulation analysis. In: IEEE Symposium on Computers and Communications, ISCC 2008, pp. 1096–1102, July 2008
4. Archer, A., Bateni, M., Hajiaghayi, M., Karloff, H.: Improved approximation algorithms for prize-collecting steiner tree and tsp. SIAM J. Comput. **40**(2), 309–332 (2011)
5. Archetti, C., Speranza, M., Vigo, D.: Vehicle routing problems with profits. Technical report WPDEM2013/3, University of Brescia (2013)
6. Arkin, E., Mitchell, J.S.B., Narasimhan, G.: Resource-constrained geometric network optimization. In: Symposium on Computational Geometry, pp. 307–316 (1998)
7. Arkin, E.M., Hassin, R., Levin, A.: Approximations for minimum and min-max vehicle routing problems. J. Algorithms **59**(1), 1–18 (2006)
8. Ausiello, G., Bonifaci, V., Laura, L.: The online prize-collecting traveling salesman problem. Inf. Process. Lett. **107**(6), 199–204 (2008)
9. Ausiello, G., Leonardi, S., Marchetti-Spaccamela, A.: On salesmen, repairmen, spiders, and other traveling agents. In: Bongiovanni, G., Petreschi, R., Gambosi, G. (eds.) CIAC 2000. LNCS, vol. 1767, pp. 1–16. Springer, Heidelberg (2000)
10. Awerbuch, B., Azar, Y., Blum, A., Vempala, S.: Improved approximation guarantees for minimum-weight k-trees and prize-collecting salesmen. SIAM J. Comput. 277–283 (1995)
11. Balas, E.: The prize collecting traveling salesman problem. Networks **19**(6), 621–636 (1989)
12. Bansal, N., Blum, A., Chawla, S., Meyerson, A.: Approximation algorithms for deadline-tsp and vehicle routing with time-windows. In: Proceedings of the Thirty-Sixth Annual ACM Symposium on Theory of Computing, STOC 2004, pp. 166–174. ACM, New York (2004)
13. Bienstock, D., Goemans, M.X., Simchi-Levi, D., Williamson, D.: A note on the prize collecting traveling salesman problem. Math. Program. **59**(1), 413–420 (1993)
14. Blum, A., Chawla, S., Karger, D., Lane, T., Meyerson, A., Minkoff, M.: Approximation algorithms for orienteering and discounted-reward tsp. In: Proceedings of 44th Annual IEEE Symposium on Foundations of Computer Science, pp. 46–55, Oct 2003
15. Chekuri, C., Korula, N., Pál, M.: Improved algorithms for orienteering and related problems. ACM Trans. Algorithms **8**(3), 23:1–23:27 (2012)
16. Chen, K., Har-Peled, S.: The euclidean orienteering problem revisited. SIAM J. Comput. **38**(1), 385–397 (2008)
17. Coene, S., Spieksma, F.C.R.: Profit-based latency problems on the line. Oper. Res. Lett. **36**(3), 333–337 (2008)
18. Eksioglu, B., Vural, A.V., Reisman, A.: Survey: the vehicle routing problem: a taxonomic review. Comput. Ind. Eng. **57**(4), 1472–1483 (2009)
19. Even, G., Garg, N., Könemann, J., Ravi, R., Sinha, A.: Covering graphs using trees and stars. In: Arora, S., Jansen, K., Rolim, J.D.P., Sahai, A. (eds.) RANDOM/APPROX 2003. LNCS, vol. 2764, pp. 24–35. Springer, Heidelberg (2003)

20. Feillet, D., Dejax, P., Gendreau, M.: Traveling salesman problems with profits. Transp. Sci. **39**(2), 188–205 (2005)
21. Golden, B.L., Levy, L., Vohra, R.: The orienteering problem. Naval Res. Logistics **34**, 307–318 (1987)
22. Grossglauser, M., Tse, D.: Mobility increases the capacity of ad hoc wireless networks. IEEE/ACM Trans. Netw. **10**(4), 477–486 (2002)
23. Gupta, P., Kumar, P.R.: The capacity of wireless networks. IEEE Trans. Inf. Theory **46**(2), 388–404 (2000)
24. Hochbaum, D.S., Pathria, A.: Analysis of the greedy approach in problems of maximum k-coverage. Naval Res. Logistics **45**(6), 615–627 (1998)
25. Jea, D., Somasundara, A., Srivastava, M.: Multiple controlled mobile elements (data mules) for data collection in sensor networks. In: Prasanna, V.K., Iyengar, S.S., Spirakis, P.G., Welsh, M. (eds.) DCOSS 2005. LNCS, vol. 3560, pp. 244–257. Springer, Heidelberg (2005)
26. Kansal, A., Rahimi, M., Kaiser, W.J., Srivastava, M.B., Pottie, G.J., Estrin, D.: Controlled mobility for sustainable wireless networks. In: IEEE Sensor and Ad Hoc Communications and Networks (SECON 2004) (2004)
27. Khani, M.R., Salavatipour, M.R.: Improved approximation algorithms for the min-max tree cover and bounded tree cover problems. Algorithmica **69**(2), 443–460 (2014)
28. Kim, D., Uma, R., Abay, B., Wu, W., Wang, W., Tokuta, A.: Minimum latency multiple data mule trajectory planning in wireless sensor networks. IEEE Trans. Mob. Comput. **13**(4), 838–851 (2014)
29. Laporte, G.: The vehicle routing problem: an overview of exact and approximate algorithms. Eur. J. Oper. Res. **59**(3), 345–358 (1992)
30. Ma, M., Yang, Y.: Data gathering in wireless sensor networks with mobile collectors. In: IEEE International Symposium on Parallel and Distributed Processing, IPDPS 2008, pp. 1–9. IEEE (2008)
31. Nagarajan, V., Ravi, R.: Approximation algorithms for distance constrained vehicle routing problems. Networks **59**(2), 209–214 (2012)
32. Pillac, V., Gendreau, M., Guéret, C., Medaglia, A.L.: A review of dynamic vehicle routing problems. Eur. J. Oper. Res. **255**(1), 1–11 (2013)
33. Shah, R.C., Roy, S., Jain, S., Brunette, W.: Data mules: modeling a three-tier architecture for sparse sensor networks. In: IEEE SNPA Workshop, pp. 30–41 (2003)
34. Somasundara, A., Kansal, A., Jea, D., Estrin, D., Srivastava, M.: Controllably mobile infrastructure for low energy embedded networks. IEEE Trans. Mob. Comput. **5**(8), 958–973 (2006)
35. Somasundara, A.A., Ramamoorthy, A., Srivastava, M.B.: Mobile element scheduling with dynamic deadlines. IEEE Trans. Mob. Comput. **6**(4), 395–410 (2007)
36. Vansteenwegen, P., Souffriau, W., Oudheusden, D.V.: The orienteering problem: a survey. Eur. J. Oper. Res. **209**(1), 1–10 (2011)
37. Vincze, Z., Vida, R.: Multi-hop wireless sensor networks with mobile sink. In: CoNEXT 2005: Proceedings of the 2005 ACM Conference on Emerging Network Experiment and Technology, pp. 302–303. ACM Press, New York (2005)
38. Wu, F.J., Tseng, Y.C.: Energy-conserving data gathering by mobile mules in a spatially separated wireless sensor network. Wirel. Commun. Mob. Comput. **13**(15), 1369–1385 (2013)
39. Yu, W., Liu, Z.: Vehicle routing problems with regular objective functions on a path. Naval Res. Logistics (NRL) **61**(1), 34–43 (2014)

Limitations of Current Wireless Scheduling Algorithms

Magnús M. Halldórsson, Christian Konrad, and Tigran Tonoyan[⊠]

ICE-TCS, Reykjavik University, Reykjavik, Iceland
{mmh,christiank,tigran}@ru.is

Abstract. We consider the problem of scheduling wireless links in the physical model, where we seek a partition of a given a set of wireless links into the minimum number of subsets satisfying the signal-to-interference-and-noise-ratio (SINR) constraints. We consider the two families of approximation algorithms that are known to guarantee $O(\log n)$ approximation for the scheduling problem, where n is the number of links. We present network constructions showing that the approximation ratios of those algorithms are no better than logarithmic, both in n and in Δ, where Δ is a geometric parameter – the ratio of the maximum and minimum link lengths.

Keywords: Wireless scheduling · Oblivious power · Lower bound

1 Introduction

The task of the MAC layer in TDMA-based (time-division multiple access) wireless networks is to determine which nodes can communicate in which time-frequency slot. A scheduler aims to optimize criteria involving throughput and fairness. This requires obtaining effective spatial reuse while satisfying the interference constraints. We treat the fundamental *scheduling* problem of partitioning a given set of communication links into the fewest possible feasible sets.

We adopt the *SINR model* of communication, where signal decays as it travels and a transmission is successful if its strength at the receiver exceeds the accumulated signal strength of interfering transmission by a sufficient (technology determined) factor. Considerable progress has been made in recent years in elucidating essential algorithmic properties of the SINR model (e.g., [1,8,9,13,24,27,29,31,34]). Early work on the scheduling problem includes [5,6,10]. Gupta and Kumar [16] proposed the geometric version of SINR and initiated average-case analysis of network capacity known as scaling laws. NP-completeness has been shown for scheduling with different forms of power control: none [14], limited [28], and unbounded [32]. Moscibroda and Wattenhofer [34] initiated worst-case analysis in the SINR model.

Although the standard analytic assumption that signal decays polynomially with the distance traveled is far from realistic [33,36], it has been shown that

© Springer International Publishing Switzerland 2015
P. Bose et al. (Eds.): ALGOSENSORS 2015, LNCS 9536, pp. 71–84, 2015.
DOI: 10.1007/978-3-319-28472-9_6

results obtained with that assumption can be translated to the setting of arbitrary measured signal decay [3,15], as well as to Rayleigh fading model [7].

The scheduling problem has been considered both for fixed *oblivious* power assignments (where the power chosen for a link depends only on the link itself) and for arbitrary power control. This paper addresses algorithms dealing with both scenarios.

Finding constant-factor approximation to the scheduling problem has proved challenging. However, there are several approaches giving logarithmic approximation.

I. One way of achieving $O(\log n)$ approximation is by *greedy or first-fit* algorithms, where the links are processed in a non-decreasing order of length and are assigned to the first slot where they "fit" [22,26,29,30]. The approximation ratio of such algorithms is usually obtained by arguing that such algorithms provide constant factor approximation to the *capacity problem*, where the goal is to select a large subset of links that can successfully transmit in the same time slot. It has also been shown that such algorithms perform well for some randomly generated network instances [2].

II. Another approach giving $O(\log n)$-approximation (for fixed power assignments) is to use randomization: links try to transmit in each time slot with certain transmission probabilities, and each link is assigned to the slot where its transmission succeeded [21,31]. It is known that appropriate choice of the probabilities guarantees an $O(\log n)$ approximation (w.h.p.).

III. Yet another approach is to divide the links into groups of nearly equal length and schedule each group separately. Following this approach, numerous $O(\log \Delta)$-approximation results have been argued [12,14,18], where Δ is the ratio between the longest and shortest link length.

The only known algorithms to achieve less than $O(n)$ approximation for scheduling are from the first two families. The only known constant-factor approximation algorithms for scheduling are obtained in the case of the *linear power scheme* [23,37]. In a recent paper, we presented an $O(\log^* \Delta)$-approximation algorithm for scheduling with power control [25], but here too, the approximation factor is only $O(n)$ in terms of n.

The optimum number of slots for scheduling has also been approximated through *interference measures* [19,21,31]. However, no measure is known to give better than $O(\log n)$ approximation. It is also not evident how to efficiently compute such measures.

Many variants of the scheduling problem are known to be NP-hard but we are not aware of any significant inapproximability result.

Our Results. While for the third family of algorithms it is easy to find examples attaining the approximation ratio $\Omega(\log \Delta)$, we are not aware of constructions showing that the approximation ratio of $O(\log n)$ of the first and second families cannot be improved. For the second family of algorithms, a construction has been presented in [21] for which the output of the algorithm is a $\Theta(\log n)$-approximation, but their construction does not exclude that $\Theta(\log n)$ is only additive, and it is asked in [21] whether another analysis of the algorithm could give a smaller approximation ratio.

We show that algorithms of families I and II achieve (including the algorithms for both fixed power assignments and optimized powers) no better than $\Omega(\frac{\log \Delta}{\log \log \Delta})$ (in terms of only Δ) or $\Omega(\frac{\log n}{\log \log n})$ (in terms of only n) approximation even in one dimensional networks. Our constructions are obtained by modeling sets of links after certain graphs that are "hard" instances for graph coloring algorithms similar to the families I and II.

These results suggest that new methods are needed for obtaining better than logarithmic approximation for scheduling. Note, however, that our constructions do not work for the uniform power assignment, where all links use equal power.

2 Model and Definitions

Communication Links. Consider a set L of n *links*, numbered from 1 to n. Each link i represents a unit-demand communication request from a sender s_i to a receiver r_i - point-size wireless transmitter/receivers located in a metric space with distance function d. We denote $d_{ij} = d(s_i, r_j)$ the distance from the sender of link i to the receiver of link j, $l_i = d(s_i, r_i)$ the *length* of link i and $d(i, j)$ the minimum distance between $\{s_i, r_i\}$ and $\{s_j, r_j\}$. We let $\Delta(L)$ denote the ratio between longest and shortest link lengths in L, and drop L when clear from context.

SINR Feasibility and the Scheduling Problem. In the *physical model (or SINR model)* of communication [35], a transmission of a link i is successful iff

$$\mathcal{S}_i > \beta \cdot \left(\sum_{j \in S \setminus \{i\}} \mathcal{I}_{ji} + N \right), \tag{1}$$

where $\mathcal{S}_i$ denotes the received signal power of link i, $\mathcal{I}_{ji}$ denotes the interference power on link i caused by link j, $N \geq 0$ is a constant denoting the ambient noise, $\beta \geq 1$ is the minimum SINR (Signal to Interference and Noise Ratio) required for a message to be successfully received and S is the set of links transmitting concurrently with link i. If $P : L \to \mathbb{R}_+$ is the power assignment of links – $P(i)$ defines the transmission power of the sender node s_i, then $\mathcal{S}_i = \frac{P(i)}{l_i^\alpha}$ and $\mathcal{I}_{ji} = \frac{P(j)}{d_{ji}^\alpha}$, where $\alpha \in (2, 6)$ is the path-loss exponent.

A set L of links is called *P-feasible* if (1) holds for each link $i \in L$ when using power assignment P. A collection of sets is P-feasible if each set in the collection is P-feasible. When transmission power is also subject to optimization, we call a set of links (collection) simply *feasible* if there is a power assignment P such that the set (collection) is P-feasible.

The *scheduling* problem with a fixed power scheme P is to partition a given set L into the minimum number of P-feasible subsets (or *slots*).

The *scheduling problem with power control* is to partition a given set L into the minimum number or feasible subsets.

For simplicity of constructions, we assume henceforth that $N = 0$. The bounds can be adapted for the case of positive noise by scaling power levels.

Affectance: Fixed Power Assignments and Power Schemes. Following [26], we define the *affectance* $a_P(i, j)$ of link j by link i under power assignment P by

$$a_P(i, j) = \frac{\mathcal{I}_{ij}}{\mathcal{S}_j} = \frac{P(i)l_j^\alpha}{P(j)d_{ij}^\alpha}.$$

We let $a_P(j, j) = 0$ and extend a_P additively over sets: $a_P(S, i) = \sum_{j \in S} a_P(j, i)$ and $a_P(i, S) = \sum_{j \in S} a_P(i, j)$. It is readily verified (recall that we assumed $N = 0$) that a set of links S is P-feasible if and only if $a_P(S, i) \leq 1/\beta$ for all $i \in S$.

We will be particularly interested in *power schemes* P_δ of the form $P_\delta(i) = C \cdot l_i^{\delta\alpha}$, where C is constant for the given network instance. These are called *oblivious power assignments* because the power level of each link depends only on a local information - the link length. Examples of such power schemes are *uniform* power scheme (P_0), *linear* power scheme (P_1) and *mean* power scheme ($P_{1/2}$) [11].

It will be useful to note that for any power scheme P_δ, $a_{P_\delta}(i, j) = \left(\frac{l_i^\delta l_j^{(1-\delta)}}{d_{ij}}\right)^\alpha$.

Affectance: Power Control. For scheduling with power control, the following definition of affectance is used [29]: $a(i, j) = \left(\frac{l_i}{d(i, j)}\right)^\alpha$.

Similarly as before, we define $a(i, i) = 0$ and extend a additively to sets. As shown in [29], if the following condition holds for any link $i \in S$ with a sufficiently small constant γ (depending on α and β), then set S is feasible: $a(S_i^-, i) < \gamma$, where S_i^- denotes the subset of links in S that are no longer than link i: $S_i^- = \{j \in S : l_j \leq l_i\}$.

3 Lower Bounds for First-Fit Algorithms

3.1 Scheduling with Fixed Power Schemes

The *first-fit* algorithm considered in [26] was originally used for the uniform power scheme, but applies also to other oblivious power schemes [22]. The algorithm is a simple greedy procedure, where one starts with empty slots in a fixed order, then the links are processed in *non-decreasing order by length* and a link i is assigned to the first slot S such that $a_P(S, i) + a_P(i, S) < \gamma$ for a given constant γ. One may also generalize the acceptance condition with $f[a_P(S, i), a_P(i, S)] < \gamma$, where $f : \mathbb{R}_+ \times \mathbb{R}_+ \to \mathbb{R}$ is a decreasing function of both arguments that goes to 0 when both arguments go to 0. Below, NDFirstFit will refer to such an algorithm.

The family of hard network instances for the first-fit algorithm is inspired by a well known tree construction called *binomial trees* (in relation to binomial heaps), that has been used to obtain lower bounds for first-fit algorithms for graph coloring [4,17]. For any given power scheme P_δ with $\delta \in (0,1)$, we construct a family L_R of network instances on the real line using binomial trees as a model, where no pair of links corresponding to adjacent nodes in the tree can be in the same P_δ-feasible set together, but are otherwise spatially well separated. We show that while L_R can be scheduled in constant number of slots using P_δ, NDFirstFit gives only an $\Omega(\log n)$ approximation in terms of n and $\Omega(\frac{\log \Delta}{\log \log \Delta})$ approximation in terms of Δ.

Theorem 1. *Let $\delta \in (0,1)$. For each $n_0 > 0$, there is a set of $n > n_0$ links L_R on the real line such that NDFirstFit achieves no better than $\Omega(\log n) = \Omega(\frac{\log \Delta}{\log \log \Delta})$ approximation for scheduling with P_δ.*

Binomial Trees. We use the following family of trees, known as *binomial trees*, as a model for our construction. The rooted tree T_R, $(R \geq 0)$, is constructed recursively, as follows. T_0 consists of a single root vertex. For $R \geq 0$, the tree T_{R+1} is obtained from T_R by adding a new child vertex to the root, then adding a copy of T_R, by identifying its root node with the new child. For example, T_1 consists of two nodes connected by an edge and T_2 consists of a root vertex that has two children and one "grandchild". Note that the number of vertices in T_R is $n = 2^R$. Let us call the set of leaves of T_R *layer R*. For $t = R-1, R-2, \ldots, 1$, *layer t* denotes the set of leaves of the tree that remains after removing layers $R, R-1, \ldots, t+1$. Thus, T_R has $R+1$ layers and the root is in layer 0. Further, each layer t contains 2^{t-1} vertices, except layer 0, which contains one vertex (the root). Note also that each layer t node has exactly one child from each of layers $R, R-1, \ldots, t+1$.

Remark. "Layer" should not be confused with "level", i.e. the set of vertices at a given distance from the root. Note that a layer contains links from many levels.

The Network Instance. Let us fix an integer $R > 0$ and $\delta \in (0,1)$. For simplicity of the argument, we assume that $\beta = 1$. We model the set L_R of links after the tree T_R. Each link $i \in L_R$ corresponds to a vertex v_i of the tree and all the links are arranged on the real line. See Fig. 1 for an example. The links

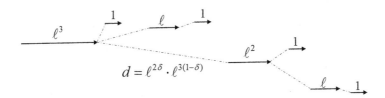

Fig. 1. The set L_R with $R = 3$.

corresponding to layer t vertices have length ℓ^{R-t}, where $\ell = cR^{1/\gamma} = c\log^{1/\gamma} n$, $\gamma = \alpha \cdot \min\{\delta, 1 - \delta\}$ and $c > 1$ is a large enough constant to be determined below. For instance, the "root link" has length ℓ^R and the "leaves" have length 1. Note also that $\Delta(L_R) = \ell^R$. Links are numbered arbitrarily, from 1 to n. The link corresponding to the root (leaves) of the tree is called *root link* (*leaf* links). Link j is called the *parent* of link i (and i is a *child* of link j) if v_j is the parent of vertex of v_i. Link j is called a *left sibling* of a link i (and i is called a *right sibling* of j) if $l_j < l_i$ and v_i and v_j have the same parent in the tree. We will place the parent and all left siblings of each link i to the left from link i (no two links occupy intersecting intervals in our construction). The *descendants* of link i are the links corresponding to the set of nodes of the subtree rooted at v_i.

The root is placed with its sender on the origin and the receiver at the coordinate ℓ^R. Assume links i and j are so that the node corresponding to i is the parent of the node of j in the tree (hence, $l_i > l_j$). Moreover, assume that the placement of link i has already been determined. Then we place link j so that $s_j = r_i + d_{ji} = r_i + l_i^{1-\delta}l_j^{\delta}$ and $r_j = s_j + l_j$. Such placement guarantees that any two links corresponding to adjacent tree nodes cannot be in the same P_{δ}-feasible set (recall that $\beta = 1$).

Analysis. The goal of the analysis below is to show that the set L_R can be scheduled in two sets, each corresponding to a color class in a proper 2-coloring of tree T_R. This is achieved by taking the constant c large enough. Having this, it will follow immediately that NDFirstFit schedules L_R into R slots, by allocating a separate slot for each layer of links (as in the increasing order of length, links come layer-by-layer), and, by the construction, no two layers can be in the same slot. This yields an approximation lower bound $\Omega(R) = \Omega(\log n) = \Omega(\log \Delta / \log \log \Delta)$.

Let us give a taste of the analysis below, by considering the affectance between a chain of three links; it will also give an idea why the construction cannot be implemented for the uniform power scheme (i.e. when $\delta = 0$). Let i, j, k be such that i is the parent of j and j is the parent of k. The placement of the links is such that the distance between e.g. j and k is $d_{kj} = l_j^{1-\delta}l_k^{\delta}$, meaning that $a_{P_{\delta}}(k,j) = \left(\frac{l_j^{1-\delta}l_k^{\delta}}{d_{kj}}\right)^{\alpha} = 1$, so j and k cannot coexist in the same feasible set, nor can i and j. On the other hand, the distance between i and k is at least $d_{ki} > d_{ji} = l_i^{1-\delta}l_j^{\delta}$. Hence, $a_{P_{\delta}}(k,j) < \left(\frac{l_i^{1-\delta}l_k^{\delta}}{l_i^{1-\delta}l_j^{\delta}}\right)^{\alpha} = \left(\frac{l_k}{l_j}\right)^{\alpha\delta} = \ell^{-\delta\alpha}$, which can be made arbitrarily small by taking large constant c (note that this property does not hold in the case of uniform power scheme, as $\delta = 0$).

Analysis: Link Placement. First we show that if the constant c is large enough then all the left siblings of a link i appear to the left of i. Note that if a link i has length ℓ^s, then its descendants constitute an instance L_s (i.e. correspond to the tree T_s). We start by computing the diameter $d(L_s)$ of L_s for any $s > 0$, i.e. the distance from the left-most node to the rightmost node of L_s.

Proposition 1. *If $\ell \geq 2$ then, for any $s > 0$, $\ell^s < d(L_s) < 4\ell^s$.*

Proposition 2. *If link j is a left sibling of link k then link j and its descendants are placed to the left from link k, provided that the constant c is large enough.*

Proof. Let link i be the parent of links j, k and assume w.l.o.g. that $l_i = \ell^p$, $l_j = \ell^t$ and $l_k = \ell^{t+1}$. We want link k to appear to the right of the whole "subtree" of links rooted at link j (i.e. descendants of link j); namely, $d(r_i, s_k) > d(r_i, s_j) + 2\ell \cdot d(L_t)$. Since i is the parent of j and k, we have, by definition, $d(r_i, s_k) = \ell^{p \cdot (1-\delta)} \cdot \ell^{(t+1) \cdot \delta}$ and $d(r_i, s_j) = \ell^{p \cdot (1-\delta)} \cdot \ell^{t \cdot \delta}$. Thus, due to the bound $d(L_t) < 4\ell^t$, it suffices to have: $\ell^{(1-\delta)p+\delta(t+1)} > \ell^{(1-\delta)p+\delta t} + 8\ell^{t+1}$ or

$$\ell^{(1-\delta)p+\delta t}(\ell^\delta - 1) > 8\ell^{t+1}.$$

Recall that $p \geq t + 2$ as link i is strictly longer than its children. Thus, assuming $\ell \geq 2^{1/\delta}$, the requirement above boils down to $\ell^{t-\delta+2} > 8\ell^{t+1}$, and thus to $\ell > 8^{\frac{1}{1-\delta}}$ which holds if the constant c is large enough.   □

Analysis: Affectance. Note that by the claim above, if links i and j are such that $l_i > l_j$ and j is not a descendant of link i then $d(i, j) > 2l_i$.

Let us fix a link i with $l_i = \ell^p$ and let L_R^i denote the set of all links in L_R except the parent and the children of link i. We show that the affectance of link i by links in L_R^i can be made arbitrarily small if the constant c is large enough.

We split L_R^i into the following subsets: D_i - the descendants of i, A_i - links that are longer than i, B_i - links that are shorter than i, *excluding the descendants of link i*, and E_i – links of length l_i. Recall that $\gamma = \alpha \min\{\delta, 1 - \delta\}$. While the following three claims follow relatively easily from the construction, the last one requires more care.

Proposition 3. $a_{P_\delta}(E_i, i) + a_{P_\delta}(i, E_i) = O(c^{-\alpha})$.

Proposition 4. $a_{P_\delta}(A_i, i) + a_{P_\delta}(i, A_i) = O(c^{-\gamma})$.

Proposition 5. $a_{P_\delta}(D_i, i) + a_{P_\delta}(i, D_i) = O(c^{-\gamma})$.

Proposition 6. $a_{P_\delta}(B_i, i) + a_{P_\delta}(i, B_i) = O(c^{-\gamma})$.

Proof. Let B_i^q denote the set of length ℓ^q links in B_i. We collect the links of B_i^q into disjoint subsets $S_1, S_2, \ldots$ by "climbing" from link i towards the root, as follows. Suppose we are at link i. The set S_1 contains the links of B_i^q that appear in the interval between link i and its parent; the distance between each link in S_1 and link i is at least $2l_i = 2\ell^p$, and the number of such links is at most 2^{p-q}. S_2 contains the links of B_i^q that are descendants of the first right sibling of link i; the distance between each link in S_2 and link i is at least $2\ell^{p+1}$ and the number of such links is at most 2^{p-q+1}. S_3 contains the links that are descendants of the second right sibling of link i; the links in S_3 are at a distance at least $2\ell^{p+2}$ and the number of links in S_3 is at most $2\ell^{p+3}$. After collecting the descendants of all right siblings of link i, we move to the parent of link i and repeat the same procedure. This process is carried on until reaching the root. Thus, we get the following bound on the affectance by the links in B_i^q:

$$a_{P_\delta}(B_i^q, i) < \sum_{t \geq 0} 2^{p+t-q} \left(\frac{\ell^{q\delta} l_i^{1-\delta}}{2\ell p + t} \right)^\alpha < 2^{-\alpha}(\ell/2)^{-\delta\alpha(p-q)} \sum_{t \geq 0} (\ell/2)^{-t\alpha}.$$

The last sum is clearly bounded by 2, given that $\ell > 4$. Thus, we have that $a_{P_\delta}(B_i^q, i) < (\ell/2)^{-\delta\alpha} < (c/2)^{-\delta\alpha}/R$. The first part of the claim follows because there are at most R different sets B_i^q for a fixed link i. The second part follows by a symmetric argument.   □

Analysis: Conclusion. Thus, we have that the affectance of each link i from all other links except its parent and children is small. In order to schedule the set L_R into two slots, it is enough to place each link separately from its parent and children. Two slots are sufficient because of the tree topology.

Now let us see what happens when we run NDFirstFit on the set L_R. Note that when processing the links in a non-decreasing order of length we will process all links of layer $t + 1$ before the first link of layer t. Let S_t be the set of layer-t links. We have, by the results above, that $a_{P_\delta}(S_t, i) + a_{P_\delta}(i, S_t) < \gamma$ for any constant γ and link $i \in S_t$. Thus, NDFirstFit puts the set S_R in the first slot. Each link in S_{R-1} conflicts with a link in S_R (i.e. its child link), so the next slot will consist of the set S_{R-1}. In general, each layer t link conflicts with a link from *each* of layers $t + 1, t + 2, \ldots, R$, so by the reasoning above, NDFirstFit schedules each layer in a separate slot (no two layers can occupy the same slot), taking $R + 1$ slots in total for scheduling L_R. Thus, the approximation ratio of NDFirstFit is $\Omega(R) = \Omega(\log n)$. Also, recall that $\Delta(L_R) = \ell^R = cR^{R/\gamma}$, so $R = \Omega(\frac{\log \Delta}{\log \log \Delta})$.

Note that the approximation ratio is multiplicative, as we can just multiply each link k times (i.e. replace it with its k identical copies), for any $k = poly(n)$. Then the optimum number of slots required will be $2k$, while NDFirstFit will schedule the set into $\Omega(k \cdot R) = \Omega(k \cdot \log kn) = \Omega(k \cdot \frac{\log \Delta}{\log \log \Delta})$ slots.

3.2   Scheduling with Power Control

In this section, we let NDFirstFitPC denote the first-fit scheduling algorithm (with power control) of [29]. Here also, one starts with empty slots in a fixed order, then the links are processed in non-decreasing order by length and a link i is assigned to the first slot S such that $a(S, i) < \gamma$ for a small constant γ. It is known that NDFirstFitPC achieves $O(\log n)$ approximation [29]. Using similar constructions as in the case of fixed power schemes, we prove the following.

Theorem 2. *For each $n_0 > 0$, there is a set of $n > n_0$ links L_R on the real line such that NDFirstFitPC achieves no better than $\Omega(\log n) = \Omega(\frac{\log \Delta}{\log \log \Delta})$-approximation for scheduling with power control.*

The Construction. For the sake of simplicity, we assume in this section that $\beta > 2^\alpha$: the lower bounds can be straightforwardly adapted for any constant $\beta > 1$. The construction in this case is similar to the one for fixed power schemes and is modeled after trees T_R. Let us fix an integer $R > 0$. Each link i corresponds to a vertex v_i of the tree and all the links are arranged on the real line. The links corresponding to layer t vertices have length ℓ^{R-t}, where $\ell = cR^{1/\alpha} = c \log^{1/\alpha} n$ and $c > 1$ is a large enough constant. The placement of links is similar to the construction for fixed power schemes: each child link j of a link i is placed so that $s_j = r_i + d(i, j) = r_i + l_i$ and $r_j = s_j + l_j$, assuming link i has been placed. It is known that if $\beta > 2^\alpha$, then the minimum distance between any two links that are in the same slot must be greater than the length of the smaller one [25, Theorem 4]; hence, link placement as above guarantees that any two links corresponding to adjacent tree nodes cannot be in the same slot.

Analysis. As before, it can be shown that if the constant c is large enough then all the left siblings of a link i appear to the left of link i.

Proposition 7. *If link j is a left sibling of link k then link j and its descendants are placed to the left from link k at a distance at least $l_k/2$ from link k, provided that $\ell > 6$.*

Note that by the claim above, if links i and j are such that $l_i > l_j$ and j is not a descendant of link i then $d(i, j) > l_i/2$.

For any link i, let $\underline{L}_R^i$ denote the set of links in L_R that are *not longer* than link i, excluding the children of link i. Note that we do not need to consider links longer than link i. The following claim is proved in a similar way as the analogous results for fixed powers. In fact, the bounds here are stronger because we have similar spacing between links as before, while the numerators of affectance are smaller as we consider only the affectance by smaller links.

Proposition 8. $a(\underline{L}_R^i, i) = O(c^{-\alpha})$.

The rest of the analysis is identical to the case of fixed powers.

4   Lower Bounds for Uniform Randomized Algorithms

Next we consider a generalization of the distributed algorithm presented in [31]. In this algorithm, the sender nodes of the links act in synchronous rounds and each sender node transmits with probability p_i or waits with probability $1 - p_i$ in round i, where p_i is the same for all links (but may change across the rounds). Once the transmission succeeds in round i, the node is silent in subsequent rounds.

Our construction in this case is modeled after a complete $\log^b n$-ary tree with n nodes, where $b > 0$ is a constant, and is loosely based on [20, Theorem 6]. In [20], a similar lower bound is obtained for coloring interval graphs using randomized distributed algorithms. It is worth to mention, however, that the analysis of the algorithm here is done on sparser graphs (trees of cliques) than in [20] (intersection graphs of laminar sets of intervals), but fortunately the argument can be adapted.

The main challenge is to construct a family of network instances that are structurally similar to $\log^b n$-ary trees. Namely, links correspond to adjacent nodes in the tree are not P_δ-feasible together, but are otherwise well separated.

Theorem 3. *Let* $\delta \in (0,1)$ *and the probabilities* $p_i (i = 1, 2, \dots)$ *be fixed. For each* $n_0 > 0$, *there is a set of* $n > n_0$ *links* L *on the real line s.t. the randomized algorithm that uses probabilities* $p_i (i = 1, 2, \dots)$ *yields only a* $\Omega(\frac{\log n}{\log \log n}) = \Omega(\frac{\log \Delta}{\log \log \Delta})$ *approximation for scheduling with power scheme* P_δ, *w.h.p.*

The Construction. We assume in this section that $\beta = 1$. We start with the description of a preliminary set S of links simulating a rooted complete $\log^b n$-ary tree over a set of n/M nodes, where $b > 1$ is a constant to be chosen and $M = O(n^\epsilon)$ is a parameter with $\epsilon \in (0,1)$ a constant. The main construction is a simple extension of S. We will often mix the terminology of links and trees, e.g. by speaking of children of links, hoping it does not cause confusion. We split the tree into *levels*, where the root is at level 0 and the nodes at (tree-) distance t from the root constitute level t. Note that the number of nodes at level t is $\log^{tb} n$; hence, the number of levels is $k = \Theta\left(\frac{\log(n/M)}{\log \log n}\right) = \Theta\left(\frac{\log n}{\log \log n}\right)$. For each $t \geq 0$, the level-t links have uniform length ℓ_t. We set

$$\ell_t = c\ell_{t+1} \log^{d(t+1)} n \tag{2}$$

for large enough constants $c, d > 0$. We describe the placement of links on the real line level by level, starting from level 0, which contains a single link i. We set $s_i = 0$, $r_i = s_i + \ell_0$, as shown in Fig. 2. The children of link i have length ℓ_1. We place the $\log^b n$ child links of length ℓ_1 inside the interval occupied by the link i, so that for any children j, k, it holds that (see Fig. 2):

1. $d(j,k) \geq e\ell_1$ for constant $e > 1$,
2. $d(r_j, r_i) \geq \ell_0^{1-\delta} \ell_1^\delta / 2$,
3. $d(s_j, r_i) \leq \ell_0^{1-\delta} \ell_1^\delta$,
4. $d(s_i, s_j) \geq \ell_0 / 2$.

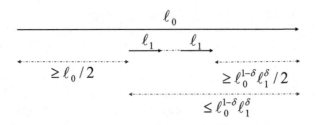

Fig. 2. The first step of the construction of Theorem 3.

This completes the first step of the construction of S. The idea behind the constraints above is to guarantee the following properties: 1. the set of links at the same level is feasible, 2. the children interfere with the parent, 3. the grandchildren do not interfere with their grandparent, 4. the parent does not interfere the children (or their descendants). At the second step, we construct the children of level-1 links in a similar manner, and continue this process until having n/M links. As shown below, the length ratios defined by (2) ensure that the construction is correct and, in particular, that no link can be in the same feasible set as any of its children. On the other hand, we prove that the affectance of any level-t link by all other links, except level-$t-1$ and level-$t+1$ links, is small. This implies that the set S can be scheduled in a constant number of slots using the power scheme P_δ.

In order to complete the construction, we replace each link in S with its M identical copies. Let L denote this set of links. Note that $|L| = n$. Note that the optimum number of slots for scheduling L is at least M, as different copies of the same link should be placed in different slots. Using the properties of set S, we show that $\Theta(M)$ slots suffice.

Analysis: Properties of Set S. We start by proving the properties of the set S. The first question to address is: what are the requirements on the lengths of links for satisfying the constraints (1-4)? The first three constraints will hold if $\ell_0^{1-\delta}\ell_1^\delta/2 > 3e\ell_1 \log^b n$, which holds if $\delta < 1$ and the constants c, d in (2) are large enough. The fourth constraint requires: $\ell_0 - \ell_0^{1-\delta}\ell_1^\delta > \ell_0/2$, which holds if $\ell_0 > 2^{1/\delta}\ell_1$. Thus, choosing the constants in (2) large enough guarantees the constraints (1-4).

Now let us show that S can be scheduled in a constant number of slots. Let S_t denote the set of level-t links for $t \geq 0$.

Proposition 9. *If the constants c, d in (2) are large enough, then for any level-t link i ($t \geq 0$), it holds that $a_{P_\delta}(T, i) < 1$, where $T = S \setminus (S_{t-1} \cup S_{t+1})$.*

Proof. First, let us bound the affectance by links from S_t. Recall that those links have equal lengths and their mutual distances are at least $e\ell_t$, by the first constraint of the construction. Thus we have:

$$a_{P_\delta}(S_t, i) < 2\sum_{r\geq 1}\left(\frac{\ell_t^\delta \ell_t^{1-\delta}}{re\ell_t}\right)^\alpha = O(e^{-\alpha}),$$

where the factor of 2 accounts for links on both sides of i. The last term can be made arbitrarily small if constant c is large enough, because $\alpha > 1$.

Now, let us fix an $s > t + 1$. The number of level-s links is $|S_s| = \log^{sb} n$. The distance from each level-s link to r_i is at least $\ell_t^{1-\delta}\ell_{t+1}^\delta/2$, by construction. Thus, we have:

$$a_{P_\delta}(S_s, i) \leq |S_s|\frac{\ell_s^{\delta\alpha}\ell_t^{(1-\delta)\alpha}}{(\ell_t^{1-\delta}\ell_{t+1}^\delta/2)^\alpha} = 2^\alpha|S_s|\left(\frac{\ell_s}{\ell_{t+1}}\right)^{\delta\alpha}.$$

Since the number of levels is $O(\log n)$, it is enough to have $a(S_s, i)\frac{e'}{\log n}$ for any constant $e' > 0$, which is provided if

$$\ell_{t+1} > 2^{1/\delta}(e'|S_s|\log n)^{1/(\delta\alpha)}\ell_s = 2^{1/\delta}e'^{1/(\delta\alpha)}\log^{(sb+1)/(\alpha\delta)} n\ell_s.$$

The last inequality holds if we set $d \geq 2b/(\alpha\delta)$ and $c \geq 2^{1/\delta}e'^{1/(\delta\alpha)}$ in (2).

Next, let us consider a layer $s < t-1$ for $t > 0$. Recall that the distance from each link of L_s to r_i is at least $\ell_s/2$, by construction. The affectance by S_s can be bounded as follows: $a_{P_\delta}(S_s, i) \leq |S_s|\frac{\ell_s^{\delta\alpha}\ell_t^{(1-\delta)\alpha}}{(\ell_s/2)^\alpha} < 2^\alpha|S_s|\left(\frac{\ell_t}{\ell_s}\right)^{\delta\alpha}$.

Hence, again, we can easily get $a_{P_\delta}(S_s, i) < \frac{e'}{\log n}$ for any constant $e' > 0$, by tuning the constants c and d in (2). This yields the claim. □

Thus, the set S can be scheduled into two feasible slots, by taking the union of the odd-numbered levels in one slot and the union of the even-numbered levels in another one. This directly implies that the set L can be scheduled in $2M$ slots.

Analysis: The Lower Bound. It remains to prove that for any sequence $p_i \in (0,1)$, $i = 1, 2\ldots$, the randomized algorithm using the probabilities p_i will schedule L in $\Omega(kM) = \Omega(M \cdot \frac{\log n}{\log\log n}) = \Omega(M \cdot \frac{\log\Delta}{\log\log\Delta})$ slots, where the last equality holds because $\Delta = poly(n)$. To that end, it will be more convenient to analyze the algorithm in terms of a *conflict graph* G corresponding to L, rather than the set of links itself. The graph G is constructed by replacing each vertex of a complete $\log^b n$-ary tree on n/M vertices with a M-clique, where the cliques corresponding to two adjacent vertices form a $2M$-clique. Obviously, $\chi(G) = 2M$. Level-t vertices in G are the vertices corresponding to level-t vertices in the tree. Let the probabilities p_i, $i = 1, 2, \ldots$ be fixed. We consider the following variant of the algorithm with relaxed constraints on transmissions. In round i, each remaining vertex v of G selects itself with probability p_i and is removed from the graph in this round if it selects itself and no neighbor is selected. Lower-bounding the runtime of this algorithm for G implies a similar bound for the original algorithm running on the set L of links, because any feasible set in L corresponds to an independent set in G, i.e. we essentially neglect some part of the interference when dealing with G, which can only make the algorithm use less slots. The argument below is similar to the proof of [20, Theorem 6]. The main idea is to show that whatever the values of p_i are, the algorithm will remove the vertices level by level, starting from the last level vertices. In particular, it will take $\Theta(M)$ steps to start making essential impact on the next level. This gives the desired bound $\Omega(k \cdot M)$.

Let T_t denote the first time step when the size of a level-t M-clique is halved. Let H_t denote the event that for all $s \leq t$, the size of the smallest level-s clique is at least $(1 - 1/\log n)M$ before iteration $T_{t+1} + 1$.

Proposition 10. *Consider $0 \leq t < k$. Suppose that $T_{t+1} < M \log n$. Then*

$$\mathbb{P}[H_t] = 1 - O(n^{-\frac{M}{130\log n}+1}).$$

Observe that given the event H_t, the difference between the times T_{t+1} and T_t is at least $M/4$ if n is large enough. Indeed, H_t implies that in round T_{t+1}, the size of each clique in levels $t, t-1, \ldots, 0$ is at least $3M/4$, and in order for a clique of size $3M/4$ to become less than $M/2$, at least $M/4$ rounds must pass. Thus, $\mathbb{P}[T_t - T_{t+1} \geq M/4] \geq \mathbb{P}[H_t] = 1 - O(n^{-\frac{M}{130\log n}+1})$ holds for each fixed t. By the union bound, the probability that the event $T_{t+1} - T_t \geq M/4$ is violated for at least one t is at most $O(k \cdot n^{-\frac{M}{130\log n}+1}) = O(n^{-\frac{M}{130\log n}+2})$. Thus, if $M > 130c\log n$ (recall that $M = \Theta(n^\epsilon)$), then with probability $1 - O(n^{2-c})$, it takes at least $k \cdot M/4 = \Omega(M\log n/\log\log n)$ steps until all the vertices of the graph are removed. This completes the lower bound argument.

References

1. Avin, C., Emek, Y., Kantor, E., Lotker, Z., Peleg, D., Roditty, L.: SINR diagrams: convexity and its applications in wireless networks. J. ACM **59**(4), 18 (2012)
2. Belke, L., Kesselheim, T., Koster, A.M.C.A., Vöcking, B.: Comparative study of approximation algorithms and heuristics for SINR scheduling with power control. In: Bar-Noy, A., Halldórsson, M.M. (eds.) ALGOSENSORS 2012. LNCS, vol. 7718, pp. 30–41. Springer, Heidelberg (2013)
3. Bodlaender, M., Halldórsson, M.M.: Beyond geometry: towards fully realistic wireless models. In: PODC (2014)
4. Caragiannis, I., Fishkin, A.V., Kaklamanis, C., Papaioannou, E.: A tight bound for online colouring of disk graphs. Theor. Comput. Sci. **384**(2–3), 152–160 (2007)
5. Chafekar, D., Kumar, V., Marathe, M., Parthasarathy, S., Srinivasan, A.: Cross-layer latency minimization for wireless networks using SINR constraints. In: Mobihoc (2007)
6. Cruz, R.L., Santhanam, A.: Optimal routing, link scheduling, and power control in multi-hop wireless networks. In: INFOCOM (2003)
7. Dams, J., Hoefer, M., Kesselheim, T.: Scheduling in wireless networks with Rayleigh-fading interference. In: SPAA (2012)
8. Daum, S., Gilbert, S., Kuhn, F., Newport, C.: Broadcast in the ad hoc SINR model. In: Afek, Y. (ed.) DISC 2013. LNCS, vol. 8205, pp. 358–372. Springer, Heidelberg (2013)
9. Dinitz, M.: Distributed algorithms for approximating wireless network capacity. In: INFOCOM (2010)
10. ElBatt, T., Ephremides, A.: Joint scheduling and power control for wireless ad-hoc networks. In: INFOCOM (2002)
11. Fanghänel, A., Kesselheim, T., Räcke, H., Vöcking, B.: Oblivious interference scheduling. In: PODC, August 2009
12. Fu, L., Liew, S.C., Huang, J.: Power controlled scheduling with consecutive transmission constraints: complexity analysis and algorithm design. In: INFOCOM. IEEE (2009)
13. Goussevskaia, O., Halldórsson, M.M., Wattenhofer, R.: Algorithms for wireless capacity. IEEE/ACM Trans. Netw. **22**(3), 745–755 (2014)
14. Goussevskaia, O., Oswald, Y.A., Wattenhofer, R.: Complexity in geometric SINR. In: MobiHoc (2007)
15. Gudmundsdottir, H., Ásgeirsson, E.I., Bodlaender, M., Foley, J.T., Halldórsson, M.M., Vigfusson, Y.: Measurement based interference models for wireless scheduling algorithms. In: MSWiM (2014)

16. Gupta, P., Kumar, P.R.: The capacity of wireless networks. IEEE Trans. Inf. Theor. **46**(2), 388–404 (2000)
17. Gyárfás, A., Lehel, J.: On-line and first fit colorings of graphs. J. Graph Theor. **12**(2), 217–227 (1988)
18. Halldórsson, M.M.: Wireless scheduling with power control. ACM Trans. Algorithms **9**(1), 7 (2012)
19. Halldórsson, M.M., Holzer, S., Mitra, P., Wattenhofer, R.: The power of non-uniform wireless power. In: SODA, pp. 1595–1606 (2013)
20. Halldórsson, M.M., Konrad, C.: Distributed algorithms for coloring interval graphs. In: Kuhn, F. (ed.) DISC 2014. LNCS, vol. 8784, pp. 454–468. Springer, Heidelberg (2014)
21. Halldórsson, M.M., Mitra, P.: Nearly optimal bounds for distributed wireless scheduling in the SINR model. In: Aceto, L., Henzinger, M., Sgall, J. (eds.) ICALP 2011, Part II. LNCS, vol. 6756, pp. 625–636. Springer, Heidelberg (2011)
22. Halldórsson, M.M., Mitra, P.: Wireless capacity with oblivious power in general metrics. In: SODA (2011)
23. Halldórsson, M.M., Mitra, P.: Wireless capacity and admission control in cognitive radio. In: INFOCOM (2012)
24. Halldórsson, M.M., Mitra, P.: Wireless connectivity and capacity. In: SODA (2012)
25. Halldórsson, M.M., Tonoyan, T.: How well can graphs represent wireless interference? In: STOC (2015)
26. Halldórsson, M.M., Wattenhofer, R.: Wireless communication is in APX. In: Albers, S., Marchetti-Spaccamela, A., Matias, Y., Nikoletseas, S., Thomas, W. (eds.) ICALP 2009, Part I. LNCS, vol. 5555, pp. 525–536. Springer, Heidelberg (2009)
27. Jurdzinski, T., Kowalski, D.R., Rozanski, M., Stachowiak, G.: On the impact of geometry on ad hoc communication in wireless networks. In: PODC (2014)
28. Katz, B., Volker, M., Wagner, D.: Energy efficient scheduling with power control for wireless networks. In: WiOpt (2010)
29. Kesselheim, T.: A constant-factor approximation for wireless capacity maximization with power control in the SINR model. In: SODA (2011)
30. Kesselheim, T.: Approximation algorithms for wireless link scheduling with flexible data rates. In: Epstein, L., Ferragina, P. (eds.) ESA 2012. LNCS, vol. 7501, pp. 659–670. Springer, Heidelberg (2012)
31. Kesselheim, T., Vöcking, B.: Distributed contention resolution in wireless networks. In: Lynch, N.A., Shvartsman, A.A. (eds.) DISC 2010. LNCS, vol. 6343, pp. 163–178. Springer, Heidelberg (2010)
32. Lin, H., Schalekamp, F.: On the complexity of the minimum latency scheduling problem on the Euclidean plane. arXiv preprint 1203.2725 (2012)
33. Maheshwari, R., Jain, S., Das, S.R.: A measurement study of interference modeling and scheduling in low-power wireless networks. In: SenSys (2008)
34. Moscibroda, T., Wattenhofer, R.: The complexity of connectivity in wireless networks. In: INFOCOM (2006)
35. Rappaport, T.S.: Wireless Communications: Principles and Practice, 2nd edn. Prentice Hall, Upper Saddle River (2002)
36. Son, D., Krishnamachari, B., Heidemann, J.: Experimental study of concurrent transmission in wireless sensor networks. In: SenSys (2006)
37. Tonoyan, T.: On some bounds on the optimum schedule length in the SINR model. In: Bar-Noy, A., Halldórsson, M.M. (eds.) ALGOSENSORS 2012. LNCS, vol. 7718, pp. 120–131. Springer, Heidelberg (2013)

Deterministic Rendezvous with Detection Using Beeps

Samir Elouasbi$^{(\boxtimes)}$ and Andrzej Pelc

Département d'informatique, Université du Québec en Outaouais,
Gatineau, QC J8X 3X7, Canada
{elos02,pelc}@uqo.ca

Abstract. Two mobile agents, starting at possibly different times from arbitrary nodes of an unknown network, have to meet at some node. Agents move in synchronous rounds. They have different positive integer labels. Each agent knows its own label but not the label of the other agent. In traditional formulations of the rendezvous problem, meeting is accomplished when agents get to the same node in the same round. We seek a more demanding goal, called *rendezvous with detection*: agents must become aware that the meeting is accomplished, simultaneously declare this and stop. This awareness depends on how agents communicate. We use two variations of a very weak communication model, called the *beeping model*, introduced in [8]. In each round an agent either listens or beeps. In the *local beeping model*, an agent hears a beep if it listens in this round and if the other agent is at the same node and beeps. In the *global beeping model*, an agent hears a *loud* (resp. a *soft*) beep if it listens in this round and if the other agent is at the same node (resp. at another node) and beeps.

We first present a deterministic algorithm of rendezvous with detection working, even for the local beeping model, in an arbitrary unknown network in time polynomial in the size of the network and in the length of the smaller label (i.e., in the logarithm of this label). However, in this algorithm, agents spend a lot of energy: the number of moves that an agent must make, is proportional to the time of rendezvous. It is thus natural to ask if *bounded-energy agents*, i.e., agents that can make at most c moves, for some integer c, can always achieve rendezvous with detection as well, in bounded size networks. We prove that the answer to this question is positive, even in the local beeping model but, perhaps surprisingly, this ability comes at a steep price of time: the meeting time of bounded-energy agents is *exponentially* larger than that of unrestricted agents. By contrast, we show an algorithm for rendezvous with detection in the global beeping model that works for bounded-energy agents (in bounded-size networks) as fast as for unrestricted agents.

Keywords: Algorithm · Rendezvous · Detection · Synchronous · Deterministic · Network · Graph · Beep

A. Pelc—Supported in part by NSERC discovery grant 8136 – 2013 and by the Research Chair in Distributed Computing of the Université du Québec en Outaouais.

P. Bose et al. (Eds.): ALGOSENSORS 2015, LNCS 9536, pp. 85–97, 2015.
DOI: 10.1007/978-3-319-28472-9_7

1   Introduction

The Background and the Problem. Two mobile agents, starting at arbitrary, possibly different times from arbitrary nodes of an unknown network, have to meet at some node of it. This task is known as rendezvous [2]. The network is modeled as a simple undirected connected graph, and agents move in synchronous rounds: in each round an agent can either stay at the current node or move to one of its neighbors. Hence in each round an agent is at a specific node. Agents are mobile entities with unlimited memory; from the computational point of view they are modeled as Turing machines. In applications, these entities may represent mobile robots navigating in a labyrinth or in corridors of a building, or software agents moving in a communication network. The purpose of meeting might be to exchange data previously collected by the agents at nodes of the network, or to coordinate future network maintenance tasks, for example checking functionality of websites or of sensors connected in a network.

Agents have different labels which are positive integers. Each agent knows its own label, but not the label of the other agent. Agents do not know the topology of the network. They do not know the starting node or activation time of the other agent. They cannot mark the visited nodes in any way. Each agent appears at its starting node at the time of its activation by the adversary.

We seek rendezvous algorithms that do not rely on the knowledge of node labels, and can work in anonymous networks as well (cf. [2]). The importance of designing such algorithms is motivated by the fact that, even when nodes are equipped with distinct labels, agents may be unable to perceive them because of limited sensory capabilities, or nodes may refuse to reveal their labels to agents, e.g., due to security or privacy reasons. On the other hand, we assume that edges incident to a node v have distinct labels in $\{0, \dots, d-1\}$, where d is the degree of v. Thus every undirected edge $\{u, v\}$ has two labels, which are called its *port numbers* at u and at v. Port numbering is *local*, i.e., there is no relation between port numbers at u and at v. An agent entering a node learns the port of entry and the degree of the node. Note that, in the absence of port numbers, rendezvous is usually impossible, as all ports at a node look identical to an agent and the adversary may prevent the agent from taking some edge incident to the current node.

In traditional formulations of the rendezvous problem, meeting is accomplished when agents get to the same node in the same round. We want to achieve a more demanding goal, called *rendezvous with detection*: agents must become aware that the meeting is accomplished, simultaneously declare this and stop. This awareness depends on how an agent can communicate to the other agent its presence at a node. We use two variations of the *beeping model* of communication. In each round an agent can either listen, i.e., stay silent, or beep, i.e., emit a signal. In the *local beeping model*, an agent hears a beep in a round if it listens in this round and if the other agent is at the same node and beeps. In the *global beeping model*, an agent hears a *loud* beep in a round if it listens in this round and if the other agent is at the same node and beeps, and it hears a *soft* beep in a round if it listens in this round and if the other agent is at some other node and beeps. The beeping model was introduced in [8] for vertex coloring and used in [1] to solve the MIS problem. In the variant from [1,8] the beeping entities were nodes rather than agents, and beeps of a node were heard at adjacent nodes. The beeping model is widely applicable, as it makes small demands on communicating devices, relying only on carrier sensing. In the case of mobile agents, the local and the global beeping models are applicable in different settings. The local model is applicable even for agents having very weak transmissions capabilities, limiting reception of a beep to the same node. The global model is applicable for more powerful agents, that can beep sufficiently strongly to be heard in the entire network, and having a listening capability of differentiating a beep emitted at the same node from a beep emitted at a different node.

It should be noted that our local beeping model is an extremely weak way of communication between agents: they can communicate only when residing simultaneously at the same node, they cannot hear when they beep, and messages are the simplest possible. In fact, as mentioned in [8], beeps are an even weaker way of communicating than using one-bit messages, as the latter ones allow three different states (0,1 and no message), while beeps permit to differentiate only between a signal and its absence. Clearly, without any communication, rendezvous with detection is impossible, as agents cannot become aware of each other's presence at a node. Notice also that in the global beeping model it would not be possible to remove the distinction between hearing a loud beep when the beeping agent is at the same node and hearing a soft beep when the beeping agent is at a different node. Indeed, the same strength of beep reception would make it impossible for an agent A to inform the other agent B of the presence of A at the same node, and hence rendezvous with detection would be impossible. The global beeping model is at least as strong as the local one, in the sense that any algorithm of rendezvous with detection working in the local model works also in the global model, by simply ignoring soft beeps. We will see that the converse is not true.

For a given network, the execution time of an algorithm of rendezvous with detection, for agents with given labels starting in given rounds from given initial positions, is the number of rounds from the activation of the later agent to the declaration of rendezvous. For a given class of networks, the *time* of an algorithm of rendezvous with detection is its worst-case execution time, over all networks in the class, all initial positions, all pairs of distinct labels and all starting times.

Our Results. Our first result answers the basic question: Is it possible to achieve rendezvous with detection in arbitrary networks, and if so, how fast it can be done? We present a deterministic algorithm of rendezvous with detection working, even for the local beeping model, in an arbitrary unknown network in time polynomial in the size of the network and in the length of the smaller label (i.e., in the logarithm of this label). We do not assume the knowledge of any upper bound on the size of the network. The time complexity of our algorithm matches that of the fastest, known to date, rendezvous algorithm without detection, constructed in [21].

However, in this algorithm, agents spend a lot of energy: the number of moves of an agent is proportional to the time of rendezvous. On the other hand, in many applications, e.g., when agents are mobile robots, they are battery-powered devices, and hence the energy that an agent can spend on moves is limited. It is thus natural to ask if *bounded-energy agents*, i.e., agents that can make at most c moves, for some integer c, can always achieve rendezvous with detection as well. This is impossible for some networks of unbounded size. Hence we rephrase the question: Can bounded-energy agents always achieve rendezvous with detection in bounded-size networks? We prove that the answer to this question is positive, even in the local beeping model but, perhaps surprisingly, this ability comes at a steep price of time: the meeting time of bounded-energy agents is *exponentially* larger than that of unrestricted agents. By contrast, we show an algorithm for rendezvous with detection in the global beeping model that works for bounded-energy agents (in bounded-size networks) as fast as for unrestricted agents. Since algorithms for bounded-energy agents can work only for networks of bounded size, in these algorithms we assume knowledge of some such upper bound. Due to space constraints, proofs will appear in the journal version of the paper.

Related Work. The vast literature on rendezvous can be divided according to the mode in which agents move (deterministic or randomized) and the environment where they move (a network modeled as a graph or a terrain in the plane). An extensive survey of randomized rendezvous in various scenarios can be found in [2], cf. also [3,16]. Rendezvous of two or more agents in the plane has been considered e.g., in [12,13].

Our paper is concerned with deterministic rendezvous in networks, surveyed in [19]. In this setting a lot of effort has been dedicated to the study of the feasibility of rendezvous, and to the time required to achieve this task, when feasible. For instance, deterministic rendezvous with agents equipped with tokens used to mark nodes was considered, e.g., in [17]. Time of deterministic rendezvous of agents equipped with unique labels was discussed in [10,21]. Memory required by the agents to achieve deterministic rendezvous has been studied in [4,14] for trees and in [9] for general graphs. In [18] the authors studied tradeoffs between the time of rendezvous and the total number of edge traversals by both agents until the meeting.

Apart from the synchronous model used in this paper, several authors have investigated asynchronous rendezvous in the plane [7,12,13] and in network environments [5,11]. In the latter scenario the agent chooses the edge which it decides to traverse but the adversary controls the speed of the agent. Under this assumption rendezvous in a node cannot be guaranteed even in very simple graphs, and hence the rendezvous requirement is relaxed to permit the agents to meet inside an edge.

2   Preliminaries

In the rest of the paper the word "graph" means a simple connected undirected graph modeling a network. The *size* of a graph is the number of its nodes.

In this section we recall two procedures known from the literature, that will be used as building blocks in our algorithms. The aim of the first procedure is graph exploration, i.e., visiting all nodes of a graph by a single agent. The procedure, called $EXP(m)$, is based on universal exploration sequences (UXS) [15], and follows from the result of Reingold [20]. Given any positive integer m, it allows the agent to visit all nodes of any graph of size at most m, starting from any node of this graph, using $R(m)$ edge traversals, where R is some polynomial. After entering a node of degree d by some port p, the agent can compute the port q by which it has to exit; more precisely $q = (p + x_i) \mod d$, where x_i is the corresponding term of the UXS.

The second procedure, due to Ta-Shma and Zwick [21], guarantees rendezvous (without detection) in an arbitrary graph. Below we briefly sketch this procedure, which will be used in our algorithm of rendezvous with detection for unrestricted agents.

Let $\mathbb{Z}^+$ denote the set of positive integers and let $\mathbb{Z}^*$ denote the set of integers greater or equal than -1. For any positive integer L, Ta-Shma and Zwick define a function $\Phi_L : \mathbb{Z}^+ \times \mathbb{Z}^+ \times \mathbb{Z}^* \longrightarrow \mathbb{Z}^*$. The function Φ_L is *applied* by an agent with label L in a graph G at a node v of G as follows. Let $v_0 = v$ and let v_1 be the node adjacent to v_0, such that the edge $\{v_0, v_1\}$ has port number 1 at v_0. Suppose that nodes $v_0, v_1, \ldots, v_{t-1}$ are already constructed, so that v_{i+1} either equals v_i or is adjacent to v_i. The node v_t is defined as follows. In the case when $v_{t-1} = v_{t-2}$ and the degree of v_{t-1} is d, then $v_t = v_{t-1}$ if $\Phi_L(t, d, -1) = -1$; if $\Phi_L(t, d, -1) = q \geq 0$ then v_t is the node adjacent to v_{t-1} such that the port number at v_{t-1} corresponding to edge $\{v_{t-1}, v_t\}$ is q. In the case when $v_{t-1} \neq v_{t-2}$, the port number at v_{t-1} corresponding to edge $\{v_{t-1}, v_{t-2}\}$ is p and the degree of v_{t-1} is d, then $v_t = v_{t-1}$ if $\Phi_L(t, d, p) = -1$; if $\Phi_L(t, d, p) = q > 0$ then v_t is the node adjacent to v_{t-1} such that the port number at v_{t-1} corresponding to edge $\{v_{t-1}, v_t\}$ is q. Hence the application of function Φ_L at node v defines an infinite walk of the agent with label L in the graph G. This walk starts at v and in each round t the agent either stays at the current node or moves to an adjacent node by a port determined by the function Φ_L on the basis of the degree of the current node and of the port by which the agent entered it. A round t is called *active* for the agent if $v_t \neq v_{t-1}$ and it is called *passive* if $v_t = v_{t-1}$.

The following result, proved in [21], guarantees rendezvous without detection in polynomial time, if two agents apply functions Φ_L corresponding to their labels, in an unknown graph.

Theorem 1. *There exists a polynomial P in two variables, with the following property. Let G be an n-node graph and consider two agents with distinct labels L_1, L_2 respectively, starting at nodes v and w of the graph in rounds $t_1 \geq t_2$. Let $t \geq t_1$ and let ℓ be the smaller label. If agent with label L_i applies function Φ_{L_i} at its starting node, for $i = 1, 2$, then agents are simultaneously at the same node in some round of the time interval $[t, t + P(n, \log \ell)]$. Moreover, rendezvous occurs in a round which is active for one of the agents and passive for the other. The same property remains true if one of the agents is inert and the other agent applies its function Φ_{L_i}.*

3 Rendezvous with Detection of Unrestricted Agents

In this section we describe and analyze an algorithm of rendezvous with detection which works for unrestricted agents, i.e., for agents that can spend an arbitrary amount of energy on moves. It works even for the weaker of our two models, i.e., for the local beeping model. Our algorithm uses the following procedure which describes an infinite walk of an agent with label L, based on the above described application of the function Φ_L.

Procedure `Beeping walk`. Consider an agent with label L starting at node v of a graph G. Let W be the walk resulting from the application of Φ_L in graph G at node v. Each round of W is replaced by 2 consecutive rounds as follows. If round t of W is passive, i.e., $v_t = v_{t-1}$, then this round is replaced by two rounds in which the agent stays at v_t and listens. If round t of W is active, i.e., $v_t \neq v_{t-1}$, then this round is replaced by the following two rounds: in the first of these rounds the agent goes to v_t and beeps, and in the second of these rounds the agent stays at v_t and listens.

We now describe our algorithm for rendezvous with detection. It is executed by each agent. Note that the execution of procedure `Beeping walk`, called by the algorithm, depends on the label of the agent.

Algorithm. RV-with-detection
Perform procedure `Beeping walk` **until** you hear a beep
Let s be the round number when you first hear a beep
(counted since your wake-up)
Stay inert forever
Beep in round $s + 1$ listen in round $s + 2$
If you hear no beep in round $s + 2$ **then**
      declare rendezvous in round $s + 3$ and stop
else
      listen in round $s + 3$, declare rendezvous in round $s + 4$, and stop.

We now show that Algorithm `RV-with-detection` correctly accomplishes rendezvous with detection and works in time polynomial in the size of the graph and in the logarithm of the smaller label. The agent that starts later will be called the *later* agent and the other one the *earlier* agent. If agents start simultaneously, these qualifiers are attributed arbitrarily.

Theorem 2. *Consider two agents with distinct labels L_1, L_2 respectively, starting at nodes v and w of an n-node graph in possibly different rounds. Let ℓ be the smaller label. If both agents execute Algorithm* `RV-with-detection`, *then they meet and simultaneously declare rendezvous in time $O(P(n, \log \ell))$, i.e., polynomial in n and in $\log \ell$, after the start of the later agent.*

4 Rendezvous with Detection of Bounded-Energy Agents

In this section we study rendezvous with detection of agents that can perform a bounded number of moves. Let c be a positive integer. A *c-bounded agent* is defined as an agent that can perform at most c moves. (Notice that we do not restrict the number of beeps; indeed, the amount of energy required to make a move is usually so much larger than the amount of energy required to beep that ignoring the latter seems to be a reasonable approximation of reality in many applications.) Can c-bounded agents, for some integer c, perform rendezvous with detection in arbitrary graphs? The answer to this question is, of course, negative, even if detection is not required. For any integer c, c-bounded agents starting at distance larger than $2c$ cannot meet because at least one of them would have to make more than c steps. Even if we assume that the initial distance between the agents is 1, meeting of c-bounded agents is impossible in some graphs. Indeed, consider two n-node stars whose centers are linked by an edge, with agents starting at the centers of the stars. In the worst case, at least one of the agents must make at least $n - 1$ steps before meeting (to find the connecting edge), which is impossible for c-bounded agents, when n is large.

Thus, we rephrase the question: Can c-bounded agents always achieve rendezvous with detection in bounded-size graphs? More precisely, for any integer n, does there exist an integer c, such that c-bounded agents can achieve rendezvous with detection in all graphs of size at most n? (Notice that, for example, Algorithm RV-with-detection cannot be used here. In this algorithm, the number of steps performed by an agent with label L is proportional to $P(n, \log L)$, and hence, even when the size n of the graph is bounded, this number can be arbitrarily large.) The answer to our question turns out to be positive, even in the local beeping model. Below we describe an algorithm that performs this task.

4.1   Bounded-Energy Agents in the Local Beeping Model

Our algorithm uses the following procedure, for an integer parameter n.

Procedure Beeping exploration (n). Let $EXP(n)$ be the procedure described in Sect. 2 that permits exploration of all graphs of size at most n. Replace each round r of $EXP(n)$ by three consecutive rounds as follows. If in round r of $EXP(n)$ the agent takes port p to move to node w, then in the first of the three replacing rounds the agent takes port p to move to w and beeps, and in the second and third of the replacing rounds it stays at w and listens.

Hence, in each of the three rounds replacing a round r of $EXP(n)$, the agent is at the same node in Procedure Beeping exploration (n) as it is in Procedure $EXP(n)$ in round r.

We now describe our algorithm for rendezvous with detection of bounded-energy agents, executed by an agent with label L in a graph of size at most n. Recall that $R(n)$ is the execution time of $EXP(n)$. The idea of the algorithm is the following. Its main block consists of two executions of Procedure Beeping exploration (n) between which a long *waiting period* is inserted, during which the agent is silent (it listens) and inert. The length of this period depends on the label of the agent. We will prove that, regardless of the delay between the starting times of the agents, an entire execution of Procedure Beeping exploration (n) of one of the agents must either fall within the waiting period of the other agent, or must be executed after both executions of this procedure by the other agent. This main block of the algorithm executed by a given agent is interrupted in one of the two cases: either when (a) the agent hears a beep during its waiting period or after completing its main block, or when (b) it hears beeps in two consecutive rounds during one of the executions of Procedure `Beeping exploration` (n). In case (a) the agent responds by beeps in two consecutive rounds, declares rendezvous in the next round and stops. In case (b) it declares rendezvous in the next round and stops.

Below we give the pseudo-code of the algorithm executed by an agent with label L in a graph of size at most n. During the executions of Procedure `Beeping exploration` (n), a boolean variable *waiting* is set to false, and during the waiting period and after the second execution of Procedure `Beeping exploration` (n) it is set to true. We use a boolean valued function `condition` which takes the variable *waiting* as input, and returns, after each round, the boolean value of the expression (*waiting* **and** you hear a beep) **or** ($\neg$*waiting* **and** you hear beeps in two consecutive rounds)

Algorithm. `Bounded-energy-RV-with-detection`

waiting := false
Perform the following sequence of actions in consecutive rounds
and verify the value of `condition` in each round
until the first round when `condition` becomes true
      Perform Procedure `Beeping exploration` (n)
      waiting := true
      Stay inert for $6L \cdot R(n)$ rounds and listen
      waiting := false
      Perform Procedure `Beeping exploration` (n)
      waiting := true
      Stay inert forever and listen
s := the round number when `condition` becomes true
      (counted since your wake-up)
if *waiting* **then**
      beep in rounds $s + 1$ and $s + 2$
      declare rendezvous in round $s + 3$ and stop
else
      declare rendezvous in round $s + 1$ and stop.

Theorem 3. *For any positive integer constant n there exists a positive integer c, such that Algorithm `Bounded-energy-RV-with-detection` can be executed by c-bounded agents in any graph of size at most n. If two such agents with distinct labels execute this algorithm in such a graph, then they meet and simultaneously declare rendezvous in time $O(\ell^*)$ after the start of the later agent, where ℓ^* is the larger label.*

It is interesting to compare the time sufficient to complete the task of rendezvous with detection, given by Algorithm `RV-with-detection` for unrestricted

agents, with the time given by Algorithm `Bounded-energy-RV-with-detection`
for bounded-energy agents. This comparison is meaningful on the class of graphs
for which both types of agents can achieve rendezvous with detection, i.e., for
graphs of bounded size. Consider the class C_n of graphs of size at most n,
for some constant n, and consider c-bounded agents for some integer c large
enough to achieve rendezvous with detection on the class C_n using Algorithm
`Bounded-energy-RV-with-detection`. By Theorem 2, unrestricted agents can
accomplish rendezvous with detection in time $O(P(n, \log \ell))$, i.e., since n is constant, in time *polylogarithmic in the smaller label*. By contrast, by Theorem 3,
bounded-energy agents can accomplish rendezvous with detection in time $O(\ell^*)$,
i.e., *linear in the larger label*. It is natural to ask if this exponential gap in time,
due to energy restriction, is unavoidable. The following lower bound shows that
the answer to this question is yes. In fact, this lower bound holds even for the
two-node graph, even with simultaneous start of the agents, and even for rendezvous without detection.

Theorem 4. *Let c be a positive constant. In the local beeping model, the time
of rendezvous on the two-node graph of c-bounded agents with labels from the set
$\{1, \ldots, M\}$ is $\Omega(\sqrt[c]{M})$.*

Theorem 4 implies that in the local beeping model, any rendezvous algorithm
for bounded-energy agents must have time at least $\Omega(\sqrt[c]{\ell^*})$, where ℓ^* is the larger
label and c is some constant. Theorems 2, 3 and 4 imply the following corollary.

Corollary 1. *Rendezvous with detection of bounded-energy agents is feasible in
the class of bounded-size graphs in the local beeping model, but its time must
be exponentially larger than the best time of rendezvous with detection of unrestricted agents in this class of graphs.*

4.2   Bounded-Energy Agents in the Global Beeping Model

Our final result shows that in the stronger of our two models, i.e., the global
beeping model, the lower bound on time proved in Theorem 4 does not hold
anymore. In fact, we show that in this model, bounded-energy agents can meet
with detection in the class of bounded-size graphs in time logarithmic in the
smaller label. We will also prove that this time is optimal even in the two-node
graph.

The high-level idea of the algorithm is to first break symmetry between the
agents in time logarithmic in the smaller label, without making any moves,
using the possibility of hearing the beeps of the other agent, wherever it is in the
graph. Then one of the agents remains idle and the other agent finds it using a
bounded number of moves. Correct declaration of rendezvous is possible due to
the distinction between hearing *loud* and *soft* beeps.

The main conceptual difference between Algorithm `RV-with-detection`
(that works even for the local model) and our present algorithm for the global
model is that the former breaks symmetry between agents *while they move*, which

results in the number of moves polynomial in two parameters: n (which can be neglected for bounded n) *and in the logarithm of the smaller label.* Thus the total number of moves can be arbitrarily large even for bounded-size networks, and hence impossible for bounded-energy agents. By contrast, our present algorithm uses the possibility of hearing the other agent regardless of its position in the graph to break symmetry between the agents *with no moves,* then fixes one agent and uses an exploration algorithm for the other agent, instead of using a rendezvous algorithm. In bounded-size networks this makes a crucial difference because, unlike rendezvous, exploration can be performed using a bounded number of moves, and hence can be executed by bounded-energy agents.

We first define the following transformations of the label L of an agent. Let $(c_1 \ldots c_k)$ be the binary representation of the label L. Let $T_1(L)$ be the binary sequence $(01c_1c_1c_2c_2 \ldots c_kc_k01)$, and let $T_2(L)$ be the result of replacing each bit 0 of $T_1(L)$ by the string (00) and each bit 1 by the string (10). Note that the length of the binary string $T_2(L)$ is $2(2k+4) \in O(\log L)$.

The following procedure, executed by an agent with label L and called upon the activation of the agent, does not involve any moves and permits to break symmetry between any two agents with different labels.

Procedure. Symmetry-breaking

Let $T_2(L) = (d_1 \ldots d_s)$
$i := 1$
repeat in consecutive rounds **until** you hear a beep
       if $(i \leq s$ **and** $d_i = 1)$ **then** beep **else** listen
       $i := i + 1$
Let r be the round when you first hear a beep (counted since your wake-up)
if you beeped in round $r - 1$ **then**
       declare round $r + 1$ *red*
       $role := waiting$
else
       beep in round $r + 1$
       declare round $r + 2$ *red*
       $role := walking$;
if the beep you heard was *loud* **then**
       declare rendezvous in the *red* round and stop;

Lemma 1. *Upon completion of Procedure* Symmetry-breaking, *both agents declare the same round to be* red. *For one of the agents round* red *is the next round after it heard a beep for the first time, and this agent sets* role := waiting. *For the other agent round* red *is two rounds after it heard a beep for the first time, and this agent sets* role := walking. *The round declared* red *is* $O(\log \ell)$ *rounds after the activation of the later agent, where ℓ is the smaller label.*

We will also use a modified version of Procedure Beeping-exploration, described at the beginning of this section, for an integer parameter n.

Procedure `Modified-beeping-exploration` (n). Let $EXP(n)$ be the procedure described in Sect. 2 that permits exploration of all graphs of size at most n. Replace each round r of $EXP(n)$ by two consecutive rounds as follows. If in round r of $EXP(n)$ the agent takes port p to move to node w, then in the first of the two replacing rounds the agent takes port p to move to w and beeps, and in the second replacing round it stays at w and listens.

Hence, in each of the two rounds replacing a round r of $EXP(n)$, the agent is at the same node in Procedure `Modified-beeping-exploration` (n) as it is in Procedure $EXP(n)$ in round r.

Below we give the pseudo-code of the algorithm executed by an agent with label L in a graph of size at most n.

Algorithm. `Fast-bounded-energy-RV-with-detection`

Perform Procedure `Symmetry-breaking`
if $role = waiting$ **then**
      stay idle and listen **until** you hear a *loud* beep
      let t be the round when you first hear a *loud* beep
      (counted since your wake-up)
      beep in round $t + 1$, declare rendezvous in round $t + 2$, and stop
else
      perform Procedure `Modified-beeping-exploration` (n)
      starting in round *red*
      until you hear a *loud* beep
      let t be the round when you first hear a *loud* beep
      (counted since your wake-up)
      declare rendezvous in round $t + 1$, and stop.

Theorem 5. *For any positive integer constant n there exists a positive integer c, such that Algorithm* `Fast-bounded-energy-RV-with-detection` *can be executed by c-bounded agents in any graph of size at most n, in the global beeping model. If two such agents with distinct labels execute this algorithm in such a graph, then they meet and simultaneously declare rendezvous in time $O(\log \ell)$ after the start of the later agent, where ℓ is the smaller label. This time is optimal, even in the two-node graph.*

5   Conclusion

We presented three algorithms of rendezvous with detection. The first two of them work even in the local beeping model: one for unrestricted agents in arbitrary graphs, and the other for bounded-energy agents in bounded-size graphs. We showed that in the latter case the meeting time of bounded-energy agents must be exponentially larger than the best time of rendezvous with detection of unrestricted agents. More precisely, in order to meet in bounded-size graphs, bounded-energy agents must use time polynomial in the larger label, while unrestricted agents can meet in time polylogarithmic in the smaller label. The third algorithm works for bounded-energy agents only in the global beeping model, but it is much faster: it enables such agents to perform rendezvous with detection in bounded-size graphs in time logarithmic in the smaller label, which is optimal.

Rendezvous with detection may be considered as a preprocessing procedure for other important tasks in graphs. One of them is the task of constructing a map of an unknown graph by an agent. It is well known that this task cannot be accomplished by a single agent operating in a graph, if it cannot mark nodes (e.g., a single agent cannot learn the size of an oriented ring). For the same reason it cannot be accomplished by two non-communicating agents, as they would not be aware of the presence of each other, and thus each of them would act as a single agent. By contrast, our algorithms of rendezvous with detection in the beeping model can serve, with a simple addition, to achieve map construction by the agents: the algorithm working for arbitrary agents can be used to accomplish this task in arbitrary graphs, and the algorithms working for bounded-energy agents can be used to accomplish this task in bounded-size graphs. This addition can be described as follows. Note that, in all our algorithms, at the time when agents declare rendezvous, symmetry between them is broken: in the case of algorithms in the local model, one of the agents heard two beeps at the meeting node, and the other agent heard only one beep, and in the case of the algorithm in the global model, one of the agents has *role* set to *waiting* and the other to *walking*. Hence agents can start simultaneously the following procedure in the round after rendezvous declaration. The first agent stays inert and acts as a stationary token, beeping in every second round, while the second agent silently executes exploration with a stationary token (at the end of which it acquires the map of the graph), cf. e.g., [6], replacing each exploration round by two rounds, in the first of which it moves as prescribed and in the second it stays inert. Beeps of the inert agent allow the circulating silent agent to recognize the token at each visit and complete exploration and map construction. At the end of the exploration, the second agent is with the first one and can inform it of the end of the exploration by beeping in the last round, in which the first agent is silent (listens). Then the roles of the agents may change to allow the previously inert agent to acquire the map in its turn. (Note that an agent cannot efficiently communicate the already acquired map due to the restrictive communication model.)

References

1. Afek, Y., Alon, N., Bar-Joseph, Z., Cornejo, A., Haeupler, B., Kuhn, F.: Beeping a maximal independent set. In: Peleg, D. (ed.) Distributed Computing. LNCS, vol. 6950, pp. 32–50. Springer, Heidelberg (2011)
2. Alpern, S., Gal, S.: The Theory of Search Games and Rendezvous. International Series in Operations Research and Management Science. Kluwer Academic Publisher, Dordrecht (2002)
3. Anderson, E., Fekete, S.: Two-dimensional rendezvous search. Oper. Res. **49**, 107–118 (2001)

4. Baba, D., Izumi, T., Ooshita, F., Kakugawa, H., Masuzawa, T.: Space-optimal rendezvous of mobile agents in asynchronous trees. In: Patt-Shamir, B., Ekim, T. (eds.) SIROCCO 2010. LNCS, vol. 6058, pp. 86–100. Springer, Heidelberg (2010)
5. Bampas, E., Czyzowicz, J., Gąsieniec, L., Ilcinkas, D., Labourel, A.: Almost optimal asynchronous rendezvous in infinite multidimensional grids. In: Lynch, N.A., Shvartsman, A.A. (eds.) DISC 2010. LNCS, vol. 6343, pp. 297–311. Springer, Heidelberg (2010)
6. Chalopin, J., Das, S., Kosowski, A.: Constructing a map of an anonymous graph: applications of universal sequences. In: Lu, C., Masuzawa, T., Mosbah, M. (eds.) OPODIS 2010. LNCS, vol. 6490, pp. 119–134. Springer, Heidelberg (2010)
7. Cieliebak, M., Flocchini, P., Prencipe, G., Santoro, N.: Distributed computing by mobile robots: gathering. SIAM J. Comput. **41**, 829–879 (2012)
8. Cornejo, A., Kuhn, F.: Deploying wireless networks with beeps. In: Lynch, N.A., Shvartsman, A.A. (eds.) DISC 2010. LNCS, vol. 6343, pp. 148–162. Springer, Heidelberg (2010)
9. Czyzowicz, J., Kosowski, A., Pelc, A.: How to meet when you forget: log-space rendezvous in arbitrary graphs. Distrib. Comput. **25**, 165–178 (2012)
10. Dessmark, A., Fraigniaud, P., Kowalski, D., Pelc, A.: Deterministic rendezvous in graphs. Algorithmica **46**, 69–96 (2006)
11. Dieudonné, Y., Pelc, A., Villain, V.: How to meet asynchronously at polynomial cost. In: Proceedings of the 32nd ACM Symposium on Principles of Distributed Computing (PODC 2013), pp. 92–99 (2013)
12. Flocchini, P., Prencipe, G., Santoro, N., Widmayer, P.: Gathering of asynchronous robots with limited visibility. Theor. Comput. Sci. **337**, 147–168 (2005)
13. Flocchini, P., Santoro, N., Viglietta, G., Yamashita, M.: Rendezvous of two robots with constant memory. In: Moscibroda, T., Rescigno, A.A. (eds.) SIROCCO 2013. LNCS, vol. 8179, pp. 189–200. Springer, Heidelberg (2013)
14. Fraigniaud, P., Pelc, A.: Delays induce an exponential memory gap for rendezvous in trees. ACM Trans. Algorithms, 9 (2013). Article 17
15. Koucký, M.: Universal traversal sequences with backtracking. J. Comput. Syst. Sci. **65**, 717–726 (2002)
16. Kranakis, E., Krizanc, D., Morin, P.: Randomized Rendez-Vous with limited memory. In: Laber, E.S., Bornstein, C., Nogueira, L.T., Faria, L. (eds.) LATIN 2008. LNCS, vol. 4957, pp. 605–616. Springer, Heidelberg (2008)
17. Kranakis, E., Krizanc, D., Santoro, N., Sawchuk, C.: Mobile agent rendezvous in a ring. In: Proceedings of the 23rd International Conference on Distributed Computing Systems (ICDCS 2003), pp. 592–599 (2003)
18. Miller, A., Pelc, A.: Time versus cost tradeoffs for deterministic rendezvous in networks. In: Proceedings of 33rd Annual ACM Symposium on Principles of Distributed Computing (PODC 2014), pp. 282–290 (2014)
19. Pelc, A.: Deterministic rendezvous in networks: a comprehensive survey. Networks **59**, 331–347 (2012)
20. Reingold, O.: Undirected connectivity in log-space. J. ACM **55**, 1–24 (2008)
21. Ta-Shma, A., Zwick, U.: Deterministic rendezvous, treasure hunts and strongly universal exploration sequences. In: Proceedings of 18th ACM-SIAM Symposium on Discrete Algorithms (SODA 2007), pp. 599–608 (2007)

Minimizing Total Sensor Movement for Barrier Coverage by Non-uniform Sensors on a Line

Robert Benkoczi[1], Zachary Friggstad[2], Daya Gaur[1], and Mark Thom[1] (✉)

[1] Mathematics and Computer Science, University of Lethbridge,
Lethbridge, AB T1K 3M4, Canada
{benkoczi,gaur,thom}@cs.uleth.ca
[2] Department of Computing Science, University of Alberta,
Edmonton, AB T6G 2E8, Canada
zacharyf@cs.ualberta.ca

Abstract. Barrier coverage is a cost effective approach for intruder detection applications. It relies on monitoring the perimeter, or barrier, around the area of interest by placing sensors at appropriate locations on the barrier. In this paper we consider the problem of barrier coverage of a line segment by moving sensors along the line containing the segment. We extend the results existing in the literature by considering the case of non-uniform sensors placed at initial positions that do not overlap with the interval of interest.

1 Introduction

One of the fundamental applications of wireless sensor networks is to provide coverage of an area of interest. When the purpose for coverage is surveillance, a cost effective approach is to monitor the perimeter of the area in order to detect intruders. This type of coverage is called *barrier coverage.* Kumar *et al.* [8,9] were among the first to investigate barrier coverage problems. Their work has motivated a large number of contributions ranging from density estimates of random deployments [1] to relaxations of coverage requirements that are suitable for the study of localized algorithms [2], to name just a few examples.

For large scale applications such as border protection, achieving full coverage by randomly deploying sensors may be too expensive. To reduce the number of sensors needed for barrier coverage, Czyzowicz *et al.* [4,5] considered the barrier coverage problem with mobile or relocatable sensors. The idea is that, once the initial deployment is complete, the deployed sensors can adjust their position to attain coverage and thus the deployment of additional sensors is not needed. Mobility also allows flexibility in choosing areas along the border that need coverage if full coverage is not an option.

R. Benkoczi and D. Gaur—These authors acknowledges the support for this research received from an NSERC Discovery Grant.

Z. Friggstad—This research was undertaken, in part, thanks to funding from the Canada Research Chairs program.

© Springer International Publishing Switzerland 2015
P. Bose et al. (Eds.): ALGOSENSORS 2015, LNCS 9536, pp. 98–111, 2015.
DOI: 10.1007/978-3-319-28472-9_8

The optimization problems defined by Czyzowicz *et al.* [4,5] are one dimensional problems where the barrier is modeled by a line segment and sensors are initially located on the line containing the segment. The goal is to compute new positions for a subset of the sensors so that every point in the target line segment is within the sensing range of at least one sensor. Several objective functions have been studied: minimizing the maximum distance (MinMax) traveled by one sensor [5], minimizing the number (MinNum) of sensors moved [10], and minimizing the total (MinSum) travel distance [4].

Problem MinMax was shown by Chen *et al.* in [3] to be solved in time $O(n^2 \log n)$, where n represents the number of sensors. More efficient algorithms are possible if the sensors are uniform (they have the same covering range). Czyzowicz *et al.* [5] distinguish two cases for the barrier problem with uniform mobile sensors: complete coverage or $R \geq L$ and incomplete coverage or $R < L$, where R represents the sum of covering ranges of all sensors and L represents the length of the interval to be covered. They give exact algorithms with time complexity $O(n)$ for the case of $R < L$ and $O(n^2)$ for the case $R \geq L$. For the later case, they show that the running time can be improved at the expense of solution quality. They give a $(1+\epsilon)$ fully polynomial time approximation scheme (FPTAS) with time complexity sub-quadratic in n and a 2-approximation with time complexity linear in n. An exact algorithm with time complexity $O(n \log n)$ for uniform sensors in the case $R \geq L$ was later proposed by Chen *et al.* [3].

Two dimensional analogues of MinMax and MinSum are considered in [6], where sensors of arbitrary ranges are located at arbitrary points in the plane and both Euclidean and rectilinear distances are considered. Barriers are modeled as line segments in the plane, and variants of the MinMax and MinSum problems are established according to the number of barriers, whether multiple barriers are oriented parallel or perpendicular to one another and whether sensors may move arbitrarily or are restricted to move to the closest point on a barrier. Dobrev *et al.* [6] develop exact algorithms for solving the MinMax and MinSum problems for the 1 barrier and k parallel barrier restricted movement cases, and show the NP-hardness of MinMax and MinSum in all other cases.

Surprisingly, the combinatorial structure of the MinSum problem is not completely understood yet. Czyzowicz *et al.* [4] proved the NP-hardness for the general problem with non-uniform sensing ranges. We note that their proof constructs a MinSum instance where the initial position for some of the sensors is inside the target line segment. The proof also indicates that constant factor approximations for the general MinSum problem are not possible unless P = NP. Except for this inapproximability result, the only restricted instances solved are those with uniform sensors, for which an exact algorithm with time complexity $O(n)$ for the case of $R < L$ and an $O(n^2)$ exact algorithm for the case $R \geq L$ are possible.

We note that in contrast to our approach and those of [5,8], local algorithms for barrier coverage problems have been studied, in [7].

Our Contributions: We extend the set of instances for the MinSum barrier problem with movable sensors that can be solved by investigating restrictions on another parameter of the problem: the initial positions of the sensors. We prove that the problem remains NP-hard even when the initial position of the sensors is such that the sensing areas do not intersect the target interval. We give a fully polynomial time approximation scheme (FPTAS) based on dynamic programming for MinSum with non-uniform sensor ranges when the initial positions of the sensing intervals are on one side of the target interval. We then modify the dynamic programming algorithm to process intervals in reverse order and extend the FPTAS to instances where the sensors lie on both sides of the target interval.

In light of the negative result of Czyzowicz et al. [4] concerning the implausibility of constant factor approximations for the general case, and of the fact that the natural greedy algorithm for MinSum has an $\Omega(n)$ performance ratio (details omitted), our result is best in some sense.

1.1   Notation and Problem Definitions

Suppose we have a region $\mathcal{R}$ in the geometric plane with a simple, closed boundary. We are given a set of n wireless sensors S_i located at predesignated points x_i. Each sensor S_i has a range of detection specified by $r_i > 0$, and triggers a central alarm upon detecting movement at a point x with Euclidean distance $d(x_i, x) \leq r_i$.

A *barrier coverage* of a region is any placement of sensors in the plane to detect intruders across a maximal portion of the region's boundary. An optimization problem in this setting is to minimize the total distance traveled by all sensors such that every point on the boundary is within the sensing range of some sensor. We refer to the problem as the *Minimum Sum Distance*, or *MinSum*, problem as defined in [4,5].

We consider the one-dimensional version of the *MinSum* problem in which the barrier is represented by a line segment. Sensors are represented by points on the line, and sensor movements are restricted to the line. As argued in the introduction, we study a new restriction of the MinSum barrier coverage problem where the sensor ranges do not intersect the target line segment but the ranges of the sensors are arbitrary. We call this problem the one-dimensional *DisjointMinSum* problem.

We adapt the notation from [5]. The barrier is a closed interval $I = [0, L]$ on the real line, for some $L > 0$. Each sensor S_i is specified by a single point x_i on the real line, and its sensing range is given by interval $I(S_i, x_i) = [x_i - r_i, x_i + r_i]$. After moving S_i by m_i units, the resulting range of S_i is $I(S_i, x_i + m_i) = [x_i + m_i - r_i, x_i + m_i + r_i]$. The sign of m_i determines the direction of movement on the line.

The *DisjointMinSum* problem is defined for n sensors with initial positions $x_1 \leq x_2 \leq \ldots \leq x_n$ where $x_i - r_i > L$ or $x_i + r_i < 0$ for all $1 \leq i \leq n$ as

$$\min \left\{ \sum_{1 \leq i \leq n} |m_i| \right\} \text{ subject to } [0, L] \subseteq \bigcup_{i=1}^{n} I(S_i, x_i + m_i). \tag{1}$$

2   NP-completeness Results

2.1   NP-hardness of *LeftDisjointMinSum*

In order to motivate the development of an FPTAS, we show that *LeftDis-jointMinSum* is NP-complete by reducing from the Partition problem, which is defined in the following way. Given a sequence of positive integers $a_1, \ldots, a_n$ and an integer B such that $\sum_{1 \le i \le n} a_i = 2B$, find a subset of integers S such that $\sum_{a_i \in S} a_i = B$.

From the instance $(\{a_i\}_{1 \le i \le n}, B)$, we create a barrier from the line segment $[0, B]$, and for each integer $a_i > 0$, we create a sensor with range $r_i = a_i/2$ and distance $d_i = |x_i| - r_i \ge 0$ from 0, $x_i < 0$, for x_i determined later in the reduction. Let S be any subset of sensors whose total range covers all of the barrier. Using the formula for the cheapest movement covering the barrier completely, we have

$$c(S) = |S| \cdot B - \sum_{i=1}^{|S|}(|S| - i) \cdot a_{s_i} + \sum_{i=1}^{|S|} d_{s_i} \tag{2}$$

where the sequence $\{s_i\}_{1 \le i \le |S|}$ indexes the intervals of S, and is again ordered so that $r_{s_1} \ge r_{s_2} \ge \ldots \ge r_{s_{|S|}}$.

We choose d_i in order to ensure that, for any covering sensor subsets S and S', $l(S) > l(S')$ implies $c(S) > c(S')$. An optimal algorithm for *DisjointMinSum* then gives an optimal algorithm for the Partition problem. We need only apply the optimal algorithm for *DisjointMinSum* to the reduction of the Partition problem instance and check that the solution has total length B.

To that end, we decide the values of d_i. Suppose S, S' are covering subsets of sensors satisfying $l(S) > l(S')$. Using the cost formula, we get

$$c(S) - c(S') = (|S| - |S'|) \cdot B - \sum_{i=1}^{|S|}(|S| - i) \cdot a_{s_i}$$

$$+ \sum_{i=1}^{|S'|}(|S'| - i) \cdot a_{s'_i} + \sum_{i=1}^{|S|} d_{s_i} - \sum_{i=1}^{|S'|} d_{s'_i}$$

$$= B \sum_{i=1}^{|S|}\left(d_{s_i}^B + 1 - \frac{|S| - i}{B} \cdot a_{s_i}\right)$$

$$- B \sum_{j=1}^{|S'|}\left(d_{s'_j}^B + 1 - \frac{|S'| - j}{B} \cdot a_{s'_j}\right)$$

where we define $d_i^B = d_i/B$. Then $c(S) - c(S') > 0$ is equivalent to

$$\sum_{i=1}^{|S|}\left(d_{s_i}^B + 1 - \frac{|S| - i}{B} \cdot a_{s_i}\right) > \sum_{j=1}^{|S'|}\left(d_{s'_j}^B + 1 - \frac{|S'| - j}{B} \cdot a_{s'_j}\right)$$

Let $d_i^B = (2B + 1) \cdot a_i - 1 > 0$. In particular, we have

$$\sum_{i=1}^{|S|} d_{s_i}^B + 1 - \frac{|S| - i}{B} \cdot a_{s_i} = \sum_{i=1}^{|S|} (2B + 1 - \frac{|S| - i}{B}) \cdot a_{s_i}$$
$$> \sum_{i=1}^{|S|} 2B \cdot a_{s_i}$$

where the inequality holds by the following argument. $|S| - i < |S| \le B$, and so $1 - (|S| - i)/B > 0$. With the assumptions $a_{s_1} > 0$ and $|S| \ge 1$, we establish strict inequality.

From $l(S) - l(S') \ge 1$, we get that

$$\sum_{i=1}^{|S|} 2B \cdot a_{s_i} - \sum_{j=1}^{|S'|} 2B \cdot a_{s'_j} = 2B \cdot (l(S) - l(S')) \ge 2B$$

It is easy to see that

$$2B \ge \sum_{j=1}^{|S'|} \left(1 - \frac{|S'| - j}{B}\right) \cdot a_{s'_j}$$

giving $c(S) > c(S')$, combined with the earlier inequalities.

3   Approximation Schemes

3.1   An FPTAS for *LeftDisjointMinSum*

We devise an FPTAS for the problem when the initial sensor placements lie entirely to one side of the barrier. By symmetry, we suppose that all sensors lie to the left of the barrier. We refer to this restricted problem as *LeftDisjointMinSum*.

We note that in an optimum solution, no two sensors that are moved to form the barrier will overlap; they are packed from L leftward. Specifically, if S is an optimum solution and if the sensors $s_1, \ldots, s_{|S|}$ are indexed in increasing order of their final position, then $x_{s_1} + r_{s_1} + m_{s_1} = L$, $x_{s_2} + r_{s_2} + m_{s_2} = L - 2r_{s_1}$, $x_{s_3} + r_{s_3} + m_{s_3} = L - 2r_{s_1} - 2r_{s_2}$, and so on. This scheme of packing, along with the following order preservation lemma, is key in structuring the dynamic programming algorithm that is the basis of our FPTAS. The proof is found in the appendix.

Lemma 1. *Let the sensors be indexed so that $r_1 \ge r_2 \ge \ldots \ge r_n$. Then an optimum solution consists of a subset of sensors ordered in their final positions according to this indexing scheme in such a way that*

$$x_i + m_i < x_j + m_j \text{ if } m_i > 0, \ m_j > 0, \text{ and } r_i \le r_j.$$

where m_i represents the distance moved by the i-th sensor in the optimum solution.

Theorem 1. *There is an FPTAS for problem LeftDisjointMinSum with running time $O(\frac{n^5}{\epsilon^2})$.*

We briefly discuss some intuition before presenting the algorithm and proof. If L and all the values r_i are integers then it is easy to get a pseudo-polynomial time exact solution. For various integers x and i, we could simply compute the cheapest solution that covers the range $[x, L]$ using the i longest intervals using dynamic programming. However, this is inefficient for instances with large values.

However, we cannot simply scale the r_i values: if we scaled them down then the optimum solution may no longer be feasible in the scaled instance and if we scale them up then the solution we find may no longer be feasible when we revert to the original r_i values. Instead, we use a different approach where the dynamic programming table is indexed with a budget that must pay for the cost of the partial solutions, and we scale this budget.

Proof (of Theorem 1). We assume that the sensors are ordered in decreasing order of range: $r_1 \geq r_2 \geq \ldots \geq r_n$. We describe a recurrence that can be used to determine the exact optimum solution. Following this, we discuss how to modify this approach to find near optimum solution in poly$(n, 1/\epsilon)$ time.

For any $0 \leq i \leq n$ and any value $z \geq 0$, we let $f^*(i, z)$ be the smallest value such that we can cover the interval $[f^*(i, z), L]$ using only the first i sensors (according to the sorted order) with total cost at most z.

Note that $f^*(i, z)$ satisfies the following recurrence relation.

$$f^*(i, z) = \max\{\min\{f^*(i - 1, z), g^*(i, z)\}, 0\}$$

where for $i > 0$ and $z \geq 0$ we let

$$g^*(i, z) := \min_{0 \leq x \leq z} \{f^*(i - 1, z - x) - 2r_i : |f^*(i, z - x) - r_i - x_i| \leq x\}.$$

Intuitively, this is capturing the idea that either the solution witnessing $f(i, z)$ uses S_i (in which case the min in g is "guessing" how much is spent from z in moving S_i) or else it does not (meaning $f(i, z) = f(i-1, z)$). We will not actually use this recurrence in our final algorithm, so we will not prove these claims.

The cost OPT of the optimal complete coverage obtained from among all n sensors is then

$$OPT = \min_{z \geq 0}\{z \mid f^*(n, z) \leq 0\}$$

We now present our FPTAS. We first find a value Z that coarsely approximates the optimum solution. This will be used to discretize the cost of building partial solutions. Specifically, Z is cost of the coverage returned by the following greedy algorithm.

1. Sort the sensors in ascending order of $|x_i|$.
2. For each sensor, compute the length-greedy coverage among all sensors of lesser or equal barrier distance $|x_i|$, if one exists. A length-greedy coverage moves the sensor of greatest range r_i to the rightmost position of the barrier, and packs further sensors to the rightmost uncovered spot by next greatest r_i.
3. Once all length-greedy coverages have been computed, choose the coverage of least cost and call this cost Z.

The proof of the lemma is found in the appendix.

Lemma 2. $OPT \leq Z \leq n \cdot OPT$.

For $\epsilon > 0$, let $\zeta = \epsilon Z / (n(n+1))$. We slightly modify the recurrence for f^*. For integers $0 \leq i \leq n$ and $0 \leq k \leq \lceil n^2/\epsilon \rceil$ we define values $f(i, k\zeta)$ and $g(i, k\zeta)$ recursively by

$$f(i, k\zeta) = \min\{f(i-1, k\zeta), f(i-1, (k-1)\zeta), g(i, k\zeta)\}$$

where

$$g(i, k\zeta) = \min_{1 \leq c \leq k} \{f(i-1, (k-c)\zeta) - 2r_i$$
$$\text{s.t. } |f(i-1, (k-c)\zeta) - r_i - x_i| \leq c\zeta\}$$

We consider $f(i, k\zeta) = L$ if $i = 0$ or $k = 0$. These f-values approximate the true f^* values in the following sense.

Lemma 3. For every $0 \leq i \leq n$ and every integer $0 \leq k$, there is a barrier for the interval $[f(i, k\zeta), L]$ that uses the first i sensors and has movement cost at most $k \cdot \zeta$.

Conversely, for $x \geq 0$ suppose there is a barrier for an interval $[x, L]$ that uses the first i sensors with total movement cost at most z. Let k be such that $k \cdot \zeta \geq z$. Then $f(i, (k+i)\zeta) \leq x$.

The proof of this lemma can be found in the appendix. Given this claim, the cost of the optimum solution is approximated by $k^*\zeta$ where

$$k^* = \min\{k\zeta \mid 0 \leq k \leq \frac{Z}{\zeta} + n + 1 \text{ and } f(n, k\zeta) \leq 0\} \tag{3}$$

To see this, let $k'\zeta$ be such that $OPT - \zeta < k'\zeta \leq OPT$. Note that $(k' + n + 1)\zeta \leq OPT + (n+1)\zeta \leq Z + (n+1)\zeta$ so k' is considered in the min in (3). Furthermore, $f(i, (k' + n + 1)) \leq 0$ by the second part of Lemma 3 and the fact that $OPT < (k' + 1)\zeta$.

Now, k^* is the smallest number such that $f(n, k^*\zeta) \leq 0$ so $k^* \leq k' + n + 1$. By the first part of the claim, there is in fact a barrier for $[0, L]$ with cost at most $k^*\zeta$ so $OPT \leq k^*\zeta$. Overall, we have computed a value k^* satisfying

$$OPT \leq k^*\zeta \leq (k' + n + 1)\zeta \leq OPT + (n+1)\zeta$$
$$= OPT + (n+1)\frac{\epsilon Z}{n(n+1)} \leq (1 + \epsilon) \cdot OPT$$

where the last bound uses $Z \leq n \cdot OPT$. An actual solution with value at most $(1+\epsilon) \cdot OPT$ can be recovered in a standard way by examining how the dynamic programming table is constructed.

The pseudo-code for this FPTAS is presented in Algorithm 1. Each table entry $f(i, k\zeta)$ calculated by this recurrence has $0 \leq i \leq n$ and $0 \leq k \leq n^2/\epsilon + n$ so the number of table entries is $O(n^3/\epsilon)$. Furthermore, there are $O(n^2/\epsilon)$ values considered in the min defining g, so calculating all table entries takes $O(n^5/\epsilon^2)$ time.

Algorithm 1. The FPTAS for *LeftDisjointMinSum*

1: **procedure** LEFTFPTAS(x, r, L, ϵ)
2:     $Z \leftarrow$ a value in $[OPT, n \cdot OPT]$ computed as described above
3:     $\zeta = \epsilon Z/(n(n+1))$
4:     $n \leftarrow$ # of sensors
5:     **for** k from 0 to $\ldots Z/\zeta + n + 1$ **do**
6:         $f[0, k\zeta] \leftarrow L$
7:     **for** i from 1 to n **do**
8:         **for** k from 0 to $Z/\zeta + n + 1$ **do**
9:             $g[i, k\zeta] \leftarrow +\infty$

10:             **for** c from 0 to k **do**
11:                 **if** $|f[i-1, k\zeta - c\zeta] - r_i - x_i| \leq c\zeta$ **then**
12:                     endpt $\leftarrow f[i-1, k\zeta - c\zeta] - 2r_{i+1}$
13:                     $g[i, k\zeta] \leftarrow \min(g[i-1, k\zeta], \text{endpt})$

14:             $f[i, k\zeta] \leftarrow \min(f[i-1, k\zeta], g[i, k\zeta])$
15:             **if** $k > 0$ **then**
16:                 $f[i, k\zeta] \leftarrow \min(f[i, k], f[i, (k-1)\zeta])$
17:     **return** $\min\{k\zeta : k \leq Z/\zeta + n + 1, f(n, k\zeta) \leq 0\}$

3.2 The FPTAS for *DisjointMinSum*

We now use Algorithm 1 to give an FPTAS on *DisjointMinSum*. As with the *LeftDisjointMinSum* case, we can assume that the sensors are tightly packed. We also note the following properties for later use in the FPTAS. The proofs are in the appendix.

Lemma 4. *For any minimum-cost complete barrier coverage of a* DisjointMin-Sum *instance, we may suppose without loss of generality that there exists a point* $x^* \in [0, L]$ *such that the interval* $[0, x^*)$ *is covered only by sensors* S_i *with* $x_i < 0$ *and* $(x^*, L]$ *is covered only by sensors* S_i *with* $x_i > L$.

Another property we will use is the absence of "overhang" on one of the sides of the optimum coverage. A sensor S_i overhangs from the right side of a barrier coverage if $x_i + m_i - r_i < L < x_i + m_i + r_i$ and similarly when hanging from the left side.

Lemma 5. *There exists an optimum barrier coverage in which there is either no left overhang or no right overhang.*

Theorem 2. *There is an FPTAS for the* DisjointMinSum *problem with running time* $O(\frac{n^7}{\epsilon^3})$.

To get an FPTAS, it would suffice to guess this middle point x^* and then use our FPTAS for *LeftDisjointMinSum* on the interval $[0, x^*]$ using sensors S_i with $x_i < 0$, and then using the corresponding FPTAS on the interval $[x^*, L]$ using sensors S_i with $x_i > L$. However, there could be too many possible midpoints to guess.

Refining this idea, we choose a small list of guesses for x^* such that some value in this list, say $\tilde{x}$, satisfies $\tilde{x} \leq x^*$. We remain able to compute a covering $[\tilde{x}, L]$ using sensors S_i with $x_i > L$ whose total cost is close to that of the covering the optimum uses over $[x^*, L]$. The *LeftDisjointMinSum* sub-problem on $[0, \tilde{x}]$ is then solved using our previous FPTAS.

Proof (of Theorem 2). The FPTAS is based on computing values $f_R(i, z)$, which minimizes the leftmost endpoint of the best coverage purchasable using budget z, packing from the right of L using the i *shortest* sensors that lie to the right of the barrier. If we think of $f_R^*(i, z)$ as starting at L and growing positively to the left from L, it is given as the recurrence relation

$$f_R(i, z) = \max\{f_R^*(i-1, z), g_R^*(i, z)\},$$

where

$$g_R^*(i, z) = \max_{x \in [0, z]} \{f_R^*(i-1, z-x) - 2r_i :$$
$$x_i - f_R^*(i-1, z-x) + r_i \leq x\}$$

We will solve a discretized version of this recurrence $f_R^*(i, z)$. For each x-value of the form $f_R(i, z)$ for this discretized version, we will produce for each $f_R(i, z)$ a coverage of the interval $[0, f_R(i, z)]$ using only the sensors that lie to the left of 0 using our FPTAS for *LeftDisjointMinSum*. The cheapest coverage obtained from this process will have a performance ratio of $(1 + \epsilon)OPT$, where OPT is the cost of an optimum two-sided uncrossed coverage that has no overhang on at least one side.

To discretize f_R, we start with a coarse estimate of the optimum solution. Let Z be determined by the following procedure. For a sensor i, let α_i be the distance from x_i to the nearest endpoint $\{0, L\}$. Try all n guesses (and keep the cheapest solution found) for the largest value α_i for sensors S_i used in the optimum solution. Move all of these sensors S_j with $\alpha_j \leq \alpha_i$ to their nearest endpoint. For the proper guess α_i, the total movement so far is at most n times the total movement used by the optimum to move sensors from their start points to the endpoints of the interval.

Next, guess the number i_L of these sensors that the optimum uses from the left-sensors. Say the i_L longest of these sensors have total length ℓ. Greedily cover $[0, \ell]$ using these sensors, and then greedily cover $[\ell, L]$ using the longest right-sensors that were moved to L. The movement cost is at most the total movement of sensors when moving within $[0, L]$ used by the optimum.

If we let Z denote the total movement in this solution, then we have $OPT \leq Z \leq n \cdot OPT$. Let $\zeta = \epsilon Z/(n(n+1))$ as before. We will discretize the indices of f_R^* over integer multiples of ζ, as in the FPTAS for *LeftDisjointMinSum*. For $0 \leq i \leq n$ and integers $k \geq 0$ we compute the following values recursively.

$$f_R(i, k\zeta) = \max\{f_R(i-1, k\zeta), f_R(i-1, (k-1)\zeta), g_R(i, k\zeta)\}$$

where

$$g_R(i, k\zeta) = \max_{1 \leq c \leq k} \{f_R(i - 1, (k - c)\zeta) + 2r_i$$

$$x_i - \mid f_R(i - 1, (k - c)\zeta) + r_i \leq c\zeta\}$$

and

$$f_R(0, k\zeta) = 0 \text{ for all } k$$

where, as before, we iterate over k.

As with the previous FPTAS, we have the following claim.

Claim. For every $0 \leq i \leq n$ and every integer $0 \leq k$, there is a barrier for the interval $[f_R(i, k\zeta), L]$ that uses the shortest i sensors to the right of L and has movement cost at most $k \cdot \zeta$.

Conversely, for $x \geq 0$ suppose there is a barrier for an interval $[x, L]$ that uses the shortest i sensors to the right of L with total movement cost at most z. Let k be such that $k \cdot \zeta \geq z$. Then $f_R(i, (k + i)\zeta) \leq x$.

The proof is essentially identical, so it is omitted from this extended abstract.

Let x^* denote the midpoint in the optimum solution, so $[0, x^*]$ is covered by sensors with $x_i < 0$ and $[x^*, L]$ is covered by sensors with $x_i > L$. Let OPT_L and OPT_R denote the movement in this optimum solution coming from sensors on the left and right, respectively.

Let $k_R^*\zeta$ be the smallest integer multiple of ζ that is at least $OPT_R + (n_R + 1)\zeta$. Then $f_R(n_r, k_R^*\zeta) \leq x^*$. We also know there is a covering of $[f_R(n_r, k_R^*\zeta), L]$ with cost at most $k_R^*\zeta$ by the claim.

Our final algorithm is to approximate each *LeftDisjointMinSum* problem formed by the left-lying sensors for each interval of the form $[0, f_R(n_R, k'\zeta)]$ for some $0 \leq k' \leq Z + n\zeta$. When $k' = k_R^*$, the cost of covering $[0, f_R(n_R, k_R^*\zeta)]$ is at most $(1 + \epsilon) \cdot OPT_L$ because we know the interval $[0, x^*]$ can be covered with cost OPT_L and $f_R(n_R, k_R^*\zeta) \leq x^*$. The total cost of the cheapest solution that covers $[0, L]$ by breaking it into two sub-intervals around some value of the form $f_R(n_R, k'\zeta)$ is at most

$$(1 + \epsilon) \cdot (1 + \epsilon)OPT_L + OPT_R + (n_R + 1)\zeta$$
$$\leq (1 + \epsilon) \cdot OPT + (n + 1)\zeta$$
$$\leq (1 + 2\epsilon) \cdot OPT$$

The pseudo-code for this procedure is summarized in Algorithm 2. It is called with the values Z, ζ described above. It is presented at a bit of a higher-level than Algorithm 1, but the details behind filling out the tables via dynamic programming are similar. It only computes a value $k\zeta$ such that $OPT \leq k\zeta \leq (1 + \epsilon) \cdot OPT$, but it is easy to recover a solution with cost at most $k\zeta$ by examining the dynamic programming tables in the standard way.

Overall, computing the f_R-values takes $O(n^5/\epsilon^2)$ time. For each of the $O(n^2/\epsilon)$ values of the form $f_R(n_R, k'\zeta)$, we use the FPTAS for the resulting *LeftDisjoint-MinSum* instance for a total running time of $O(n^7/\epsilon^3)$.

Algorithm 2. The FPTAS for *DisjointMinSum*

1: **procedure** DISJOINTMINFPTAS(x, r, L)
2:     $Z \leftarrow$ a value in $[OPT, n \cdot OPT]$ computed by the procedure described above
3:     $\zeta \leftarrow \epsilon Z/(n(n+1))$
4:     $n_R \leftarrow$ # of right-sensors
5:     $x^L, r^L \leftarrow$ the sub-lists of x, r for the left-sensors
6:     find all f_R values using dynamic programming.
7:     best $\leftarrow +\infty$
8:     **for** k_R from 0 to $Z/\zeta + n + 1$ **do**
9:         left $\leftarrow$ LEFTFPTAS($x^L, r^L, f(_R n_R, k_R \zeta)$)
10:        best $\leftarrow$ min(best, left $+ k_R \zeta$)
11:    **return** best

The above FPTAS works under the assumption that there is an optimal two-sided barrier coverage with no right overhang. We run the symmetric procedure under the assumption that there is no left overhang, and return the cheaper of the two coverages.

4   Conclusion and Open Problems

In this paper we consider the one dimensional barrier coverage problem of a line segment by mobile sensors. The objective is to minimize the total distance travelled by the sensors (MinSum). The problem was proposed in [4]. Unlike the related MinMax problem for which efficient exact algorithms exist [3,5], MinSum is NP-hard when sensors have arbitrary covering ranges [4]. The hardness proof also shows that it is unlikely that constant factor approximation algorithms for MinSum with arbitrary covering ranges exist.

We give the first algorithm (an FPTAS) for MinSum with arbitrary covering ranges when the sensors' covering ranges do not intersect the target line segment and show that this version is also NP-hard. We also mention that the natural greedy algorithm for the general MinSum problem has an $\Omega(n)$ approximation factor.

Our results motivate several new directions of research. Since the natural greedy algorithm for MinSum problem has an $\Omega(n)$ approximation factor, are there any $o(n)$ approximations for the general MinSum problem? A natural generalization of the notion of coverage is k-coverage, where every point in the target area must be within the sensing range of at least k sensors. In this setting, both MinSum and MinMax problems are completely open.

A   Appendix

A.1   Proofs from Sect. 3

Proof (of Lemma 1). Let S be a collection of sensors covering the barrier $[0, L]$. If the sensors of S are ordered from right to left and labeled accordingly by the indices $s_1, s_2, \ldots s_{|S|}$, then the cost of the barrier is

$$c(S) = |S| \cdot L - \sum_{i=1}^{|S|} (|S| - i) \cdot 2r_{s_i} + \sum_{i=1}^{|S|} d_{s_i} \qquad (4)$$

where L, r_i and d_i assume the definitions given above. If S is fixed, then $c(S)$ is minimized by maximizing $\sum_{i=1}^{|S|} (|S| - i) \cdot 2r_{s_i}$. This is accomplished by re-ordering the index so that $r_{s_1} \geq r_{s_2} \geq \ldots \geq r_{s_{|S|}}$.

Proof (of Lemma 2). We have $OPT \leq Z$ because Z corresponds to the movement in a feasible solution. To see that $Z \leq n \cdot OPT$, let $X = \max_{i \in S^*} |x_i|$ where S^* is the set of sensors moved in an optimum solution. Let $OPT = OPT_1 + OPT_2$ where $OPT_1 = \sum_{i \in S^*} |x_i|$. Think of OPT_1 as the cost of moving the optimum intervals from their start positions to 0 and OPT_2 as the cost of moving them from 0 to their final positions.

Consider the iteration of the greedy algorithm that uses sensors i with $|x_i| \leq X$. An upper bound on the cost of moving them from their start positions to 0 is $n \cdot X \leq n \cdot OPT_1$ and an upper bound on moving them further to their final destinations is OPT_2. This is because greedily moving by length is optimum if all start positions are 0. Therefore, $Z \leq n \cdot OPT_1 + OPT_2 \leq n \cdot OPT$.

Proof (of Lemma 3). We prove the first statement of Lemma 3 by induction on i, with the case $i = 0$ being trivial. So, suppose $i > 0$. If $f(i, k\zeta) = f(i - 1, k\zeta)$ then there is nothing to prove. Otherwise, let c be such that $f(i - 1, k\zeta) = f(i - 1, (k - c)\zeta) - 2r_i$ and $|f(i - 1, (k - c)\zeta) - r_i - x_i| \leq c\zeta$. By induction, we can cover $[f(i - 1, (k - c)\zeta), L]$ using the first $i - 1$ intervals with cost at most $(k - c) \cdot \zeta$. Extending this to a cover of $f(i, k\zeta)$ costs at most $c\zeta$, which proves this part of the claim.

The second statement is also proved by induction on i, with the case $i = 0$ again being clear. So, let $i > 0$ and let x be such that the first i sensors can cover $[x, L]$ with total movement cost at most z. Say S' is the collection of sensors used in this cover. If $S_i \notin S'$, then by induction we have $f(i - 1, (k + i - 1)\zeta) \leq x$ so $f(i, (k + i - 1)\zeta) \leq x$ as well. But $f(i, (k + i)\zeta) \leq f(i, (k + i - 1)\zeta)$ also holds (by the recurrence) so $f(i, (k + 1)\zeta) \leq x$ as required.

So, suppose $S_i \in S'$. Without loss of generality (by Lemma 1), we may assume that S_i is the leftmost sensor in the cover of $[x, L]$ and that S_i moves $m_i := x + r_i - x_i$ to its position in this cover (this will be positive, otherwise we are saying sensor i was moved left, in which case it can be discarded from S' to get an even cheaper solution). Let c be such that $c\zeta \leq m_i < (c + 1) \cdot \zeta$, so the remaining sensors in $S' - \{S_i\}$ have total movement at most $z - m_i \leq (k - c)\zeta$ in this cover. Now, $S' - \{S_i\}$ covers $[x + 2r_i, L]$ with cost at most $x - m_i$ so $f(i - 1, (k - c + i - 1)\zeta) \leq x + 2r_i$ by induction. Because $|f(i - 1, (k - c + i - 1)\zeta) - r_i - x_i| \leq x + r_i - x_i = m_i < (c + 1) \cdot \zeta$, then by the recurrence for g (when the min indexes with $c + 1$) we have $f(i, (k + i)\zeta) \leq x$.

Proof (of Lemma 4). If we assume a complete barrier coverage containing consecutive sensors S_i, S_j such that $x_i < 0$, $x_j > L$, and $x_i + m_i > x_j + m_j$, then the paths the sensors travel can be "uncrossed", so that S_i takes the former place of S_j and vice versa. Since the distance traveled by either sensor is only decreased by uncrossing, it follows that there exists some optimum barrier coverage without crossed sensors.

Proof (of Lemma 5). We can eliminate overhang on one side of any complete barrier coverage as follows. Suppose we have a coverage with the uncrossed property, so that there exists a unique x^* with n_L sensors originally positioned to the left of the barrier covering $[0, x^*]$ and n_R sensors originally positioned to the right covering $[x^*, L]$. Suppose without loss of generality that $n_L \geq n_R$ and that overhang is present on both sides of the barrier. Each sensor in the coverage can be shifted contiguously to the left until the rightmost point covered by the sensor overhanging $[0, L]$ on the right is shifted to L. Since $n_L \geq n_R$, the reduction in cost required to move the left-side barriers to their new positions is no less than the added cost of moving the right-side barriers further to the left. Therefore, we obtain a coverage without right overhang at equal or lesser cost. A symmetric argument works in the case that $n_L \leq n_R$.

References

1. Balister, P., Bollobas, B., Sarkar, A., Kumar, S.: Reliable density estimates for coverage and connectivity in thin strips of finite length. In: Proceedings of the 13th Annual ACM International Conference on Mobile Computing and Networking, MobiCom 2007, pp. 75–86. ACM, New York (2007)
2. Chen, A., Kumar, S., Lai, T.H.: Designing localized algorithms for barrier coverage. In: Proceedings of the 13th Annual ACM International Conference on Mobile Computing and Networking, MobiCom 2007, pp. 63–74. ACM, New York (2007)
3. Chen, D.Z., Gu, Y., Li, J., Wang, H.: Algorithms on minimizing the maximum sensor movement for barrier coverage of a linear domain. Discrete Comput. Geom. **50**(2), 374–408 (2013)
4. Czyzowicz, J., Kranakis, E., Krizanc, D., Lambadaris, I., Narayanan, L., Opatrny, J., Stacho, L., Urrutia, J., Yazdani, M.: On minimizing the sum of sensor movements for barrier coverage of a line segment. In: Nikolaidis, I., Wu, K. (eds.) ADHOC-NOW 2010. LNCS, vol. 6288, pp. 29–42. Springer, Heidelberg (2010)
5. Czyzowicz, J., Kranakis, E., Krizanc, D., Lambadaris, I., Narayanan, L., Stacho, L., Urrutia, J., Yazdani, M.: On minimizing the maximum sensor movement for barrier coverage of a line segment. In: Proceedings of 8th International Conference on Ad Hoc Networks and Wireless, pp. 22–25 (2002)
6. Dobrev, S., Durocher, S., Eftekhari, M., Georgiou, K., Kranakis, E., Krizanc, D., Narayanan, L., Opatrny, J., Shende, S., Urrutia, J.: Complexity of barrier coverage with relocatable sensors in the plane. In: Spirakis, P.G., Serna, M. (eds.) CIAC 2013. LNCS, vol. 7878, pp. 170–182. Springer, Heidelberg (2013)
7. Eftekhari, M., Kranakis, E., Krizanc, D., Morales-Ponce, O., Narayanan, L., Opatrny, J., Shende, S.: Distributed algorithms for barrier coverage using relocatable sensors. In: Proceedings of the 2013 ACM Symposium on Principles of Distributed Computing, PODC 2013, pp. 383–392. ACM, New York (2007)
8. Kumar, S., Lai, T., Arora, A.: Barrier coverage with wireless sensors. Wirel. Netw. **13**(6), 817–834 (2007)

9. Kumar, S., Lai, T.H., Arora, A.: Barrier coverage with wireless sensors. In: Proceedings of the 11th Annual International Conference on Mobile Computing and Networking, pp. 284–298. ACM (2005)
10. Mehrandish, M., Narayanan, L., Opatrny, J.: Minimizing the number of sensors moved on line barriers. In: 2011 IEEE Wireless Communications and Networking Conference (WCNC), pp. 653–658. IEEE (2011)

A Comprehensive and Lightweight Security Architecture to Secure the IoT Throughout the Lifecycle of a Device Based on HIMMO

Oscar Garcia-Morchon$^{(\boxtimes)}$, Ronald Rietman, Sahil Sharma, Ludo Tolhuizen, and Jose Luis Torre-Arce

Philips Group Innovation, Research, Eindhoven, The Netherlands
{oscar.garcia,ronald.rietman,sahil.sharma,
ludo.tolhuizen,jose.luis.torre.arce}@philips.com

Abstract. Smart objects are devices with computational and communication capabilities connected to the Internet forming the so called Internet of Things (IoT). The IoT enables many applications, for instance outdoor lighting control, smart energy and water management, or environmental sensing in a smart city environment. Security in such scenarios remains an open challenge due to the resource-constrained nature of devices and networks or the multiple ways in which opponents can attack the system during the lifecycle of a smart object. This paper firstly reviews security and operational goals in an IoT scenario inspired in a smart city environment. Then, we present a comprehensive and lightweight security architecture to secure the IoT throughout the lifecycle of a device. Our solution relies on the lightweight HIMMO scheme – a novel key pre-distribution scheme that is both collusion resistance and efficient – as the building stone enabling not only efficient resource-wise but also advanced and scalable IoT protocols and architectures. Our design and analysis show that our HIMMO-based security architecture can be easily integrated in existing communication protocols such as IEEE 802.15.4 or OMA LWM2M providing a number of advantages that existing solutions cannot provide both performance and operation-wise.

Keywords: Lightweight · Key distribution · Security architecture · Internet of Things

1   Introduction

The ubiquitous connection of devices to the Internet, the Internet of Things (IoT), will account for more than a third of the total Internet connections by 2018, according to the Cisco M2M Devices Forecast 2013–2018 [1]. These devices will be deployed in multiple scenarios including smart homes, healthcare, or smart cities. In each of these environments, multiple applications are enabled: the IoT in a smart city can mean connecting city infrastructure such as water meters, environmental sensors, or lighting infrastructure to automate and improve city flows and functionality.

© Springer International Publishing Switzerland 2015
P. Bose et al. (Eds.): ALGOSENSORS 2015, LNCS 9536, pp. 112–128, 2015.
DOI: 10.1007/978-3-319-28472-9_9

Privacy and security are still two of the technical issues that remain unsolved. The reason is that solutions created for traditional computer networks do not suit the IoT performance, operational and security requirements. For instance, public-key cryptography allows any pair of devices to setup a secure channel or enables accountability. However, it is computationally expensive and requires the exchange of long keys, which has a negative impact during IoT operation. Symmetric cryptography is lightweight but it does not scale, in particular regarding key distribution and management, and this can lead to a lower security level. For instance, if a wireless network relies on a single key, then the capture of a single device will break down the whole system. Finally, IoT scenarios involve both traditional and new threats and security protocols and primitives should be adapted to address them. In particular, a security solution for the IoT should be secure and technically feasible (e.g., performance wise) not only during operation but during the whole lifecycle of a device.

In this context, this paper firstly reviews these operational and security goals in the context of the lifecycle of a smart object deployed in a smart city scenario. We then propose a comprehensive security architecture that addresses both goals during the lifecycle of a smart device. Our security architecture relies on the recently introduced and lightweight HIMMO scheme [2]. In this paper we show that HIMMO not only provides good performance, but that it also enables very attractive features from the point of view of deployment, operation, and maintenance. Our solution relies on an infrastructure of Trusted Third Parties (TTPs) for the management of the HIMMO security domains, device credentials and keying materials. This facilitates a secure manufacturing process and distribution of HIMMO keying materials. The easy integration of HIMMO with standard protocols such as LWM2M or IEEE 802.15.4 and simple extensions of these protocols allows us to ensure efficient and secure network access and device registration with a back-end server. Secure network operation is further achieved in the sense of full collusion resistance and easy and lightweight management of device credentials.

The rest of the paper is organized as follows: Sect. 2 describes the use cases for IoT in a smart city scenario, introduces relevant communication protocols, and analyzes operational and security goals for the IoT. Section 3 reviews the HIMMO scheme. Section 4 details the proposed security architecture according to the typical lifecycle of a smart object. Section 5 describes the implementation and evaluation of the Section main building blocks of our security architecture. Section 6 discusses how our architecture addresses the identified operational and security goals and compares it with related work. Section 7 concludes this paper and points out future work.

2   Background

2.1   Use Cases

The IoT in the context of smart cities enables services such as outdoor lighting control, water control, smart energy networks, and environmental sensing [3].

For instance, an environmental sensing network based on the IoT can be realized by means of a wireless mesh network that enables the communication between sensors: the environmental sensors would use the mesh network to reach a border router that would further forward the messages containing the gathered measurements towards a back-end system in charge of device and data management. Here, the devices constitute one of the key enablers of these services.

However, the limited amount of resources, e.g., regarding energy or communication, has heavily limited and limits the functionality that these systems can offer. Security is specially affected by this since most cryptographic primitives or protocols are just too heavy for many use cases so that a designer is confronted with two options: (i) using strong security, but that negatively affects the operation of the system and the way a user interacts with the system, or (ii) using weaker security, in order to not affect the expected system operation.

In this context, it is important to note that devices follow a lifecycle [4] that is to be considered in order to build a security architecture. This lifecycle includes several phases: manufacturing, bootstrapping, commissioning, and operational. In each of these phases, the system should remain secure while operating as expected.

2.2   Relevant Protocols

The above architecture in which a device communicates over a mesh network with a back-end system can be realized by means of multiple communication protocols. For instance, 6LoWPAN [5]/IEEE 802.15.4 [6] networks enable IP connectivity in a mesh network. Network connectivity from the border router to the back-end can be based on a cellular link. End to end communication can be based on OMA LWM2M, in which application data is exchanged between client and server by means of CoAP [7] and the end to end communication is secured by means of DTLS [8] using pre-shared keys, raw public-keys or certificates.

While protocols are available, security primitives are not optimal. For instance, IEEE 802.15.4 networks often rely on a network or system wide symmetric-key. The performance of DTLS and its cipher modes lack either flexibility or performance in many cases.

2.3   Operational and Performance Goals

The above scenarios have very specific operational and performance needs. A security architecture for the IoT should address all of them.

Because of the resource-constrained nature of the devices, security solutions should achieve good performance (**O-1**) regarding energy consumption, bandwidth requirements, number of round trips, memory needs, and CPU usage.

The deployment of an IoT network is usually done in several phases. It should thus be possible to add devices to a running system in a simple way (**O-2**).

Scalability is a key requirement so that a very high number of devices and back-end servers can be easily supported (**O-3**). In order to ensure this scalability, it is required to enable easy management of device's credentials and attributes in both centralized and distributed communication patterns (**O-4**).

Another important requirement refers to the easy integration with existing communication protocols and architectures (**O-5**). This facilitates adoption and allows for a smooth transition by avoiding the costs related to a change in the technology of the network.

Finally, solutions should fit the use cases not only during the operational phase but throughout the whole lifecycle of a device (**O-6**). In the context of IoT this also means that the deployed solutions have to remain secure during a long period of time (10, 20, or even 30 years) (**O-7**).

2.4   Attack Model and Security Goals

In addition to the above operational and performance goals, we consider an attack model in which the opponent can aim at disrupting the system operation at different stages of the lifecycle of a device. Next, we identify potential attacks and discuss security goals that aim to prevent them.

First of all, the attacker (either external or insider) can aim at compromising a root of trust, such as a Certification Authority (CA) in a Public Key Infrastructure (PKI). This would allow him to gain full control over the system. One example of this type of attack is the one suffered by DigiNotar [9]. Therefore, the first security goals refer to being resilient to the compromise of a root of trust (**S-1**) and ensuring that a single root of trust cannot monitor and control communication links (**S-2**). This last requirement should still be compatible with key escrow if required (**S-3**).

The next type of attack focuses on the manufacturing process in which the devices are configured with credentials and secret keys. If an opponent manages to get control on a manufacturing facility, then he can modify or copy this information. An important security goal is to facilitate a secure manufacturing process that prevents this (**S-4**).

The next type of attack can happen when the devices are being deployed. In this case, the back-end server might be exposed to fake devices and, in a similar way, fake servers might impersonate the actual back-end server to gain control over the devices. Authentication and authorization of device (**S-5**) and back-end server (**S-6**) is required to prevent this situation. Even with end to end authentication, the devices that are routing the information could still be exposed to a Denial of Service (DoS) attack, since they would not be able to identify by themselves the communicating parties. Thus, another goal is the prevention of DoS attacks during network access (**S-7**).

During operation, an attacker might aim at physically capturing devices to misuse their secret keys and credentials towards the back-end server or towards any other device in the network. Therefore, a key goal will be that the compromise of any number of devices does not affect the security of the whole system (**S-8**). Another related goal is to facilitate the identification and blacklisting of compromised devices (**S-9**).

Also during operation, another security goal refers to the capability of establishing a common shared key for providing further security services (**S-10**). As the devices will be on the field for many years, a security solution should provide long term security including resilience against post-quantum attacks (**S-11**). Other security goals are perfect forward secrecy (**S-12**), meaning that a session key derived from a set of long-term keys cannot be compromised if one of the long-term keys is compromised in future, and non-repudiation (**S-13**), e.g. allowing a metering device signing the energy consumption so that there is proof of the amount of consumed energy.

3   The HIMMO Scheme

HIMMO [2] is a Key Pre-Distribution Scheme (KPS) based on the HI and MMO problems. HIMMO enables any pair of devices in a system to directly agree on a pairwise symmetric key based on their identifiers and a secret key-generating polynomial. HIMMO is the first KPS that achieves collusion-resistance, making interpolation attacks by colluding nodes infeasible, while being efficient in the generation of pairwise keys. This means that a network of M devices, a total of $M(M-1)/2$ can be efficiently distributed so that each pair of devices shares a different pairwise key. Like any KPS [10], HIMMO requires a TTP and three phases can be distinguished in its operation.

Setup Phase: The TTP, upon reception of some public parameters, generates a secret root keying material that consists of the coefficients of several bi-variate symmetric polynomials: $R^{(i)}(x, y)$.

Keying Material Extraction Phase: The TTP provides each node in the system with the coefficients of the generated polynomial that arises from the addition of evaluations of the secret bi-variate symmetric polynomials in the identity of the node, for node ξ we have: G_ξ.

Key Generation Phase: A node can compute a pairwise key with any other node of the system by evaluating its keying material in the identity of the other node: $K_{\xi,\eta} = \langle\langle G_\xi(\eta)\rangle_N\rangle_{2^b}$. It can be shown that $K_{\xi,\eta}$ and $K_{\eta,\xi}$ need not be equal but within a certain range.

A detailed description of this procedure can be found in [2,11], and Appendix A.1.

In addition to the basic scheme for key agreement, HIMMO enables interesting extensions. First, HIMMO supports multiple TTPs as previously introduced by Matsumoto and Imai [10] enhancing privacy as well as improving the security of the system—compromising a sub-set of TTPs does not break the overall system [2]. Another extension refers to the capability for implicit certification and verification of credentials at the only cost of a hashing operation. This capability builds on the fact that HIMMO is based on identifiers that can be bound to any bit string by means a one way hash function. This has been used in [11] to enable the verification of credentials between client and server in a TLS connection without the need of digital certificates and it is further discussed in Section Appendix A.2.

4 Design

This section describes the proposed security architecture focusing on the security goals presented in Sect. 2.4 and the operational goals described in Sect. 2.3. In the following subsections, we detail our solution according to the lifecycle of a smart object in which we show how keying materials are managed by means of a TTP infrastructure. This TTP infrastructure can be used to enable secure manufacturing of the smart objects (Sect. 4.1). The HIMMO keying material enables secure network access of a smart object in a typical smart city scenario (Sect. 4.2). Finally, this HIMMO keying material provides a way of ensuring secure operation and credential management during normal operation (Sect. 4.3).

Figure 1 depicts the components of our security architecture including: (i) an infrastructure of roots of trust in charge of handling the HIMMO root keying materials (Sect. 4.1). (ii) A number of manufacturing units producing smart objects configured with secret keys and credentials (Sect. 4.1). (iii) An authentication process, in which a smart object registers with a back-end system in order to get credentials for access to the network (Sect. 4.2). (iv) A secure operation phase in which smart objects can securely communicate with each other, using the credentials obtained in the previous step (Sect. 4.3).

4.1 TTP Infrastructure and Smart Object Manufacturing

Our architecture relies on the HIMMO capability for working with multiple TTPs to address goals **S-1, S-2, S-3, S-4, and S-11**.

We consider multiple TTPs can generate and securely manage multiple HIMMO root keying materials. The role of these TTPs is similar to today's CA infrastructure with the obvious differences in the underlying technology.

When a new set of devices needs to be manufactured, the back-end server in charge of those devices will request a subset of the TTPs to extract HIMMO keying materials linked to some device credentials. In the following, we will assume that these credentials depend on the unique MAC address of each device η, e.g., $\eta = f(\eta' MACaddress)$. The back-end server will also determine which factories will get HIMMO keying materials from which TTPs. Next, TTP j will extract HIMMO keying material $G_\eta^j(y)$ for device η and securely send these keying materials to the corresponding manufacturing facility l. Here, each device η will receive $G_\eta^j(y)$ and update its locally aggregated keying material as $G_\eta(y) = \langle G_\eta(y) + G_\eta^j(y) \rangle_N$. In a similar way, the back-end server ξ can request at any time a new set of keying materials from those TTPs and obtain its aggregated keying material in a similar way. Note that in this process, even if an attacker manages to compromise a manufacturing facility or TTP or a combination of those, the attacker will not gain knowledge about how the whole keying material of the devices is constituted since this keying material depends on all the root keying materials issued by multiple TTPs. This is the reason why the above architecture enables secure manufacturing.

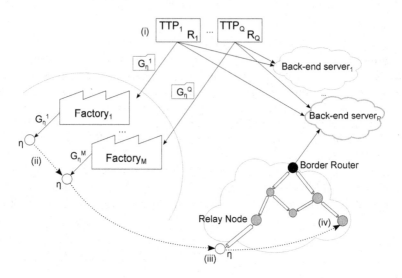

Fig. 1. Architecture overview.

4.2 The Authentication Process for Network Access

We use the HIMMO keying material to enable secure network access and device registration in a typical smart city scenario, addressing goals **S-5, S-6, S-7, S-8, S-9, S-10**. In this procedure, the joining node η aims to register with a back-end server after its verification and the back-end server aims to register the joining node after authenticating it. Although the server can verify the client, it is equally important that devices in the network can verify the authenticity of the joining node in order to prevent possible abuse that could lead to DoS attacks. The verification could be done at different levels in the network to thwart an attack as early as possible. The protocol below describes how HIMMO can be applied to a real-world scenario based on an IEEE 802.15.4 [6] mesh network using 6LoWPAN, typical example of an IoT network, where the DTLS security protocol ensures end to end security. Note that all keys described below are pairwise keys generated by means of HIMMO based on the identities, derived from their MAC addresses, assigned to the nodes during the manufacturing process.

- Step 1: A joining node η is installed and waits to hear beacons from devices in a network.
- Step 2: A neighboring node, working as relay node or a border router, regularly sends a broadcast beacon as part of the normal network traffic.
- Step 3: Upon reception of a network beacon originating, e.g., from a relay node r, node η will securely send an IEEE 802.15.4 message protected with the HIMMO key $K_{\eta,r} = \left\langle \left\langle G_\eta(r) \right\rangle_N \right\rangle_{2^b}$. The Key Identifier and Key Identifier Mode [6] can be used to indicate that HIMMO key derivation is in use.
- Step 4: The relay node will process this message protected at the IEEE 802.15.4 MAC layer, and if the verification of the message sent by η is suc-

cessful, reply with the identity of the border router. $K_{\eta,r}$ is to be cached by r for further communication.

- Step 5: Node η then will start a DTLS [8] handshake towards the back-end server through relay node r in which the communication at the IEEE 802.15.4 MAC layer is protected by $K_{\eta,r}$. Note that the relay node operates as described in [12]. Furthermore, the first DTLS message, Client Hello, includes the identity of the joining node and an authentication token created with $K_{\eta,br}$ so that the border router can verify it the identity of the joining node.
- Step 6: Node r will relay only DTLS traffic from device η. Furthermore, the border router will only forward the DTLS traffic if the verification of the authentication token is successful and the identity of the joining node is not blacklisted.
- Step 7: The back-end server gets the request from the joining node and engages in a DTLS handshake in which both joining node and back-end server mutually authenticate each other and verify their credentials. This handshake may be based on the lightweight HIMMO-based scheme proposed in [11] in which HIMMO pairwise keys are generated and credentials are verified efficiently. Upon successful establishment of the DTLS session, the joining node will receive the network key, $K_{Network}$, from the server.

Upon successful completion of this protocol, the key used to secure IEEE 802.15.4 layer communication on the link between node η and any node ξ that has joined the network is computed as $K_{\eta,\xi}^{Network} = K_{\eta,\xi} \oplus K_{Network}$, where $\oplus$ denotes bit-wise XOR.

We note that the above protocol relies on the identity-based nature of HIMMO and its properties to mutually authenticate the joining device and (i) relay node, (ii) border router, and (iii) back-end server. This allows for early detection and prevention of DoS attacks at relaying devices and at the border router. DTLS-HIMMO enables efficient operation [11]. Finally, we also remark that since the identifier, the MAC address, is linked to the keying material issued by the TTPs, attacks such as MAC spoofing are not feasible.

4.3   Secure Operation

Secure operation aims at fulfilling security goals **S-3, S-5, S-6, S-8, S-9, S-10, S-11** and all operational and performance objectives.

Firstly, we illustrate how HIMMO ensures a fully collusion resistant mesh network based on IEEE 802.15.4. IEEE 802.15.4 supports the usage of pairwise keys between devices, however, key agreement is left out of the standard. Therefore, in practice, most standards rely on a network or system-wide key so that if a single device is captured, then the whole network or system breaks down. Furthermore, compromised devices cannot be easily identified since an attacker can fake any identity so that securely updating such a network- or system-wide key is infeasible (in addition to being very costly from an operational point of view). In our security architecture, secure operation in a IEEE 802.15.4 network works as follows:

- Step 1: Note ξ will advertise its presence by means of unsecured broadcast beacon.
- Step 2: Node η will obtain a common HIMMO key with device ξ as $K_{\eta,\xi} = \langle \langle G_\eta(\xi) \rangle_N \rangle_{2^b}$. Node η will also obtain helper data $\sigma_{\eta,\xi}$. If node η has joined the network, then it will use pairwise key $K_{\eta,\xi}^{Network} = K_{\eta,\xi} \oplus K_{Network}$. If node η has not joined the network yet, then it will use pairwise key $K_{\eta,\xi}$.
- Step 3: Node η sends a secure IEEE 802.15.4 message to ξ protected with the pairwise key determined in the previous step. This message contains the MAC address of the sender (η) and $\sigma_{\eta,\xi}$ in the Key Identifier field of the 802.15.4 security header.
- Step 4: Node ξ receives the secure message from η. It generates the pairwise HIMMO key with the MAC address of η as $K_{\xi,\eta} = \langle \langle G_\xi(\eta) \rangle_N \rangle_{2^b}$ and recovers $K_{\eta,\xi}$ from $K_{\xi,\eta}$ and $\sigma_{\eta,\xi}$. Assuming that node ξ has already joined the network, then ξ knows $K_{Network}$ so that it can compute $K_{\eta,\xi}^{Network}$ and process the incoming message with both $K_{\eta,\xi}$ and $K_{\eta,\xi}^{Network}$. If the verification of the message with $K_{\eta,\xi}$ is successful, then node η is considered a joining device and only DTLS traffic for network access is allowed. Alternatively, the message is verified with $K_{\eta,\xi}^{Network}$ so that the device is identified as part of the network and normal routing and data traffic is enabled.

This solution enables pairwise keys between devices based on HIMMO so that the IEEE 802.15.4 network becomes fully collusion resistant to the physical capture of devices. Note that this represents a huge improvement with the state-of-the-art in which most deployments rely on a single network-wide or even system-wide key that represents a single point of failure. Furthermore, compromised devices can be identified since the HIMMO keying material is cryptographically bound to the device identifier and in this case it is chosen to be the MAC address of the device. This means that captured devices can be blacklisted. Note that the MAC address cannot be spoofed since it is bound to the HIMMO keying material. Finally, this method allows us to easily differentiate devices that have joined the network from devices that have not done so yet. This differentiation, however, could also be determined by indicating which key is used to protect the packet by using the Key Index field of the Key Identifier in the IEEE 802.15.4 Auxiliary Security Header (ASH) [6]. This would avoid verification of the packet with two keys in Step 4. The Key Index field (or a part of it) will indicate to node ξ which key, $K_{\eta,\xi}$ or $K_{\eta,\xi}^{Network}$, was used to secure the packet.

We further note that HIMMO cannot only be used at MAC layer to generate pairwise keys between neighboring devices, but it can also be used to enable secure communication between any pair of devices in the network/system at application layer by means of DTLS [11]. This is similar to the process described in previous section with the difference that the communication is not between the joining device and the back-end server, but between two arbitrary devices. As a consequence any pair of devices can securely agree on a common symmetric key used in DTLS and verify any other set of credentials. In addition to the security advantages, performance benefits will become clear in the next section.

5   Implementation and Evaluation

The design described in Sect. 4 addresses the needs of real-world environments in Smart Cities such as SmartSantander [3]. This section provides performance measures for the building blocks that comprise this architecture.

According to its different functionalities and resource requirements, we divide the HIMMO implementation into two modules: one that provides TTP capabilities and would likely be implemented in a server, and another one that provides node functionalities and would likely be implemented in a device. Some of these implementations are available at the HIMMO website www.himmo-scheme.com.

The TTP module allows choosing the HIMMO public parameters described in Appendix A.1. It also generates the keying materials for the network devices tailored to the word-length and architecture of each device for optimal performance. The TTP module is implemented in Java and can be easily integrated in a web service.

For the node module implementation, it is worth noting that HIMMO has been designed to enable very efficient performance in constrained devices. From Eq. (3) we observe that obtaining a symmetric key just requires the evaluation of a polynomial of degree α modulo N and taking the b least significant bits. This means that only $\alpha + 1$ modular multiplications are required to compute the key. In each multiplication, the B bit identifier multiplies the $(\alpha + 1)B + b$ bit coefficient and the result is reduced modulo N. These modular operations can be implemented in a very efficient manner for appropriate choices for N, e.g., $N = 2^{(\alpha+1)B+b} - 1$. Many implementations are possible for the node module because of the heterogeneous nature of the devices. Specialized implementations will optimize performance, sacrificing portability.

In order to evaluate the performance of the HIMMO scheme, we have implemented it on three different devices: a very resource-constrained 8-bit CPU ATMEGA128L running at 8 MHz, the 32-bit NXP LPC1769 LPCXpresso Board running at 120 MHz, and an Intel i3 3120M (64-bit), at 2.50 GHz running Xubuntu 14.04. The implementations for the NXP LPC1769 and the Intel i3 3120M are based on our flexible C library that can be easily ported to different devices and integrated in different protocols. To this end, the library can be easily configured to work with different CPU word sizes and the whole library, including the big integer arithmetic required for key generation, is written in C. On the other hand, the implementation for the ATMEGA128L is optimized in assembler so that better performance can be achieved.

For very resource constrained devices, specialized implementations are preferred: our optimized implementation for the ATMEGA128L fits in just 428 bytes of flash memory. This shows that HIMMO can fit even in very resource constrained devices or when integrated as part of the IEEE 802.15.4 logic described in the above section. We also note that the RAM consumption is linear with α since we have to keep in memory a term that is $(\alpha + 1)B + b$ bits. Tables 1 and 2 provide a brief summary of the performance of the HIMMO scheme implemented in the above CPUs.

Table 1. HIMMO performance for $B = b = 128$ as a function of α.

α		26	34	40	50
Keying material size (KB)		6.90	11.18	15.07	22.83
CPU time (msec)	Atmega128L (8-bit @ 8 MHz)	223	367	497	743
	NXP LPC1769 (32-bit @ 120 MHz)	7.46	11.82	15.74	23.48
	Intel i3 3120M (64-bit @ 2.5 GHz)	0.034	0.053	0.069	0.103

Table 2. HIMMO performance for $\alpha = 26$ as a function of $b = B$.

$b = B$		64	128	192	256
Keying material size (KB)		3.45	6.90	10.34	13.79
CPU time (msec)	Atmega128L (8-bit @ 8 MHz)	63	223	393	632
	NXP LPC1769 (32-bit @ 120 MHz)	2.55	7.46	14.93	25.16
	Intel i3 3120M (64-bit @ 2.5 GHz)	0.015	0.034	0.062	0.100

The above node module can be used to enable the functionality described in the design section. In particular, it can be easily integrated with IEEE 802.15.4 ensuring a fully collusion resistant network during operation. It can also be used to enable secure network access and registration by means of DTLS-HIMMO, achieving functionalities such as mutual authentication and verification, which today are only possible with public-key cryptography, at the speed and memory requirements of symmetric cryptography. In this context, the left part of Fig. 2 shows the time elapsed to successfully complete a DTLS handshake in different modes, we can see that HIMMO with mutual verification is almost 7 times faster than mutual verification with the ECC alternative. On the right, we can see the ratio between exchanged data and payload: for a secure interchange of 1 KB this ratio is around 40 % smaller with HIMMO compared with ECC both providing mutual authentication. More detailed performance measures, as well as a complete description of this implementation are available on [11].

6   Discussion and Comparison with Related Work

This section discuses how the proposed security architecture fulfills the goals described in Sects. 2.4 and 2.3, and compares our solution with approaches based on symmetric cryptography (Pre-Shared Key (PSK)) and on asymmetric cryptography (Public Key Cryptography (PKC)). A PSK is lightweight and very efficient but, as we can see in Table 3, it is unable to support most of the security goals identified in Sect. 2.4. Such a solution is not scalable. We might wish to share the same key with a large number of devices. This is the current approach in IEEE 802.15.4 as it does not require any additional infrastructure. However, this approach leads to a single point of failure —if the key gets compromised, the security of all the network will be broken— which is a really important

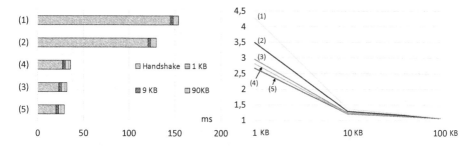

Fig. 2. (1) ECDH-ECDSA with mutual authentication, (2) ECDH-ECDSA with server authentication, (3) HIMMO with mutual verification of client's and server's credentials ($t = 5, B = 256, b = 32, \alpha = 17$), (4) HIMMO with mutual authentication ($t = 5, B = 32, b = 32, \alpha = 50$) and (5) PSK.

problem in networks where not even the physical integrity of the devices can be guaranteed. This approach also does not enable identification, authentication, or authorization of devices. On the other hand, if many (pairwise) keys are distributed, this involves huge amounts of memory, makes credential management difficult, and the addition of new devices to the network becomes more complex. In contrast, HIMMO enables the generation of pairwise keys from the identifiers assigned to the devices achieving both the simplicity of a single key approach with the advantages of having many pairwise keys. PKC solves most security goals, but is much more resource-consuming than PSK or HIMMO in terms of required CPU time, memory, bandwidth, and round-trips (e.g., see Fig. 2). In a similar way, in order to comply with security goal **S-1** —resiliency to root of trust compromise— it would be necessary to have multiple CA signing the certificates; this will increase the certificate length as well as memory usage and network overhead. As one of the main sources of energy consumption in an IoT network is the wireless radio, the battery life would also be affected. In contrast, HIMMO can achieve this property without performance penalty. The integration with OMA LWM2M would be easy, as DTLS supports PKC. However, it is not possible to use PKC in IEEE 802.15.4 without protocol modifications. Furthermore, it would imply a serious degradation in the network performance. Similarly, it could be possible to use certificates to design a joining protocol which would prevent DoS attacks, but it would result in a great increase in network overhead because of the certificate exchanges. Finally, existing PKC solutions (e.g., all PKC-based cipher-suites in (D)TLS) would break down if a quantum computer is built. HIMMO and PSK do no not offer perfect forward secrecy (**S-12**); the $*$ in the row for (**S-12**) means that PKC does offer this feature for some cipher suites, e.g. those based on DHE or ECDHE key exchange. Being symmetric schemes, HIMMO and PSK do not offer non-repudiation (**S-13**). We finally note that these architectures focus on cryptography and network security protocols. Other *complementary* approaches exist, e.g., to handle trust between devices. For instance, in [13], the authors discuss a protocol to manage how

Table 3. Comparison between different security architectures. Relative performance from most efficient (*) to least (***).

Design goals	HIMMO	PKC	PSK
O-1: Performance	*	***	*
O-2: Easy device addition to a running system	✓	✓	✓
O-3: Scalable	✓	✓	-
O-4: Easy credential management	✓	✓	-
O-5: Easy integration with existing protocols	✓	-	✓
O-6: Fits device lifecycle	✓	✓	-
O-7: Long term security	✓	✓	✓
S-1: Resilient to root of trust compromise	✓	✓	-
S-2: Single root of trust cannot monitor	✓	✓	-
S-3: Key escrow	✓	-	✓
S-4: Facilitates secure manufacturing	✓	✓	-
S-5: Device authentication and authorization	✓	✓	-
S-6: Back-end authentication and authorization	✓	✓	-
S-7: Prevents DoS attacks	✓	✓	-
S-8: Fully collusion resistance	✓	✓	-
S-9: Device identification and blacklisting	✓	✓	-
S-10: Key agreement	✓	✓	✓
S-11: Post-quantum resilience	✓	-	✓
S-12: Perfect forward secrecy	-	✓*	-
S-13: Non-repudiation	-	✓	-

devices gain trust in each other and also can lose it if a number of devices agree on it, e.g., agree on a device to be misbehaving. In this case, losing trust means that those devices can generate a revocation message against the misbehaving device so that it is excluded from the network.

7   Conclusions

The IoT is an emerging area involving many connected smart devices that need to be secured during their entire lifecycle. This problem is not solved with current security solutions: asymmetric-key cryptography is computationally hungry, difficult to integrate with protocols, and not suitable for long term deployments, while the symmetric-key option is extremely efficient, but does not scale and offers limited security properties. We have reviewed the operational and security requirements of a real-world IoT scenario and build on top of them a comprehensive and lightweight security architecture based on HIMMO. Thus, our security architecture is tailored to the requirements of IoT, its attack vectors, and resource requirements. Our solution builds on HIMMO's excellent performance —a combined HIMMO-based key agreement and credential verification handshake can be done in a few hundred milliseconds even on very resource-constrained devices— to create a scalable, operationally friendly and secure architecture. HIMMO can be easily integrated in modern protocols like IEEE 802.15.4 or DTLS, providing, at the relatively low costs of symmetric cryptographic solutions, features that before were only feasible with asymmetric cryptography. Some of the security features enabled by HIMMO are full collusion resistance, device and back-end authentication and verification, pairwise key agreement, support for multiple TTPs and key escrow, or protection against DoS attacks. Because our architecture combines great security features with low resource needs and allows easy integration in modern protocols, we believe that it is a very competitive approach to secure the IoT and related emerging areas.

A   HIMMO

HIMMO is a Key Pre-Distribution Scheme (KPS), a concept introduced by Matsumoto and Imai in 1987 [10]. Blundo et al. [14] present an elegant and efficient KPS based on symmetric polynomials. However, their KPS is prone to collusion attacks: if an attacker has compromised $\alpha + 1$ nodes, where α is the degree of the polynomial in any variable, then he can crack the complete system by using simple (Lagrange) interpolation. There was no known KPS that is both efficient and not prone to efficient attacks of multiple colluding (or compromised) nodes (see [2] for further references) until recently the HIMMO scheme solved this problem. This section reviews the operation of the HIMMO scheme that enables any pair of devices in a system to directly agree on a common symmetric-key based on their identifiers and a secret key generating polynomial as introduced in [15]. Like Blundo's scheme, HIMMO is based on symmetric polynomials, but it introduces new features to make simple interpolation attacks by colluding nodes infeasible. The underlying security principles on which HIMMO relies have been analyzed in [16,17]. Furthermore, this section describes two protocol extensions of the HIMMO scheme as described in [2].

We use the following notation: for each integer x and positive integer M, we denote by $\langle x \rangle_M$ the unique integer $y \in \{0, 1, \ldots, M-1\}$ such that $x \equiv y \bmod M$.

A.1   HIMMO Operation

Like any KPS, HIMMO requires a TTP, and three phases can be distinguished in its operation [10].

In the **setup phase**, the TTP selects positive integers B, b, m and α, where $m \geq 2$. The number B is the bit length of the identifiers that will be used in the system, while b denotes the bit length of the keys that will be generated. The TTP generates the public modulus N, an odd number of length exactly $(\alpha+1)B + b$ bits (so $2^{(\alpha+1)B+b-1} < N < 2^{(\alpha+1)B+b}$). It also randomly generates m distinct secret moduli $q_1, \ldots, q_m$ of the form $q_i = N - 2^b \beta_i$, where $0 \leq \beta_i < 2^B$

and at least one of $\beta_1, \ldots, \beta_m$ is odd. Finally, the TTP generates the secret root keying material, that consists of the coefficients of m bi-variate symmetric polynomials of degree at most α in each variable. For $1 \leq i \leq m$, the i-th root keying polynomial $R^{(i)}(x, y)$ is written as

$$R^{(i)}(x, y) = \sum_{j=0}^{\alpha} \sum_{k=0}^{\alpha} R_{j,k}^{(i)} x^j y^k$$

with $0 \leq R_{j,k}^{(i)} = R_{k,j}^{(i)} \leq q_i - 1$.

In the **keying material extraction phase**, the TTP provides each node ξ in the system, with $0 \leq \xi < 2^B$, the coefficients of the key generating polynomial G_ξ:

$$G_\xi(y) = \sum_{k=0}^{\alpha} G_{\xi,k} y^k \tag{1}$$

where

$$G_{\xi,k} = \left\langle \sum_{i=1}^{m} \left\langle \sum_{j=0}^{\alpha} R_{j,k}^{(i)} \xi^j \right\rangle_{q_i} \right\rangle_N. \tag{2}$$

In the **key generation phase**, a node ξ wishing to communicate with node η with $0 \leq \eta < 2^B$, computes:

$$K_{\xi,\eta} = \left\langle \langle G_\xi(\eta) \rangle_N \right\rangle_{2^b} \tag{3}$$

It can be shown that $K_{\xi,\eta}$ and $K_{\eta,\xi}$ need not be equal. However, as shown in Theorem 1 in [2], for all identifiers ξ and η with $0 \leq \xi, \eta \leq 2^B$,

$$K_{\xi,\eta} \in \{ \langle K_{\eta,\xi} + jN \rangle_{2^b} \mid 0 \leq |j| \leq 2m \}$$

In order to perform key reconciliation , i.e. to make sure that ξ and η use the same key to protect their future communications, the initiator of the key generation (say node ξ) sends to the other node, simultaneously with an encrypted message, information on $K_{\xi,\eta}$ that enables node η to select $K_{\xi,\eta}$ from the candidate set $C = \{ \langle K_{\eta,\xi} + jN \rangle_{2^b} \mid 0 \leq |j| \leq 2m \}$. No additional communication thus is required for key reconciliation. The key $K_{\xi,\eta}$ will be used for securing future communication between ξ and η. As an example of helper data used for key reconciliation, node ξ sends to node η the number $\sigma_{\xi,\eta} = \langle K_{\xi,\eta} \rangle_{2^s}$, where $s = \lceil \log_2(4m + 1) \rceil$. Node η can efficiently obtain the integer j such that $|j| \leq 2m$ and $K_{\xi,\eta} \equiv K_{\eta,\xi} + jN \mod 2^b$ by using that $jN \equiv K_{\xi,\eta} - K_{\eta,\xi} \equiv \sigma_{\xi,\eta} - K_{\eta,\xi}$ mod 2^s. As N is odd, the latter equation allows for determination of j. As $\sigma_{\xi,\eta}$ reveals the s least significant bits of $K_{\xi,\eta}$, only the $b - s$ most significant bits $K_{\xi,\eta}$, that is, the number $\lfloor 2^{-s} K_{\xi,\eta} \rfloor$, should be used as key.

A.2   Implicit Certification and Verification of Credentials

Implicit certification and verification of credentials is further enabled on top of the basic HIMMO scheme. A node that wants to register with the system provides the TTP with its credentials, e.g., device type, manufacturing date, etc. The TTP, which can also add further information to the node's credentials such as a unique node identifier or the issue date of the keying material and its expiration date, obtains the node's identity as $\xi = H(credentials)$, where H is a public hash function. When a first node with identity ξ wants to securely send a message M to a second node with identity η, the following steps are taken.

- Step 1: Node ξ computes a common key $K_{\xi,\eta}$ with node η. It uses the computed common key to encrypt and authenticate its credentials and message M, say $e = E_{K_{\xi,\eta}}(credentials|M)$.
- Step 2: Node ξ sends $(\xi, e, \sigma_{\xi,\eta})$ to node η, where $\sigma_{\xi,\eta}$ is helper data helping node η to find $K_{\xi,\eta}$.
- Step 3: Node η receives $(\xi', e', \sigma'_{\xi,\eta})$. Using $\sigma'_{\xi,\eta}$, it computes its common key $K_{\eta,\xi'}$ with ξ' to decrypt e' obtaining the message M and verifying the authenticity of the received message. Furthermore, it checks whether the $credentials'$ in e' correspond with ξ', that is, it validates if $\xi' = H(credentials')$.

This method not only allows for direct secure communication of message M, but also for implicit certification and verification of ξ's credentials because the key generating polynomial assigned to a node is linked to its credentials by means of H. If the output size of H is long enough, e.g., 256 bits, the input (i.e., the credentials) contains a unique node identifier, and if H is a secure one-way hash function, then it is infeasible for an attacker to find any other set of credentials leading to the same identity ξ. The fact that credential verification might be prone to birthday attacks motivates the choice for the relation between identifier and key sizes, namely, $B = 2b$. In this way, the scheme provides an equivalent security level for credential verification and key generation. The capability for credential verification enables e.g. the verification of the expiration date of the credentials (and the keying material) of a node, or verification of the access roles of the sender node ξ.

A.3   Enhancing Privacy by Using Multiple TTPs

Using multiple TTPs was introduced by Matsumoto and Imai [10] for KPS and can also be elegantly supported by HIMMO [2]. In this scheme, a number of TTPs provide a node with keying materials linked to the node's identifier during the keying material extraction phase. Upon reception, the device combines the different keying materials by adding the coefficients of the key generating polynomials modulo N. Key generation is performed as usual. This scheme enjoys two interesting properties without increasing the resource requirements of the nodes. First, privacy is enhanced since a single TTP cannot eavesdrop the communication links. In fact, all TTPs should collude to monitor the communication links. Secondly, compromising a sub-set of TTPs does not break the overall system.

References

1. Pepper, R.: The Internet of Things is Now: M2M Devices Forecast 2013–2018. IIC Annual Conference (2014). http://www.iicom.org

2. García-Morchón, O., Gómez-Pérez, D., Gutiérrez, J., Rietman, R., Schoenmakers, B., Tolhuizen, L.: HIMMO - A Lightweight, Fully Colluison Resistant Key-Predistribution Scheme. Cryptology ePrint Archive, Report 2014/698 (2014). http://eprint.iacr.org/

3. Sanchez, L., Galache, J.A., Gutierrez, V., Hernández, J.M., Bernat, J., Gluhak, A., García, T.: Smartsantander: the meeting point between future internet research and experimentation and the smart cities. In: Future Network & Mobile Summit (FutureNetw), pp. 1–8. IEEE (2011)

4. Garcia-Morchon, O., Kumar, S., Keoh, S., Hummen, R., Struik, R.: Security considerations in the ip-based internet of things. Internet-Draft draft-garcia-core-security-06, IETF Secretariat, September 2013. http://www.ietf.org/internet-drafts/draft-garcia-core-security-06.txt

5. Kushalnagar, N., Montenegro, G., Schumacher, C.: IPv6 over Low-Power Wireless Personal Area Networks (6LoWPANs): Overview, Assumptions, Problem Statement, and Goals. RFC 4919 (Informational), August 2007

6. IEEE Computer Society. IEEE Standard for Local and metropolitan area networks - Part 15.4 2011 revision: Low-Rate Wireless Personal Area Networks (LR-WPANs), September 2011

7. Shelby, Z., Hartke, K., Bormann, C.: The Constrained Application Protocol (CoAP). RFC 7252 (Proposed Standard), June 2014

8. Rescorla, E., Modadugu, N.: Datagram Transport Layer Security Version 1.2. RFC 6347 (Proposed Standard), January 2012

9. JR Prins and Business Unit Cybercrime. DigiNotar Certificate Authority breach Operation Black Tulip (2011)

10. Matsumoto, T., Imai, H.: On the key predistribution system: a practical solution to the key distribution problem. In: Pomerance, C. (ed.) CRYPTO 1987. LNCS, vol. 293, pp. 185–193. Springer, Heidelberg (1988)

11. Garcia-Morchon, O., Rietman, R., Sharma, S., Tolhuizen, L., Torre-Arce, J.L.: DTLS-HIMMO: Efficiently Securing a Post-Quantum World with a Fully-Collusion Resistant KPS, Accepted for publication at ESORICS (2015). https://eprint.iacr.org/2014/1008

12. Kumar, S., Keoh, S., Garcia-Morchon, O.: DTLS Relay for Constrained Environments. Internet-Draft draft-kumar-dice-dtls-relay-02, IETF Secretariat, October 2014. http://www.ietf.org/internet-drafts/draft-kumar-dice-dtls-relay-02.txt

13. Garcia-Morchon, O., Kuptsov, D., Gurtov, A., Wehrle, K.: Cooperative security in distributed networks. Comput. Commun. J. **36**, 1284–1297 (2013)

14. Blundo, C., de Santis, A., Herzberg, A., Kutten, S., Vaccaro, U., Yung, M.: Perfectly secure key distribution for dynamic conferences. Inf. Comput. **146**, 1–23 (1998)

15. Garcia-Morchon, O., Tolhuizen, L., Gomez, D., Gutierrez, J.: Towards full collusion resistant ID-based establishment of pairwise keys. In: Extended Abstracts of the Third Workshop on Mathematical Cryptology (WMC 2012) and The Third International Conference on Symbolic Computation and Cryptography (SCC 2012), pp. 30–36 (2012)

16. García-Morchón, O., Gómez-Pérez, D., Gutiérrez, J., Rietman, R., Tolhuizen, L.: The MMO problem. In: Proceedings of ISSAC 2014, pp. 186–193. ACM (2014)

17. García-Morchon, O., Rietman, R., Shparlinski, I.E., Tolhuizen, L.: Interpolation and approximation of polynomials in finite fields over a short interval from noisy values. Exp. Math. **23**, 241–260 (2014)

Maximizing Throughput in Energy-Harvesting Sensor Nodes

Stanley P.Y. Fung[(✉)]

Department of Computer Science, University of Leicester, Leicester, UK
pyf1@le.ac.uk

Abstract. We consider an online throughput maximization problem in sensor nodes that can harvest energy. The sensor nodes generate and forward packets, which cost energy; they can also harvest energy from the environment, but the amount of energy that can be harvested is not known in advance. We give a number of algorithms and lower bounds for the case of a single node. We consider both the general case and some types of 'non-idling' adversaries where we can get better bounds. We also consider the case of networks with multiple nodes and demonstrate that some very simple scenarios already admit no competitive algorithms.

1 Introduction

Background. Sensor networks are often deployed in areas where it is infeasible to maintain a constant energy supply to the sensor nodes. Often the nodes are equipped with batteries, and a node can only operate until its battery is exhausted. There are many research work on how to extend the useful lifetime of the sensor node or the sensor network by careful scheduling. If the sensor node is equipped with some energy-harvesting device, e.g., solar cells so it can replenish used energy, it can help make the system work longer or even indefinitely. This creates a challenge of designing algorithms that can make use of this harvested energy effectively.

The Model. We consider the scenario where each sensor node senses the environment, generates packets and sends them to a target destination. First consider a single node. The model was defined in [12]. Time consists of discrete time steps 1, 2, ... A packet j is specified by a 3-tuple $(r(j), d(j), v(j))$, which represents its release time, deadline and value. A packet with release time $r(j)$ and deadline $d(j)$ can only be sent in one of the time steps between $r(j)$ and $d(j)$, inclusive. Sending a packet costs one unit of energy. The sensor is equipped with a battery with a capacity of C, and an energy-harvesting device that may harvest some amount of energy $h(t)$ at each time step t. Let $e(t)$ denote the energy level of the battery at the beginning of time t (excluding energy harvested at this time step). A packet can only be sent if the node has sufficient energy, i.e., $e(t) + h(t) \geq 1$. The energy remaining at the next step is given by $e(t+1) = \min(C, e(t) + h(t) - x(t))$ where $x(t) = 1$ if a packet is sent at time t and $x(t) = 0$ otherwise. We assume there is no 'leak' of the battery so the

© Springer International Publishing Switzerland 2015
P. Bose et al. (Eds.): ALGOSENSORS 2015, LNCS 9536, pp. 129–141, 2015.
DOI: 10.1007/978-3-319-28472-9_10

energy level stays the same when no packets are sent. The objective is to maximize the *profit* or *weighted throughput* of the schedule, i.e., the sum of values of all packets sent.

Note that $h(t)$ is not known in advance and only become known at time t. Packet arrivals are also unknown in advance: packets with release time $r(j)$ are not known until time $r(j)$. Therefore, this is an *online* problem. We measure the performance of online algorithms using competitive analysis [1]: an online algorithm A is r-competitive if the value produced by A is always at least $1/r$ that of the optimal offline algorithm OPT over all input instances. For randomized online algorithms we use the expected value of A instead for comparison.

Generalizing from a single node, we also consider the model where nodes are connected into a network. Packets may have different sources and destinations, and each sensor node needs to forward traffic generated by other nodes as well. In our model, in each time step each node can send one packet to another node. Each packet takes one time unit to pass the link between two nodes. Thus if a packet is sent at time t in an upstream node, it appears as a packet with release time $t + 1$ in the next downstream node. Sending a packet takes one unit of energy, and we ignore the energy required to listen to or receive packets. The objective is to maximize the total value of packets reaching their destinations.

Before going any further we introduce some definitions and notations. Let $V = \max_j v(j) / \min_j v(j)$. An instance is *underloaded* if all packets can be sent by OPT, respecting deadlines and energy availability. An algorithm or an adversary (the optimal offline algorithm) is *non-idling* if, at every time step, it must send a packet as long as there is energy available and there are packets pending.

Previous Work and Our Contributions. For the case of a single node, the problem without energy considerations is known as the *unit job scheduling* problem (UJS) and was studied extensively; see [5] for a survey. The current best deterministic upper and lower bounds are 1.828 [4] and 1.618 [3,6,14] respectively while for randomized algorithms they are 1.58 [2] and 1.25 [3].

There has been a lot of work in the sensor network community on the problem of energy harvesting although most of them study the problem with somewhat different objective functions, or assume that there are knowledge of probability distributions or even complete knowledge of packet arrivals and/or energy harvesting. For example, [11] assumed that future energy harvesting is known; [8] assumed both the packet arrivals and energy replenishment follow a Poisson process. The only algorithmic, worst-case analysis without prior probability assumptions that we are aware of is [12]. It considers the case of a single node, and the authors gave deterministic upper and lower bounds of V against general adversaries. Then they turned their attention to non-idling adversaries and claimed to give a randomized algorithm that is 1.25-competitive against such adversaries. We show that this is not true even when energy is not a limiting factor. (Note: in subsequent communications [9] one of the authors stated that their 'non-idling' adversary is more restricted than just not being allowed to idle; it is not allowed to have any kind of 'reserving' of energy by scheduling fewer packets. It was not made precise what it means, but it seems to share similar spirit of the strongly non-idling adversary that we define later. In any case, we show that their upper bound is not correct even when there is unlimited energy, and in such scenarios any definition of non-idling is irrelevant since there is always no harm in moving packets earlier to those idle time steps. The authors have also since published a corrigendum [13] which gave a 2.5-competitiveness proof.) In fact we prove a general lower bound of 2 for all randomized algorithms, and a lower bound of $\Omega(\sqrt{V})$ for deterministic algorithms, against oblivious, non-idling adversaries.

As can be seen, a non-idling adversary is still very powerful. Thus we define a more restricted *strongly non-idling* adversary, and against such adversaries we prove a deterministic upper bound of 2^1 and a matching randomized lower bound.

Back to the general adversary case, we show that the correct deterministic competitive ratio should in fact be $V + 1$. We also consider the unweighted packet case and show that if packets have agreeable deadlines, i.e., packets released earlier have earlier deadlines, then the Earliest Deadline First algorithm (EDF) is 1-competitive.

Finally we consider the case of a network of nodes. When energy is not a restriction, the problem becomes the one considered by [10]. They considered the case of an *uplink tree*, where the nodes are connected into a tree and the root node is the sink, and packets can originate in any node but the destination is always the sink. This is a common scenario in sensor network applications. They showed that it is possible to achieve 1-competitiveness for unweighted, underloaded instances. For general network topologies and general source/destination pairs they gave a tight $O(P \log P)$ competitive ratio bound, where P is the maximum route length. In the case with energy we demonstrate that the problem has poor cometitive ratios even for some very simple scenarios.

Due to space constraints some proofs will only appear in the full paper.

2   Non-idling Adversary

Proposition 1. *The competitive ratio of RAND [12] is at least 1.265 against an oblivious, non-idling adversary, even when there are no energy limitations.*

In fact we show the following lower bounds for all non-idling randomized algorithms:

Theorem 1. *No non-idling randomized algorithm is better than $(2 - \frac{2}{\sqrt{V}+1})$-competitive against an oblivious, non-idling adversary. For deterministic algorithms the lower bound is $\Omega(\sqrt{V})$.*

---

[1] In [12] it was stated that the greedy algorithm is 2-competitive against non-idling adversaries, apparently as a corollary from [7] which is about UJS. However our problem is not a special case of UJS, even for strongly non-idling adversaries. We give a separate 2-competitive proof, both because of this and because of the difference in the (strongly) non-idling definitions.

Proof. Consider a setting with two packets $j_1(1, 1, \sqrt{V})$ and $j_2(1, 2, 1)$, a battery with $C = 2$ and an initial $e(1) = 2$, and no energy harvested throughout. Suppose an online randomized algorithm A sends j_1 at time 1 with probability p and j_2 with probability $1 - p$. (These are the only two possibilities as it is non-idling.) If $p \leq \frac{1+V}{2V}$, no further packets are released. A can send j_2 at time 2 if it has not already done so at time 1, so the expected profit of A, $E[A] = p(\sqrt{V} + 1) + (1 - p)(1) = 1 + p\sqrt{V}$. The optimal profit is clearly $1 + \sqrt{V}$, so the competitive ratio is at least

$$\frac{1 + \sqrt{V}}{1 + p\sqrt{V}} \geq \frac{1 + \sqrt{V}}{1 + \frac{1+V}{2V}\sqrt{V}} = \frac{2V + 2V\sqrt{V}}{2V + V\sqrt{V} + \sqrt{V}} = \frac{2V + 2\sqrt{V}}{V + 2\sqrt{V} + 1} = 2 - \frac{2}{\sqrt{V} + 1}$$

Otherwise if $p > \frac{1+V}{2V}$ then $j_3(3, 3, V)$ arrives. If A sent j_1 at time 1 then it must send j_2 at time 2 since it is non-idling, leaving no energy for j_3, whereas if it sent j_2 at time 1 then there is no pending packet to send at time 2 and so has the remaining energy to send j_3. Hence $E[A] = p(\sqrt{V} + 1) + (1 - p)(1 + V) = 1 + V - p(V - \sqrt{V})$. OPT will send j_2 at time 1 and j_3 at time 3. Note that this OPT is non-idling. The competitive ratio is therefore

$$\frac{1 + V}{1 + V - p(V - \sqrt{V})} > \frac{1 + V}{(1 + V) + \frac{1+V}{2V}(\sqrt{V} - V)} = \frac{2V}{2V + (\sqrt{V} - V)} = 2 - \frac{2}{\sqrt{V} + 1}$$

For deterministic algorithms, the proof is basically the same but p can only take on two discrete values $\{0, 1\}$. If an online algorithm A sends j_2 at $t = 1$ (i.e. $p = 0$) then no more packets arrive and the competitive ratio is $1 + \sqrt{V}$. Otherwise if j_1 is sent ($p = 1$) then j_3 arrives and the competitive ratio is $\frac{1+V}{1+V-(V-\sqrt{V})} = \Theta(\sqrt{V})$. □

3   Strongly Non-idling Adversary

The instances in Theorem 1 illustrate a curious aspect of the problem. When faced with two packets p and q with $v(p) > v(q)$ and $d(p) < d(q)$, it seems natural to give preference to p over q. Such algorithms are called *rational*. Here however, the algorithm has to send q and discard p, even when $v(p)$ is much higher than $v(q)$, in order to get good performance by saving the energy for a later packet.

To get around this, we put further restrictions on what the adversary can do. We say a packet p *dominates* another packet q if (i) $v(p) > v(q)$ and $d(p) \leq d(q)$, or (ii) $v(p) \geq v(q)$ and $d(p) < d(q)$. We call a schedule S *irrational* if there are two packets $p \notin S$ and $q \in S$, q is sent in a time step t such that $r(p) \leq t \leq d(p)$, and yet p dominates q. We call an adversary *strongly non-idling* if it is non-idling and it never returns an irrational schedule. Note that when there is no energy limitation or when non-idling is not required, this additional assumption is redundant: clearly substituting q with p gives a schedule at least as good. However, what may happen is that sending p first may mean the adversary is forced to send q later due to its non-idling property, consuming the energy that could be used for sending future high-value packets, whereas sending q first may 'kill off' p and thus save the energy. The situation in the proof of Theorem 1 would not happen in strongly non-idling adversaries: OPT would not be allowed to discard j_1.

For strongly non-idling adversaries we first show a simple deterministic lower bound of 2, then show that the greedy algorithm is 2-competitive and thus optimal.

Theorem 2. *Any deterministic algorithm is at least 2-competitive against a strongly non-idling adversary.*

Proof. Consider a setting with two packets $j_1(1, 1, 1)$ and $j_2(1, 2, 1 + \epsilon)$, where $\epsilon > 0$ is very small, a battery with $C = 2$ and an initial $e(1) = 2$, and no energy harvested throughout. Clearly OPT can send both packets, hence if an online algorithm A does not send both packets then no more packets arrive, giving a competitive ratio of at least $(2 + \epsilon)/(1 + \epsilon) \approx 2$. Otherwise A sends j_1 at $t = 1$ and j_2 at $t = 2$, consuming all energy. Then $j_3(3, 3, V)$ arrives, which A has no energy to send. OPT sends j_2 at $t = 1$, dropping j_1 which a strongly non-idling adversary can do (it would not be allowed to do so if $v(j_1) \geq v(j_2)$), and then send j_3. The competitive ratio is $\frac{1+\epsilon+V}{2+\epsilon} > 2$ for large V. □

We first define a total ordering of packets as follows. For two packets x and y, we say $x \succ y$ if (i) $v(x) > v(y)$, or (ii) $v(x) = v(y)$ and $d(x) < d(y)$, or (iii) $v(x) = v(y)$ and $d(x) = d(y)$ and $ID(x) < ID(y)$, where $ID()$ is a unique ID given to each packet for tie-breaking purposes. The algorithm $GREEDY$ works as follows: at each time step, as long as there is energy to send a packet and there is at least one pending packet, send the one that is 'largest' according to the $\succ$ ordering, i.e., the packet x such that there is no other packet with $y \succ x$.

We assume OPT and $GREEDY$ tie-break using the IDs consistently: if two packets x and y have the same values and deadlines, and $ID(x) < ID(y)$ (so $GREEDY$ favours x), then OPT would not leave x out of its schedule but include y. In addition, we can assume that if OPT sent two packets x and y, where $x \succ y$, one at time step t_1 and another at t_2, where $t_1 < t_2$, and that both packets are available during $[t_1..t_2]$, then x is sent at t_1 and y at t_2 and not the other way round. This follows from a simple exchange argument; note that this does not affect the energy levels or the non-idling requirement at any other time steps.

Theorem 3. *GREEDY is 2-competitive against a strongly non-idling adversary.*

Proof. Let G denote the schedule produced by $GREEDY$. Let $e(t)$ and $e^*(t)$ denote the energy in the battery at time t of G and OPT respectively. We prove by induction the invariant that

(Inv-E): at any time t, $e(t) \geq e^*(t)$

and at the same time describe how the packet values in OPT can be charged to those in G.

Clearly (Inv-E) is true initially. When energy is harvested the battery of G increases at least as much as that of OPT, unless the battery of G is fully charged before that of OPT in which case (Inv-E) holds anyway.

Consider a time t, and assume (Inv-E) is true up to time t. If $GREEDY$ does not send a packet at t, then clearly (Inv-E) is maintained at $t+1$. Moreover, if OPT sends a packet x then by (Inv-E) $GREEDY$ also has the energy to send packets, so the only reason that it is idle is because x has already been sent earlier. Charge x to itself in G.

Now suppose $GREEDY$ sends a packet y and OPT sends a packet x. Clearly (Inv-E) remains true at $t+1$. If $v(x) \leq v(y)$, simply charge the value of x to y. If $v(x) > v(y)$, then x must already be sent by G earlier since otherwise G would have sent it instead at this time step; charge x to itself in that earlier time step.

Finally suppose $GREEDY$ sends a packet y but OPT idles. We will show below that this can only happen if OPT has zero energy ($e^*(t) = 0$ and $h(t) = 0$). This means (Inv-E) is still maintained after this time step. No packet values from OPT need to be charged.

Consider each packet in G, it receives at most two charges, one from a future copy of itself in OPT and another from a packet sent by OPT at the same time step which has at most the same value as the packet in G. Summing over all packets in G, this shows that $GREEDY$ is 2-competitive.

We now return to prove that if $GREEDY$ sends a packet but OPT idles at time t, then OPT must have zero energy. Suppose this is not true. Then OPT must have no pending packets at t since it is non-idling. Let x_1 be the packet sent by G at t. This packet x_1 must have been sent by OPT at an earlier time $t_1 < t$, since otherwise it would be pending for OPT at t. Consider the packet x_2 sent by G at time t_1. This packet must exist, i.e., G cannot idle at t_1, because x_1 is pending, and G must have energy to send it because of (Inv-E) and the fact that OPT has energy to send x_1. Moreover, $x_2 \succ x_1$ because otherwise x_1 would be sent here instead by $GREEDY$. x_2 must be sent by OPT: otherwise if $d(x_2) \geq t$ then it would still be pending at t so OPT could not idle at t, whereas if $d(x_2) < t$ then $d(x_2) < d(x_1)$ and so x_2 dominates x_1, and thus a strongly non-idling adversary could not have discarded x_2 and schedule x_1 at t_1. Let t_2 be the time where OPT sent x_2. It must be that $t_2 < t$, since otherwise OPT would not idle at t. In fact it must be that $t_2 < t_1$: otherwise, if $t_1 < t_2 < t$ then both x_1 and x_2 have been released and have not reached their deadlines during $[t_1..t_2]$, so by our assumption OPT would have sent x_2 first (because $x_2 \succ x_1$).

We then repeat the argument: again G cannot be idle at t_2 and must send a packet x_3, because it has the energy to do so by (Inv-E) and because x_2 is pending at t_2. Moreover this means $x_3 \succ x_2 \succ x_1$. Then, x_3 must appear in OPT: if $d(x_3) \geq t$ then OPT would not idle at t; if $t_1 \leq d(x_3) < t$ then $d(x_3) < d(x_1)$ and x_3 dominates x_1 and thus OPT could not have discarded x_3 when it could schedule it at t_1; if $d(x_3) < t_1$ then $d(x_3) < d(x_2)$ and similarly x_3 dominates x_2. Furthermore it must appear in a time step t_3 where $t_3 < t_2$: it cannot appear after t since OPT could send it at t; if it appeared between t_2 and t_1 then OPT would swap x_2 and x_3; and if it appeared between t_1 and t then OPT would swap x_1 and x_3.

Continuing like this, we can build a 'chain' of x_i's. In general, let t_i be the time where OPT sent x_i, where $t_i < t_{i-1} < ... < t_1 < t$ and $x_i \succ x_{i-1} \succ ... \succ x_1$. G cannot be idle at t_i because x_i, which G sent at t_{i-1}, is pending and it has at least one unit of energy by (Inv-E). Let x_{i+1} be the packet sent by G at t_i. Moreover $x_{i+1} \succ x_i$, since otherwise G would have sent x_i instead of x_{i+1} at t_i. Then x_{i+1} must appear in OPT or else a strongly non-idling adversary must include x_{i+1} and discard one of $x_1, ..., x_i$ instead, depending on its deadline. Moreover it must appear in a time step t_{i+1} where $t_{i+1} < t_i$: it cannot appear after t since OPT could send it at t; if it appeared between t_j and t_{j-1} for some $j > 1$ then OPT would swap x_{i+1} and x_j; and if it appeared between t_1 and t then OPT would swap x_{i+1} and x_1.

This process can go on indefinitely, but there are only a finite number of time steps before t and all these time steps $t_1, t_2, ...$ and packets $x_1, x_2, ...$ are distinct. Hence we will eventually run into a contradiction.   □

In fact we give a randomized lower bound of 2, showing that randomization does not help.

Theorem 4. *No randomized algorithm is better than* $(2 - \epsilon)$-*competitive against a strongly non-idling (and oblivious) adversary.*

Proof. In the following we give a construction involving k rounds, and which shows a lower bound of $2 - \frac{1}{k+1}$, for any positive integer k. Since k can be made arbitrarily large this proves the theorem.

Fix the capacity $C = 2$. At time 1 the battery is full. Fix a large x. At each round $i \geq 1$, an *early packet* $j_i(2i - 1, 2i - 1, x^{i-1})$ and a *late packet* $k_i(2i - 1, 2i, x^{i-1} + \delta)$ arrive, where $\delta > 0$ is very small (in the following calculations we ignore δ). Also at round $i \geq 2$ a unit of energy is harvested at the beginning, i.e., $h(2i - 1) = 1$.

First consider round 1 and suppose at time 1 an online algorithm A sends j_1 with probability $1 - p_1$ and k_1 with probability p_1. If it sent j_1 first then it must send k_1 at time 2, consuming all energy, while if it sent k_1 first then j_1 expires. If $p_1 < \frac{k+1}{2k+1}$, then a *big packet* $(3, 3, x)$ arrives and no further rounds are released. The expected profit of A is $E[A] = p_1(1 + x) + (1 - p_1)(2) = p_1 x + 2 - p_1$, while OPT sends k_1 and the big packet. Hence the competitive ratio $R = \frac{1+x}{p_1 x + 2 - p_1} \approx \frac{1}{p_1} > \frac{2k+1}{k+1}$ for large x. On the other hand, if $p_1 \geq \frac{2k}{2k+1}$, then no more packets or rounds arrive. OPT gets 2 while $E[A] = p_1(1) + (1 - p_1)(2)$, hence $R = \frac{2}{2-p_1} \geq \frac{2k+1}{k+1}$. Finally, if $\frac{k+1}{2k+1} \leq p_1 < \frac{2k}{2k+1}$, we proceed to round 2.

In general, we only proceed to round i if $\frac{k+1}{2k+1} \leq p_1 p_2 .. p_j < \frac{2k+1-j}{2k+1}$ for all previous rounds $1 \leq j \leq i - 1$. Suppose we are at the beginning of round i, and two packets and one unit of energy is released. Consider the event (*):

In all previous rounds the late packets were sent immediately on arrival.

Thus none of the early packets were sent and this leaves one unit of energy (plus the one just harvested). This happens with probability $p_1 p_2 ... p_{i-1}$. Let p_i be the conditional probability that A sends k_i at time $2i - 1$, conditional on

(*) happens. In this case j_i cannot be sent, and there is one unit of energy left afterwards. And with conditional probability $1 - p_i$, again conditional on (*), j_i is sent instead, forcing k_i to be sent at the next time step and with no energy left afterwards. Finally, with the rest of probability $1 - p_1 p_2 ... p_{i-1}$, in at least one of the previous rounds both the early and the late packets were sent, meaning there is no energy left at the beginning of round i (other than the one just harvested), so only one of j_i or k_i can be sent (and one of them must be sent). We now consider three cases.

Case 1: $p_1 p_2 ... p_i < \frac{k+1}{2k+1}$. A big packet $(2i + 1, 2i + 1, x^i)$ arrives and no more rounds arrive. Since x^i is much larger than any other packet values, we only consider the value of this big packet in the profits. The only way A can send this big packet is to have (*) and also send the late packet at this round immediately on arrival; thus $E[A] = (p_1..p_i)(x^i)$. Clearly OPT can get x^i. Hence $R = \frac{1}{p_1..p_i} > \frac{2k+1}{k+1}$.

Case 2: $p_1 p_2 .. p_i \geq \frac{2k+1-i}{2k+1}$. No further packet arrives. The two packets in round i, of value x^{i-1}, dominate the profits, hence we only consider them. The only way that A can sent both of these packets is to have (*), then send the early packet j_i first; this happens with probability $p_1..p_{i-1}(1 - p_i)$. In all other scenarios, A can send one of the two packets this round. Thus $E[A] = x^{i-1}(1 + p_1..p_{i-1}(1 - p_i))$. OPT gets $2x^{i-1}$. Hence

$$R = \frac{2}{1 + p_1..p_{i-1}(1 - p_i)} = \frac{2}{1 + p_1..p_{i-1} - p_1..p_i} \geq \frac{2}{1 + \frac{2k+1-i+1}{2k+1} - \frac{2k+1-i}{2k+1}} = \frac{2k+1}{k+1}.$$

Case 3: $\frac{k+1}{2k+1} \leq p_1 p_2 .. p_i < \frac{2k+1-i}{2k+1}$. We proceed to round $i + 1$.

Since Case 3 cannot happen when $i = k$, the construction stops latest at round k and in all cases the lower bound is at least $(2k + 1)/(k + 1)$.   □

4   Unrestricted Adversary

4.1   Weighted Instances

In [12] it was shown that, against general adversaries (i.e., they can idle), any deterministic non-idling rational algorithm is V-competitive. We first show that the correct competitive ratio for any deterministic non-idling algorithms is in fact $V + 1$, rational or not.

Consider the following counterexample with two packets $j_1(1, 3, 1)$, $j_2(2, 2, V)$, battery capacity $C = 1$, initial energy $e(1) = 1$, and harvesting energy $h(3) = 1$ and $h(t) = 0$ for any other t. A non-idling algorithm must send j_1 at time 1 and then cannot send j_2. OPT would send j_2 and j_1 at time 2, 3 respectively, obtaining a profit of $V + 1$. Thus the competitive ratio is at least $V + 1$.

The following lemma is useful for a number of results later on. Given the schedules of OPT and that of an online algorithm A, we say a time step t is an

OPT-only step if *OPT* sends a packet at t, but A does not despite having at least one pending packet, because it has no energy. We call a time step A-*only* if A sends a packet, but *OPT* does not despite having at least one unit of energy.

Lemma 1. *For the k-th OPT-only step in the schedule, there must be at least k A-only steps before it in the schedule.*

Theorem 5. *Any non-idling algorithm is $(V + 1)$-competitive.*

Proof. We consider how to charge the values of packets sent by *OPT* to those by the online algorithm A. Any packet sent by *OPT* is charged to itself in A if it is also sent by A. If at time t *OPT* sends a packet x that A does not send, and A sends another packet y instead at this time step, then charge x to y. Clearly $v(x)/v(y) \leq V$. If at time t *OPT* sends x but A idles because it has no pending packets, then x must have been sent by A already and therefore its value is already charged. Thus the only remaining case is when *OPT* sends x but A idles because it has no energy to send any packet, i.e., it is an *OPT*-only step.

Suppose there are a total of k *OPT*-only steps. By Lemma 1, there are at least k A-only steps. We charge each of the k packets in these *OPT*-only steps to each of these k packets in A in A-only steps (in some arbitrary way). Again, if x is the packet in *OPT* making the charge and y is the one in A receiving it then $v(x)/v(y) \leq V$.

Each packet in A is charged by at most two packets: one which is itself, and the other either from *OPT* in the same time step, or from some *OPT*-only time steps, but not both. Thus the ratio of total charges received by a packet to the value of the packet sent by A is at most $V + 1$. This shows that A is $(V + 1)$-competitive. $\square$

We can also easily prove matching randomized upper and lower bounds of $\Theta(\log V)$:

Theorem 6. *Against unrestricted adversaries, any randomized algorithm is $\Omega(\log V)$-competitive. There exists an $O(\log V)$-competitive randomized algorithm.*

4.2   Unweighted Instances

It might appear that if packets are unweighted, *EDF* is optimal. However it is not the case: following the same example in the beginning of the previous subsection, *EDF*, or any non-idling algorithm, is not better than 2-competitive. It also follows from Theorem 5 that any non-idling algorithm is 2-competitive.

It can be observed from those examples that such 'deadline inversion' is the problem to getting optimal schedules. We formalise this by showing that for instances with *agreeable deadlines*, i.e. $d(i) < d(j)$ implies $r(i) \leq r(j)$, *EDF* is 1-competitive against unrestricted adversaries. Note that *EDF* is 1-competitive for unweighted instances against non-idling adversaries (without the agreeable deadline assumption) since neither *OPT* nor the online algorithm can idle and clearly it is best to send the packet with the earliest deadline when it is the only thing that distinguishes packets. Therefore, in a sense we can replace the requirement of a non-idling adversary with agreeable deadlines to get to 1-competitiveness. Note that agreeable deadline instances include the case where all packets have the same 'lax time' $(d(j) - r(j))$ as a special case.

Similar to Theorem 3, we use IDs as a consistent way of tie-breaking deadlines. We assume EDF prefers packets with earlier release times among those that have the same deadline, and if release times are also equal, then the one with a smaller ID. We say $x \prec y$ if $d(x) < d(y)$, or $d(x) = d(y)$ and $r(x) < r(y)$, or $d(x) = d(y)$ and $r(x) = r(y)$ and $ID(x) < ID(y)$.

We also assume OPT follows a canonical structure, in that: (i) if it sent a packet x at time t_1 before sending a packet y at time t_2, and $r(y) \leq t_1$, then it must be that $x \prec y$; (ii) OPT does not idle unnecessarily, i.e., if OPT was idle at t_1 and sends a packet x at a later time step t_2, then it must be that x cannot be moved earlier to t_1 without affecting other parts of the schedule (e.g. due to energy availability), or that simply x was not released at t_1, or that there is no energy available at t_1. Both assumptions are without loss of generality by applying standard exchange arguments.

Lemma 2. *Let $e^*(t)$ and $e(t)$ be the energy in the battery of OPT and EDF at time t respectively. Then at any time t,*

Claim 1: $e^*(t) \geq e(t)$.
Claim 2: if OPT sent a packet x at t then EDF could not send x before t.

Proof. We prove both claims together by induction on t. Both claims are obviously true for the first time step $t = 1$. It is also easy to see that Claim 1 is true for $t = 2$: it can only be falsified if OPT sent a packet at time 1 but EDF idles, but they have the same starting energy and the same set of pending packets, so EDF must also send a packet if OPT can.

Suppose Claim 1 is true for all time steps up to and including t, and Claim 2 is true for all time steps up to but excluding t. Claim 1 is true for time $t + 1$ unless OPT sends a packet x at t but EDF idles. It is also true if any idling of EDF is due to that it has no energy $(e(t) + h(t) = 0)$. But if $e(t) + h(t) > 0$, EDF will send x instead of staying idle unless x has already been sent. Hence it remains to prove that x cannot have been sent earlier in EDF, i.e., to prove Claim 2 is true at time t.

So suppose x was sent by EDF at time $t' < t$. Consider the two cases.

Case 1: OPT is idle at t'. EDF has the energy to send a packet, so $e(t') + h(t') > 0$, and applying the induction hypothesis of Claim 1 to time t', $e^*(t') \geq e(t')$. Hence $e^*(t') + h(t') > 0$ and OPT has the energy to send a packet at t'. So the only reason why x is not sent by OPT at t' must be that during $(t'..t]$, there is an *energy-critical time step*, i.e. a step s where $e^*(s) + h(s) = 1$ and a packet z is sent by OPT there, so that if x was sent at t' instead it would use up one unit of energy and z then could not be sent at s. Furthermore assume s is the earliest such energy-critical step in $(t'..t]$. We have $z \prec x$ since otherwise OPT

would swap x and z. Hence either $d(z) < d(x)$, which implies $r(z) \leq r(x)$ by the definition of agreeable deadlines, or $d(z) = d(x)$ and the definition of $\prec$ also implies $r(z) \leq r(x)$. But then z could have been sent by OPT at t' because it has energy available and because there are no other energy-critical step between t' and s. Hence there is a contradiction.

Case 2: OPT sent a packet y at t'. y must still be pending in EDF at t' by induction hypothesis on Claim 2, yet EDF chooses to send x, hence $x \prec y$. But then OPT would have swapped x and y (note that $d(x) \leq d(y)$). $\qquad\square$

Theorem 7. *EDF is 1-competitive for unweighted instances with agreeable deadlines (against unrestricted adversaries).*

Proof. Consider each packet x sent by OPT at a time step t. If EDF sends some packet at t, charge x to that packet. Otherwise, EDF idles despite the fact x is still pending (by Claim 2 of Lemma 2), so it can only be because it has no energy, i.e., it is an OPT-only step. By Lemma 1, there must be at least as many A-only steps as OPT-only steps, so pair them up arbitrarily and charge the packet values as in the proof of Theorem 5.

Any packet sent by EDF can only receive charge from one other packet: if it is an A-only step that it only receives from a packet in an OPT-only step, and if it is a step where both OPT and A send packets then it gets charged from the corresponding packet in OPT. $\qquad\square$

As a note, this automatically means that EDF is V-competitive for weighted, agreeable-deadline instances.

5   Network Topologies

Here we consider a network with more than one node. We will restrict ourselves to unweighted packets. We use the notation $(r(j), d(j), s(j), t(j))$ for a packet j where $s(j)$ and $t(j)$ are the source and destination nodes. We use $h_N(t)$ to denote the energy harvesting function for node N.

The situation is already very bad even for unweighted instances:

Proposition 2. *The competitive ratio is unbounded even for line networks and even for unweighted instances if packets have different destinations.*

Proof. Consider a line network with four nodes a, b, c, d and two packets $p_1(1, 3, a, c)$, $p_2(1, 5, a, d)$. All batteries are initially empty. We have $h_a(1) = 1$ and $h_a(t) = 0$ for $t \geq 2$, $h_b(1) = h_c(1) = 0$. Hence an online algorithm A can only send one of the two packets. If A sends p_1, then $h_b(2) = 0$, so p_1 will expire. OPT sends p_2 instead, with $h_b(3) = 1, h_c(4) = 1$. If A sends p_2 instead, then $h_b(2) = 1$ but $h_c(t) = 0$ for all t, so p_2 expires while OPT sends p_1. $\qquad\square$

Proposition 3. *EDF has an infinite competitive ratio even when all packets have the same source and destination in a line network with only three nodes.*

Proof. Consider a line network a, b, c and two packets $p_1(1, 3, a, c), p_2(1, 4, a, c)$. Again nodes have empty batteries initially, $h_a(1) = 1$ and $h_a(t) = 0$ afterwards, and $h_b(t) = 0$ for all $t \neq 3$ and $h_b(3) = 1$. *EDF* sends p_1 first, but node b has no energy at time 2 and hence p_1 expires, and node a has no energy at time 2 onwards so p_2 also expires. *OPT* sends p_2 at time 1, waits at node b at time 2 until it has energy at time 3. Thus *EDF* gets 0 while *OPT* gets 1.   □

To try to get around this, we make an additional assumption that the instance is underloaded. We note that it is quite common in the real-time systems community to consider underloaded instances. However we still have the following:

Proposition 4. *For a line network where all packets have a common destination (the sink), any non-idling algorithm is at least $(n + 1)$-competitive for unweighted and underloaded instances against unrestricted adversaries, where n is the number of nodes (excluding the sink).*

Proof. Consider a line network with n nodes (in this order) $N_0, N_1, ..., N_n$ where N_0 is the sink. Each node $N_1..N_n$ have $C = 1$, initial battery energy 0 and the following energy harvesting function: $h(1) = 1, h(t) = 0$ for $2 \leq t \leq n + 1$, and $h(t) = 1$ for $t \geq n + 2$. Packet p_0 is released to node N_n with $r(p_0) = 1$ and $d(p_0)$ very large. For each $1 \leq i \leq n$, packet p_i is released to node N_i with $r(p_i) = n + 1$ and $d(p_i) = n + i + 1$. These packets are tight, i.e., they must be forwarded immediately at every node to reach N_0 in time. A non-idling algorithm will send p_0 along the line from time 1 to n, consuming the only unit of energy at each node along the way. Then when the tight packets arrive at time $n + 1$, they cannot be forwarded immediately and hence all are lost. *OPT* withholds p_0 and stays idle up to and including time n. At time $n + 1$ it forwards each of $p_1..p_n$ by one node. Starting at time $n + 2$ all nodes have plenty of energy, so they continue to forward packets $p_1..p_n$ to the sink. Finally p_0 is sent.   □

We believe the bound is indeed tight, i.e., for underloaded instances any non-idling algorithm is $O(n)$-competitive in line networks with a common sink, or even for uplink trees where n is the total number of vertices. Note that without energy limitations EDF is 1-competitive for uplink trees, but for arbitrary non-idling algorithms it can also be as bad as $(n + 1)$-competitive. Also, it is not true that the competitive ratio may be upper bounded by the depth of the tree rather than the number of nodes: we have an example to show that any non-idling algorithm is $\Omega(n)$-competitive for an uplink tree even with a depth of 2.

6   Conclusion

Most importantly we want to get an upper bound in the case of uplink trees or at least line networks. In the single node case, it is interesting to see whether there are other ways to get non-trivial competitiveness with reasonable assumptions. The power of randomized algorithms, or algorithms that choose to idle, remain to be investigated. For example in the unrestricted adversary case, it is not clear whether it is possible to get (idling) algorithms with competitive ratio better than $V + 1$; or for non-idling algorithms, what are the upper bounds.

References

1. Borodin, A., El-Yaniv, R.: Online Computation and Competitive Analysis. Cambridge University Press, New York (1998)
2. Chin, F.Y.L., Chrobak, M., Fung, S.P.Y., Jawor, W., Sgall, J., Tichý, T.: Online competitive algorithms for maximizing weighted throughput of unit jobs. J. Discrete Algorithms **4**(2), 255–276 (2006)
3. Chin, F.Y.L., Fung, S.P.Y.: Online scheduling with partial job values: does time-sharing or randomization help? Algorithmica **37**(3), 149–164 (2003)
4. Englert, M., Westermann, M.: Considering suppressed packets improves buffer management in quality of service switches. SIAM J. Comput. **41**(5), 1166–1192 (2012)
5. Goldwasser, M.H.: A survey of buffer management policies for packet switches. SIGACT News **45**(1), 100–128 (2010)
6. Hajek, B.: On the competitiveness of online scheduling of unit-length packets with hard deadlines in slotted time. In: Proceedings of 35th Annual Conference on Information Sciences and Systems, pp. 434–438 (2001)
7. Kesselman, A., Lotker, Z., Mansour, Y., Patt-Shamir, B., Schieber, B., Sviridenko, M.: Buffer overflow management in QoS switches. SIAM J. Comput. **33**(3), 563–583 (2004)
8. Lei, J., Yates, R., Greenstein, L.: A generic model for optimizing single-hop transmission policy of replenishable sensors. IEEE Trans. Wirel. Commun. **8**(2), 547–551 (2009)
9. Li, F.: Personal communication (2014)
10. Mao, Z., Koksal, C.E., Shroff, N.S.: Optimal online scheduling with arbitrary hard deadlines in multihop communication networks. In: Proceedings of IEEE INFOCOM, pp. 2463–2471 (2013)
11. Moser, C., Brunelli, D., Thiele, L., Benini, L.: Real-time scheduling for energy harvesting sensor nodes. Real-Time Syst. **37**, 233–260 (2007)
12. Wang, H., Zhang, J.X., Li, F.: Worst-case performance guarantees of scheduling algorithms maximizing weighted throughput in energy-harvesting networks. Sustainable Comput.: Inform. Syst. **4**, 172–182 (2014)
13. Wang, H., Zhang, J.X., Li, F.: Corrigendum to worst-case performance guarantees of scheduling algorithms maximizing weighted throughput in energy-harvesting networks. Sustain. Comput.: Inf. Syst. **5**, 64 (2015)
14. Zhu, A.: Analysis of queueing policies in QoS switches. J. Algorithms **53**, 137–168 (2004)

On Verifying and Maintaining Connectivity
of Interval Temporal Networks

Eleni C. Akrida[1]([✉]) and Paul G. Spirakis[1,2]

[1] Department of Computer Science, University of Liverpool, Liverpool, UK
Eleni.Akrida@liverpool.ac.uk
[2] Computer Technology Institute and Press "Diophantus" (CTI), Patras, Greece
P.Spirakis@liverpool.ac.uk

Abstract. An interval temporal network is, informally speaking, a network whose links change with time. The term *interval* means that a link may exist for one or more time intervals, called *availability intervals* of the link, after which it does not exist (until, maybe, a further moment in time when it starts being available again). In this model, we consider continuous time and high-speed (instantaneous) information dissemination. An interval temporal network is connected during a period of time $[x, y]$, if it is connected for all time instances $t \in [x, y]$ (instantaneous connectivity). In this work, we study instantaneous connectivity issues of interval temporal networks. We provide a polynomial-time algorithm that answers if a given interval temporal network is connected during a time period. If the network is not connected throughout the given time period, then we also give a polynomial-time algorithm that returns large components of the network that remain connected and remain large during $[x, y]$; the algorithm also considers the components of the network that start as large at time $t = x$ but dis-connect into small components within the time interval $[x, y]$, and answers how long after time $t = x$ these components stay connected and large. Finally, we examine a case of interval temporal networks on tree graphs where the *lifetimes* of links and, thus, the failures in the connectivity of the network are not controlled by us; however, we can "feed" the network with extra edges that may re-connect it into a tree when a failure happens, so that its connectivity is maintained during a time period. We show that we can with high probability maintain the connectivity of the network for a long time period by making these extra edges available for re-connection using a randomised approach. Our approach also saves some cost in the design of availabilities of the edges; here, the cost is the sum, over all extra edges, of the length of their availability-to-reconnect interval.

1 Introduction and Motivation

A great variety of systems in society, technology and nature can be modelled as networks, linked with edges; from the Internet to the web of social acquaintances,

Supported in part by (i) the School of EEE and CS and the NeST initiative of the Univeristy of Liverpool, and (ii) the FET EU IP Project MULTIPLEX under contract No. 317532.

P. Bose et al. (Eds.): ALGOSENSORS 2015, LNCS 9536, pp. 142–154, 2015.
DOI: 10.1007/978-3-319-28472-9_11

from the transport network of a city to the nervous system of the human body. The structure of a network describes the several connections between the participating entities and helps us understand or predict the behaviour of dynamical systems. However, in many cases the links between the participating entities do not always remain active but change or disappear as time progresses. A *temporal network* is, informally speaking, a network that changes with time. Both traditional and modern networks, such as communication networks, social networks, transportation networks and physical systems, can be modelled as temporal.

Dynamic networks in general have been attracting attention over the past years, exactly because they model real-life applications. The study of temporal networks in particular is quite interdisciplinary, which is also reflected in literature where the object of study may have different names - temporal graphs, temporal networks, evolving graphs, time-stamped graphs etc. Kempe et al. [14] considered the single-labelled discrete-time model of temporal graphs, where every edge may become available (for use) only at a discrete moment in time, called the *label* of the edge; their main motivation was to examine how basic graph properties change in this temporal setting. In their multi-labelled model, Mertzios et al. [17] extended the model of [14] to many labels per edge and mainly examined the number of labels needed for a temporal design of a network to guarantee several graph properties with certainty. They also provided an algorithm to compute foremost time-respecting paths; in this discrete-time model, a time-respecting path is a path in which successive edges have strictly increasing time labels and a foremost time-respecting path is one that *reaches* the destination vertex at the earliest possible time. Random edge availabilities in the discrete-time model of temporal networks were first considered by Akrida et al. [1] in order to study the Expected Temporal Diameter of temporal graphs.

Assuming the availability of an edge for a whole time-interval $[t_1, t_2]$ or multiple such time-intervals, and not just for discrete moments, is a clearly natural assumption since time is indeed a continuous measure. Bui-Xuan et al. [4] consider a class of dynamic networks where the changes in the topology can be predicted in advance and in which each node and each edge comes with a list of time intervals; they give algorithms for computing foremost time-respecting paths, shortest (minimum hop count) time-respecting paths and fastest (minimum time) time-respecting paths in this model. Fleischer and Tardos [11] consider a continuous-time model of dynamic graphs and prove continuous versions of known discrete-time flow algorithms for dynamic flow problems. Fleischer and Skutella [10] also engage in the study of flows in a continuous-time model of dynamic graphs. Further related work includes $[2, 3, 5, 6, 9, 12, 13, 15, 16, 18–21]$.

1.1 Our Contribution

In this work, we restrict our attention to *continuous time* and consider systems in which only the connections between the participating entities may change, while the entities remain unchanged. So we consider networks of a fixed vertex set, each edge e of which is available over a set of time intervals $L_e = \{[t_1, t'_1], \ldots, [t_k, t'_k]\}$. Each interval indicates a period of availability of e; the unprimed times mark the start of the availability period and the primed times mark the end. This is a model that could naturally represent several systems, such as proximity networks where a link may represent that two entities have been close to each other for some extent of time, or infrastructural systems like the Internet, or even seasonal food webs where a time interval may represent the fact that one species is the main food source of another for a specific period of the year.

We give a polynomial-time deterministic algorithm that decides if a given interval temporal network is connected during a given period (cf. Sect. 3); if the network is not connected, the algorithm returns the maximal interval from the beginning of the given period during which the network stays connected. We then provide a polynomial-time algorithm that decides if a given interval temporal network has *large enough* connected components during a given time period; here, the size of the components in question is determined by a parameter provided by the user as input to the algorithm (cf. Sect. 4). Finally, we provide a probabilistic analysis of a scenario where the *lifetime* of the intervals assigned to the edges of a network on a tree graph are not designed via a deterministic process and are unknown to us; instead, the edges may fail unexpectedly and we are required to supply the network with more available edges so that, when a break in the connectivity of the network happens, we can re-connect it. We wish not to keep all these extra edges available for re-connection at all times, i.e., we wish to maintain connectivity but by paying a low cost on keeping extra edges available. Assuming that the cost of keeping additional edges available is linear to the sum of lengths of their availability intervals, we show a low cost construction. Other work for maintaining some structure or property like connectivity in probabilistic dynamic graphs includes [7,8].

2   Preliminaries

We focus here on networks, the links of which are not always available. The availability of a link is described via a set of time intervals, one set per edge.

Definition 1 (Interval Temporal Network). *Let $G = (V, E)$ be a (di)graph. An interval temporal network on G is an ordered triplet $G(L) = (V, E, L)$, where $L = \{L_e = \{[t_1, t'_1], \ldots, [t_{k_e}, t'_{k_e}]\}$, for some $k_e \in \mathbb{N}$, $t_i, t'_i \in \mathbb{R}^+, t_i < t'_i$, $i = 1, 2, \ldots, k_e : e \in E\}$ is an* assignment *of availability intervals to the edges (arcs) of G. L is called a* labelling *of G.*

The availability intervals of an edge (arc) e represent the *continuous time intervals* at which e is active. When we say that an edge (arc) is active or available during the interval $[a, b]$, for some $a, b \in \mathbb{R}$, it means that the edge exists in the network $\forall t \in \mathbb{R}^+$, $t \in [a, b]$. For the analysis throughout the paper, we assume the intervals $[t_1, t'_1], \ldots, [t_{k_e}, t'_{k_e}]$ to be disjoint.[1] Every time a change in the network happens, i.e., an edge starts or stops being available, we have changes

---

[1] We can assume this, because if an edge $e \in E$ has overlapping availability intervals, then we can consider their union as an availability interval of e.

in the topology of the network; so, in a sense, an interval temporal network can be viewed as a sequence of graphs, one after every topology change. However, representing such networks as evolving graphs, i.e., the sequence of states of the network after each change, is not as efficient. The interval representation is indeed a very compact representation of such kinds of evolving graphs.

A basic assumption that we follow is that when a message or an entity passes through an available link at time t, then it can pass through a subsequent link only at some time $t' \geq t$ and only at a time at which that link is available. However, unlike what is assumed in the discrete-time model of [1,14,17], here we consider instant information dissemination through a path of the underlying (di)graph, if the consecutive edges (arcs) are consistently labelled. In fact, our model considers very high speed of information dissemination, resembling fibre-optic communication, but the small time needed to send a message through a link is considered negligible for the analysis. Consider, for example, an interval temporal network $G(L)$ and a path p of $G(L)$ such that all edges of p are available at some time $t = t_0$; if some information starts at time t_0 from one endpoint of p, it can arrive at time t_0 to the other endpoint.

Definition 2 (Connectivity of Interval Temporal Networks). *An interval temporal network $G(L) = (V, E, L)$ is connected at a given time instance t_0 if the edges that are available at time t_0, i.e., the edges that have an availability interval which includes t_0, induce a spanning tree.*

3 Connectivity of Interval Temporal Networks During a Given Time Period

A fundamental issue for any given network, dynamic or not, is to verify if the network is connected (over time, in the dynamic case), i.e., information can travel via edges between any ordered pair of vertices in it. In this section, we consider interval temporal networks and address the issue of their connectivity.

One can think of an interval temporal network as a dynamic network, where the changes in the topology of the network happen whenever an availability interval of an edge starts or finishes, but can view it as static in between these (instantaneous) changes. Since information can travel instantaneously in interval temporal networks, for such a network to be connected over a time period, all the instances of the "static" networks that are formed during that period need to be connected.

We provide below a polynomial-time procedure to determine if a given interval temporal network is connected throughout a particular time period. Henceforth, we denote by $E(t)$ the set of edges that are available at time t, and t is *not* the finish time of the availability interval that includes t.[2]

---

[2] $E(t)$ are the edges that are available at t and do not stop being available (immediately) after time t.

Theorem 1. *There is a polynomial-time algorithm (cf. Algorithm 1) which, given an interval temporal network $G(L)$ on n vertices and numbers $x, y \in \mathbb{R}^+$, $x < y$, answers whether $G(L)$ is connected during the time period $[x, y]$, i.e., is connected for every time instance $t \in [x, y]$. If for some $a \in [x, y]$, $[x, a] \subseteq [x, y]$ is the maximal sub-period of $[x, y]$ during which $G(L)$ remains connected, then the algorithm also returns the length of that period, $a - x$.*

Algorithm 1. Connectivity of interval temporal networks

Input: A temporal network $G(L)$ of n vertices and numbers $x, y \in \mathbb{R}^+$ such that $x < y$
Output: Answer if $G(L)$ remains connected during the time interval $[x, y]$

1: **if** $E(x)$ induces a spanning tree, T, of G **then**
2:      Sort the edges in T according to the finish time of their availability interval;
        //For every edge in T, we only consider the interval that includes x
3:      Let $A = \{e_i,$ with interval $[a_i, b_i] : i = 1, \ldots, n - 1\}$ be the sorted list;
4:      **if** $b_1 \geq y$ **then**
5:          **return** "Network is connected" **and** "Duration of survival $=$" $y - x$; //If all edges in T remain available until (at least) time y
6:      **else**
7:          $E' := \{e \in E(T) : b_e = b_1\}$;   //$b_1$ is the first time instance at which T becomes disconnected. E' is the set of edges of T that stop being available at time b_1.
8:          $T := T \setminus E'$;   //T is now a forest, i.e., consist of a collection of trees
9:          Remove E' from A;
10:         Let $T_1, T_2, \ldots, T_i, i \in \mathbb{N}$ be the connected components of T;
11:         **while** T is disconnected **do**
12:             **if** $\exists j, k = 1, 2, \ldots, i : \exists e = (u, v) \in E(b_1) : u \in V(T_j) \wedge v \in V(T_k)$ **then**
13:                 Find the T_j, T_k trees of T that e connects;   //If there is an edge of G with endpoints in different connected components of T and is available at time b_1, then add it to T
14:                 Merge T_j, T_k and e into a single tree;
15:                 Update the number i of connected components of T;
16:                 Insert e in the sorted list A;
17:             **else**
18:                 **return** "Network is disconnected" **and** "Duration of survival $=$" $b_1 - x$;

19:             Break;
20:         Go to line 4
21: **else**
22:     **return** "Network is disconnected" **and** "Duration of survival $=$" $t - x$;

Description of the Algorithm. The idea behind Algorithm 1 is that $G(L)$ is connected during a period $[x, y]$ if and only if $G(L)$ has a spanning tree for every time instance in $[x, y]$.

Initially, Algorithm 1 finds a spanning tree of the input network $G(L)$ at time x. If no such tree exists, then at time x the network is disconnected and the algorithm terminates. If a spanning tree T exists at time x, then T remains connected until one (or more) of its edges stop being available. Denote by b_1

the first moment in time at which T disconnects. T consists now of a number of connected components and, in fact, T is a forest (collection of trees). The algorithm checks whether there are edges of $G(L)$ that are available at time b_1, which can be added to T and re-connect it. More specifically, the algorithm finds an edge that is available at time b_1 and has endpoints in different connected components of T. The algorithm adds that edge to T and checks if this addition re-connects it. If not, then it looks for yet another edge that is available at time b_1 and has endpoints in different connected components of (the current) T. This process continues until T is re-connected or we cannot find any more edges of $G(L)$ that are available at time b_1 and have endpoints in different connected components of T. If at any step of the process there do not exist edges that can re-connect T, then the algorithm returns that the network is disconnected. However, if we can find appropriate edges to re-connect T, then we form another spanning tree of the network, available from time b_1 onwards, and the same procedure continues. The algorithm answers that the network is connected if we form a spanning tree, all the edges of which are available until the end of the period in question, namely until time y.

Running Time. The running time of Algorithm 1 depends on the number of times that the spanning tree changes during $[x, y]$. The spanning tree can only change when one or more edges stop being available, so the above number is in general upper bounded by the total number of intervals assigned to the edges of the network:

$$M = \sum_{e \in E} |L_e|$$

Initially, to find $E(x)$ we need to look at every edge $e \in E$ and decide if x is between the start and finish time of one of e's availability intervals. Performing a binary search on the ordered set of start times *and* the ordered set of finish times of e's availability intervals, we can decide if $e \in E(x)$ in time $O(\log |L_e|)$. So, to compute $E(x)$ and check if it induces a spanning tree, we need time:

$$M' = O(\sum_{e \in E} \log |L_e|)$$

Next, time $O(n \log n)$ is required to sort the edges in T, where n is the number of vertices in the network. Then, for every time T changes, we need time M' to find the new set of available edges at the time. We need time $O(n)$ to find the connected components that can be re-connected by the addition of an available edge at the time and update T. Since we add at most $O(n)$ edges to re-connect T, the addition of all edges and the updates of T take a total of $O(n^2)$ time. Also, time $O(n)$ is required to insert the added edges in the sorted list A. Therefore, the running time of Algorithm 1 is $O(M' + n \log n + M \cdot (M' + n^2)) = O(M \cdot (M' + n^2))$.

4 Large Connected Components During a Given Time Period

In this section, we examine if, given an interval temporal network $G(L)$ of n vertices, numbers $x, y \in \mathbb{R}^+$ and a parameter $0 \leq \varepsilon \leq 1$, we can find one or more *large enough* subsets of the vertices of G which remain connected and remain large within the time interval $[x, y]$. The matter of how large we want the components to be is handled by adjusting ε, which gives us a lower bound of $\varepsilon \cdot n$ on the size of the components we are looking for. In this section, we provide an algorithm that efficiently solves the above problem. Henceforth, a "large enough" connected component will be a component of size at least $\varepsilon \cdot n$.

Notice that any connected component C of $G(L)$, at time $t = x$, that is not large enough can be omitted by any algorithm that solves the above problem. Even if the vertices of C connect with more vertices in $G(L)$ at a later moment in time within $[x, y]$, resulting in a large enough connected component C' of $G(L)$ at that time, C' is not a component that was connected throughout $[x, y]$.

Theorem 2. *There is a polynomial-time algorithm which, given an interval temporal network $G(L)$ on n vertices and numbers $x, y \in \mathbb{R}^+$, $x < y$, returns all subgraphs of G of size $\varepsilon \cdot n$, $0 \leq \varepsilon \leq 1$, that remain connected and large (i.e., is always of size at least $\varepsilon \cdot n$) during the time period $[x, y]$. If $[x, a] \subseteq [x, y]$, $a \in [x, y]$, is the maximal sub-period of $[x, y]$ during which such a component remains connected, then the algorithm also returns the length of that period, $a - x$.*

Description of the Algorithm. Algorithm 2 receives as input an interval temporal network of n vertices and an interval $[x, y]$ during which we want to check whether one or more large components of the network remain connected. The algorithm also takes a non-negative parameter ε no larger than 1. This parameter defines how large we want our components to be; more specifically, the algorithm will only look for components of size (number of vertices) at least $\varepsilon \cdot n$. The algorithm returns all those subsets of the vertices of the initial graph, if any, that remain connected (and large) during $[x, y]$. Furthermore, it returns the duration of connectivity (*survival duration*) of any large enough component that was connected at time $t = x$ but disconnects at some point in $[x, y]$.

To do so, the algorithm initially checks which connected components, if any, are large enough at time x, and ignores all the rest. Then, the algorithm treats each and every one of these large components similarly, but separately. Namely, for each one of them the algorithm finds a spanning tree T and sorts all its edges according to the finish time of their availability interval, considering only the interval that includes time x. If the same tree remains connected during $[x, y]$, then the algorithm returns the respective component. Otherwise, if the tree disconnects at a moment t_0 in time, the algorithm employs a similar process to the one used in Algorithm 1, i.e., tries to reconnect the remainder of the tree via edges that are available at t_0. If T cannot be re-connected, then the algorithm

---

Algorithm 2. Connectivity of interval temporal graphs

Input: A temporal network $G(L)$ of n vertices, numbers $x, y \in \mathbb{R}^+$ such that $x < y$
 and parameter $\varepsilon : 0 \leq \varepsilon \leq 1$
Output: All components of $G(L)$ of size $\varepsilon \cdot n$ that remain connected during the time
 interval $[x, y]$

1: Find the set $E(x)$ of available edges at time x, distinguish the connected components and delete those of size smaller than $\varepsilon \cdot n$;
2: **for** each of the remaining connected components **do**
3:    Find a spanning tree, T;
4:    $n' = |V(T)|$;
5:    Sort the edges in T according to the finish time of their availability interval;
      // For every edge in T, we only consider the interval that includes x
6:    Let $A = \{e_i$, with interval $[a_i, b_i] : i = 1, \ldots, n - 1\}$ be the sorted list;
7:    **if** $b_1 \geq y$ **then**
8:       **return** $V(T)$ **and** "Duration of survival of component $= $ " $y - x$;
9:    **else**
10:      $E' := \{e \in E(T) : b_e = b_1\}$;   // b_1 is the first time instance at which T
         becomes disconnected. E' is the set of edges of T that stop being available at
         time b_1.
11:      $T := T \setminus E'$;
12:      Remove E' from A;
13:      Let $T_1, T_2, \ldots, T_i, i \in \mathbb{N}$ be the connected components of T;
14:      **while** T is disconnected **and** $|V(T)| = n'$ **do**
15:         **if** $\exists j, k = 1, 2, \ldots, i : \exists e = (u, v) \in E(b_1) : u \in V(T_j) \wedge v \in V(T_k)$ **then**
16:            Find the T_j, T_k trees of T that e connects; //If there is an edge of G
               with endpoints in different connected components of T and is available
               at time b_1, then add it to T
17:            Merge T_j, T_k and e into a single tree;
18:            Update the number i of connected components of T;
19:            Insert e in the sorted list A;
20:         **else**
21:            **for** each connected component C of T with size smaller than $\varepsilon \cdot n$ **do**
22:               $T = T \setminus C$;
23:               **return** "Duration of survival of component $= $ " $b_1 - x$;
24:         **for** each connected component, C', of T **do**
25:            $T := C'$;
26:            $n' = |V(T)|$;
27:            Go to line 7;

---

checks the sizes of its connected components; it ignores those that are not large
enough, while "processing" the rest similarly and separately as before. For each
component that is ignored in the process, the algorithm returns the duration of
its survival, meaning how long its vertices stayed connected since time x. The
algorithm stops when there are no more components that are large enough or
when the last component stays connected until time y.

Running Time. It is easy to see that the running time of the algorithm for each separate component is the same as the running time of Algorithm 1. Since there are at most $\frac{1}{\varepsilon}$ connected components of size at least $\varepsilon \cdot n$ in $G(L)$ during $[x, y]$, the running time of Algorithm 2 is $O\left(\frac{1}{\varepsilon}(M \cdot (M' + n^2))\right)$.

5   Low Cost Maintenance of a Tree Structrure

In this section, we consider an interval temporal network on an underlying clique of n nodes, i.e., all $\binom{n}{2}$ links between nodes of the network *may* exist.

The connectivity of the network needs to be maintained at all moments in time via a tree structure, i.e., a spanning tree of the clique. Each node of the tree performs an individual application determined by the operator of the structure and each link (edge) is active (alive) during a time-interval also decided by the operator, after which the link fails. We have the liberty to provide the operator with extra edges from the clique to re-connect a spanning tree when a link fails; note here that after a new edge is added to the tree structure, the operator then assigns to it a "lifetime" interval, which is determined by the application, anew. The extra edges that we can provide come from the edges of the clique that are not currently used in the tree structure, i.e., a total of $\binom{n}{2} - (n-1)$ edges. We need to assign to every such edge e out of the $\binom{n}{2} - (n-1)$ an availability interval, I_e, so that when the tree structure becomes dis-connected, there is an appropriate such edge available to re-connect it. We call those edges *reserved* edges and the set that consist exactly of all those edges (with their availability intervals) *reservoir*, denoted by R.[3]

Definition 3 (Cost of the Reservoir). *The cost of the reservoir is defined as the sum, over all reserved edges, of the length of the edges' availability interval:*

$$c = \sum_{e \in R} |I_e|$$

Let T be the tree structure that is handled by the operator. We consider the time period between 0 and n and we assume that the breaks/failures in the connectivity of T happen once inside every consecutive time interval of length $\Delta \geq \alpha \log n$, for some $\alpha > 1.2$ (*Low-frequency-of-link-breaks assumption*). We are not able to predict when exactly the failures happen, nor are we able to foresee which link will fail next. We also assume worst case breaks in the tree topology within each Δ-interval. The trivial design of the availabilities of the reserved edges would be to make them all available throughout the considered time period $[0, n]$. However, this yields cost $c = \sum_{i=1}^{\binom{n}{2}-n+1} n \in O(n^3)$. We will

---

[3] Notice, here, the distinction between the *availability* of an edge and the *lifetime* of an edge: availability refers to the interval that we assign to a reserved edge with the purpose to re-connect the tree when it breaks, and lifetime refers to the interval that the operator assigns to an edge after it is inserted in the tree structure and is the time interval after which the respective link in the tree structure will fail.

show how to provide the network with available reserved edges with lower cost, so that the network connectivity is maintained with high probability (whp).[4] In order to re-connect the tree in the worst case of breaks in the tree topology, each reserved edge needs to have been randomly assigned to an availability interval to allow for the same probability of re-connection for all edges.

Theorem 3. *Let $\alpha \in \{x \in \mathbb{R} | x \geq 0.75\}$. If failures of the edges happen once in every consecutive $\Delta \geq \alpha \log n$ time-intervals, then there exists a reservoir of cost $O(n^2 \log n)$ that keeps a spanning tree available during $[0, n]$ whp.*

Proof. Partition the time interval $[0, n]$ into consecutive equisized sub-intervals $b_1, b_2, \ldots, b_{\frac{n}{\beta \log n}}$ of length $\beta \log n$, $\beta \in \mathbb{R}$, $0.75 \leq \beta \leq \alpha$, called *boxes*.[5] For every reserved edge $e \in R$ independently, select a box uniformly at random to be the availability interval of e. For every edge $e \in R$, the probability that e is assigned a particular box b_i, $i = 1, 2, \ldots, \frac{n}{\beta \log n}$ as its availability interval is:

$$Pr[I_e = b_i] = \frac{\beta \log n}{n}$$

Denote by m' the number of edges in R that are assigned to a particular box b_i, $i = 1, 2, \ldots, \frac{n}{\beta \log n}$, $m' = |\{e \in R : I_e = b_i\}|$. The expected value of m' is:

$$\mu = E[m'] = \frac{\beta \log n}{n} \cdot \left(\frac{n(n-1)}{2} - n + 1\right) = \frac{\beta n \log n}{2} - \frac{3\beta \log n}{2} + \frac{\beta \log n}{n}$$

By Chernoff bounds, we get that the probability that m' is *close* to the expected number of edges in a particular box b_i, $i = 1, 2, \ldots, \frac{n}{\beta \log n}$ is:

$$Pr[m' \in (1 \pm \frac{1}{2})\mu] \geq 1 - e^{-\frac{1}{4}\mu}$$

$$= 1 - e^{-\frac{1}{4} \cdot \frac{\beta \log n}{n} \cdot (\frac{n^2}{2} - \frac{n}{2} - n + 1)}$$

$$\geq 1 - \frac{1}{n^{\frac{\beta n}{16}}}, \text{ for } n \text{ large enough } (n \geq 6)$$

We now show that when a failure happens in T, we can whp find an edge in R which is available at that particular moment in time. Consider the specific box b_i that includes the time moment at which the failure in T happens. The number of edges in b_i that can re-connect T depends on where the failure happens, i.e., on the sizes[6] of the two connected components after the failure. If n_1 and n_2 are the sizes of the connected components of T after a failure, then the probability that a particular edge $e \in b_i$ can re-connect T after being added to the structure is:

---

[4] An event occurs with high probability if, for any $\gamma \geq 1$, the event occurs with probability at least $1 - \frac{c_\gamma}{n^\gamma}$, where c_γ depends only on γ.

[5] The last box is not necessarily of size exactly $\beta \log n$ but this does not affect the analysis.

[6] The size of a component is the number of its vertices.

$$Pr[e \in b_i \text{ re-connects } T] = \frac{n_1 \cdot n_2}{\frac{(n_1+n_2)\cdot(n_1+n_2-1)}{2}} \geq \frac{2n_1 n_2}{(n_1+n_2)^2}$$

The probability that no edge of b_i reconnects T after a failure is:

$$Pr[\text{no } e \in b_i \text{ re-connects } T] \leq \left(1 - \frac{2n_1 n_2}{(n_1+n_2)^2}\right)^{m'}$$

So, the probability that there is an edge in b_i that re-connects T is:

$$Pr[b_i \text{ re-connects } T] \geq 1 - \left(1 - \frac{2n_1 n_2}{(n_1+n_2)^2}\right)^{m'} \geq 1 - \left(1 - \frac{2n_1 n_2}{(n_1+n_2)^2}\right)^{\frac{3\mu}{2}}$$

In the worst case, T dis-connects into a component of size $n-1$ and a single vertex. So, we can reconnect T after a failure with probability:

$$Pr[b_i \text{ re-connects } T] \geq 1 - \left(1 - \frac{2(n-1)}{n^2}\right)^{\frac{3\beta n \log n}{4} - \frac{9\beta \log n}{4} + \frac{3\beta \log n}{2n}}$$

$$\geq 1 - \left(1 - \frac{2(n-1)}{n^2}\right)^{\frac{3}{4}\beta n \log n}$$

$$\geq 1 - \frac{1}{n^{0.9\frac{3}{2}\beta}}, \text{ for } n \text{ large enough } (n \geq 10)$$

$$= 1 - \frac{1}{n^{1.35\beta}} \xrightarrow{n \to +\infty} 1$$

We require $\beta \geq 0.75$ so that the above event happens whp. The probability that within the time period $[0, n]$, there is a box that will not re-connect T is:

$$Pr[\exists b_i, \ i=1,\dots, \frac{n}{\beta \log n} : b_i \text{ doesn't re-connect } T] \leq \sum_{i=1}^{\frac{n}{\beta \log n}} \frac{1}{n^{1.35\beta}}$$

$$= \frac{n}{\beta \log n} \cdot \frac{1}{n^{1.35\beta}} \xrightarrow{n \to +\infty} 0$$

So, we can almost surely[7] re-connect T during $[0, n]$ by employing the above random assignment of availability intervals to the reserved edges, having total cost $c = \sum_{i=1}^{\binom{n}{2}-n+1} \beta \log n \in O(n^2 \log n)$. $\square$

Conjecture. *If failures of the edges happen once in every consecutive* $\Delta \geq \alpha \log n$ *time-intervals, we conjecture that there is no reservoir of cost* $o(n^2 \log n)$ *that keeps a spanning tree available during* $[0, n]$ *whp.*

Open Problem 1. *For spanning tree breaks of frequency* $o(\log n)$ *within the time period* $[0, n]$, *the reservoir of* Theorem 3 *does not re-connect* T *whp. It remains an open question to derive a scheme that does so for breaks of so high frequency.*

Open Problem 2. *What is a low cost reservoir to maintain a spanning tree of the clique network, if the failures in the links of the tree happen randomly, e.g., if each link receives a lifetime given by the Exponential Distribution?*

---

[7] Note that increasing the size of the boxes by a constant factor, i.e., increasing the lower bound for β and α, can enforce the re-connection probability to also increase.

References

1. Akrida, E.C., Gąsieniec, L., Mertzios, G.B., Spirakis, P.G.: Ephemeral networks with random availability of links: diameter and connectivity. In: Proceedings of the 26th ACM Symposium on Parallelism in Algorithms and Architectures (SPAA) (2014)
2. Angluin, D., Aspnes, J., Diamadi, Z., Fischer, M., Peralta, R.: Computation in networks of passively mobile finite-state sensors. Distrib. Comput. **18**(4), 235–253 (2006)
3. Avin, C., Koucký, M., Lotker, Z.: How to explore a fast-changing world (cover time of a simple random walk on evolving graphs). In: Aceto, L., Damgård, I., Goldberg, L.A., Halldórsson, M.M., Ingólfsdóttir, A., Walukiewicz, I. (eds.) ICALP 2008, Part I. LNCS, vol. 5125, pp. 121–132. Springer, Heidelberg (2008)
4. Bui-Xuan, B.-M., Ferreira, A., Jarry, A.: Computing shortest, fastest, and foremost journeys in dynamic networks. Int. J. Found. Comput. Sci. **14**(2), 267–285 (2003)
5. Casteigts, A., Flocchini, P., Quattrociocchi, W., Santoro, N.: Time-varying graphs and dynamic networks. Int. J. Parallel Emergent Distrib. Syst. (IJPEDS) **27**(5), 387–408 (2012)
6. Clementi, A.E.F., Macci, C., Monti, A., Pasquale, F., Silvestri, R.: Flooding time of edge-markovian evolving graphs. SIAM J. Discrete Math. (SIDMA) **24**(4), 1694–1712 (2010)
7. Cooper, C., Klasing, R., Radzik, T.: A randomized algorithm for the joining protocol in dynamic distributed networks. Theor. Comput. Sci. **406**(3), 248–262 (2008)
8. Duchon, P., Duvignau, R.: Local update algorithms for random graphs. In: Pardo, A., Viola, A. (eds.) LATIN 2014. LNCS, vol. 8392, pp. 367–378. Springer, Heidelberg (2014)
9. Dutta, C., Pandurangan, G., Rajaraman, R., Sun, Z., Viola, E.: On the complexity of information spreading in dynamic networks. In: Proceedings of the 24th Annual ACM-SIAM Symposium on Discrete Algorithms (SODA), pp. 717–736 (2013)
10. Fleischer, L., Skutella, M.: Quickest flows over time. SIAM J. Comput. **36**(6), 1600–1630 (2007)
11. Fleischer, L., Tardos, É.: Efficient continuous-time dynamic network flow algorithms. Oper. Res. Lett. **23**(3–5), 71–80 (1998)
12. Gavoille, C., Peleg, D., Perennes, S., Raz, R.: Distance labeling in graphs. In: Proceedings of the 12th Annual ACM-SIAM Symposium on Discrete Algorithms (SODA), pp. 210–219 (2001)
13. Katz, M., Katz, N.A., Korman, A., Peleg, D.: Labeling schemes for flow and connectivity. SIAM J. Comput. **34**(1), 23–40 (2004)
14. Kempe, D., Kleinberg, J.M., Kumar, A.: Connectivity and inference problems for temporal networks. In: Proceedings of the 32nd Annual ACM Symposium on Theory of Computing (STOC), pp. 504–513 (2000)
15. Koch, R., Nasrabadi, E., Skutella, M.: Continuous and discrete flows over time - a general model based on measure theory. Math. Methods OR **73**(3), 301–337 (2011)
16. Kuhn, F., Lynch, N.A., Oshman, R.: Distributed computation in dynamic networks. In: Proceedings of the 42nd Annual ACM Symposium on Theory of Computing (STOC), pp. 513–522 (2010)
17. Mertzios, G.B., Michail, O., Chatzigiannakis, I., Spirakis, P.G.: Temporal network optimization subject to connectivity constraints. In: Fomin, F.V., Freivalds, R., Kwiatkowska, M., Peleg, D. (eds.) ICALP 2013, Part II. LNCS, vol. 7966, pp. 657–668. Springer, Heidelberg (2013)

18. Michail, O., Chatzigiannakis, I., Spirakis, P.G.: Causality, influence, and computation in possibly disconnected synchronous dynamic networks. In: Baldoni, R., Flocchini, P., Binoy, R. (eds.) OPODIS 2012. LNCS, vol. 7702, pp. 269–283. Springer, Heidelberg (2012)
19. Molloy, M., Reed, B.: Graph Colouring and the Probabilistic Method. Algorithms and Combinatorics, vol. 23. Springer, Heidelberg (2002)
20. O'Dell, R., Wattenhofer, R.: Information dissemination in highly dynamic graphs. In: Proceedings of the 2005 Joint Workshop on Foundations of Mobile Computing (DIALM-POMC), pp. 104–110 (2005)
21. Scheideler, C.: Models and techniques for communication in dynamic networks. In: Alt, H., Ferreira, A. (eds.) STACS 2002. LNCS, vol. 2285, pp. 27–49. Springer, Heidelberg (2002)

Beachcombing on Strips and Islands

Evangelos Bampas[1], Jurek Czyzowicz[2], David Ilcinkas[1(✉)], and Ralf Klasing[1]

[1] LaBRI, CNRS and University of Bordeaux, Talence, France
{evangelos.bampas,david.ilcinkas,ralf.klasing}@labri.fr
[2] Département d'informatique, Université du Québec en Outaouais,
Gatineau, Canada
Jurek.Czyzowicz@uqo.ca

Abstract. A group of mobile robots (beachcombers) have to search collectively every point of a given domain. At any given moment, each robot can be in *walking mode* or in *searching mode*. It is assumed that each robot's maximum allowed searching speed is strictly smaller than its maximum allowed walking speed. A point of the domain is searched if at least one of the robots visits it in searching mode. The Beachcombers' Problem consists in developing efficient *schedules* (algorithms) for the robots which collectively search all the points of the given domain as fast as possible.

We first consider the *online* Beachcombers' Problem, where the robots are initially collocated at the origin of a semi-infinite line. It is sought to design a schedule A with maximum *speed* S, defined as $S = \inf_\ell \frac{\ell}{t_A(\ell)}$, where $t_A(\ell)$ denotes the time when the search of the segment $[0, \ell]$ is completed under A. We consider a *discrete* and a *continuous* version of the problem, depending on whether the infimum is taken over $\ell \in \mathbb{N}^*$ or $\ell \geq 1$. We prove that the `LeapFrog` algorithm, which was proposed in [Czyzowicz et al., SIROCCO 2014, LNCS 8576, pp. 23–36 (2014)], is in fact optimal in the discrete case. This settles in the affirmative a conjecture from that paper. We also show how to extend this result to the more general continuous online setting.

For the *offline* version of the Beachcombers' Problem, we consider the *single-source* Beachcombers' Problem on the cycle, as well as the *multi-source* Beachcombers' Problem on the cycle and on the finite segment. For the *single-source* Beachcombers' Problem on the cycle, we show that the structure of the optimal solutions is identical to the structure of the optimal solutions to the two-source Beachcombers' Problem on a finite segment. In consequence, by using results from [Czyzowicz et al., ALGOSENSORS 2014, LNCS 8847, pp. 3–21 (2014)], we prove that the single-source Beachcombers' Problem on the cycle is NP-hard, and we derive approximation algorithms for the problem. For the *multi-source* variant of the Beachcombers' Problem on the cycle and on the finite segment, we obtain efficient approximation algorithms.

Part of this work was done while Jurek Czyzowicz was visiting the LaBRI as a guest professor of the University of Bordeaux. This work was partially funded by the ANR project DISPLEXITY (ANR-11-BS02-014). This study has been carried out in the frame of "the Investments for the future" Programme IdEx Bordeaux – CPU (ANR-10-IDEX-03-02).

© Springer International Publishing Switzerland 2015
P. Bose et al. (Eds.): ALGOSENSORS 2015, LNCS 9536, pp. 155–168, 2015.
DOI: 10.1007/978-3-319-28472-9_12

One important contribution of our work is that, in all variants of the offline Beachcombers' Problem that we discuss, we allow the robots to *change direction of movement* and search points of the domain on both sides of their respective starting positions. This represents a significant generalization compared to the model considered in [Czyzowicz et al., ALGOSENSORS 2014, LNCS 8847, pp. 3–21 (2014)], in which each robot had a fixed direction of movement that was specified as part of the solution to the problem. We manage to prove that changes of direction do not help the robots achieve optimality.

1 Introduction

A group of n mobile robots have to explore collectively a given one-dimensional domain. The robots may be initially collocated or dispersed in the domain. At every moment of time, a robot can be either in *walking mode* or in *searching mode*. A robot in walking mode traverses the domain with a speed not exceeding its maximal *walking speed*. A robot in searching state can travel using at most its maximal *searching speed*, which is strictly smaller than its walking speed, reflecting the fact that a searching activity is more time-consuming. Different robots may have distinct maximal walking and searching speeds. A robot can change mode, speed, and direction of movement instantaneously. There is no communication between the robots during the execution of the algorithm. In the Beachcombers' Problem, the goal is to design a schedule for the movement of all robots so that the domain is *searched* as fast as possible. A domain is said to be searched under a given schedule, if every point of the domain is visited by at least one robot in searching mode.

As pointed out in [11], where the Beachcombers' Problem was introduced, there are numerous examples in quite diverse domains in which exploration using *two-speed* robots arises as a natural model for the underlying processes. For example, *foraging* or *harvesting* a field may take longer than inadvertent walking. In computer science, *web page indexing* or *code inspection* require a more involved investigation. A common feature of these examples is that the activity of searching, or other action to be performed on the territory, takes more time than casual territory traversal. The analogy to *beachcombers* has been introduced in [11] to bring out that, e.g., a beachcomber looking for things of value performs a meticulous search of the beach, which takes significantly more time than simply walking from one point of the beach to another. Further motivation for the two-speed model can be found in [11,12].

Preliminaries and Notation. We consider searching schedules using two-speed robots in the following one-dimensional geometric domains: the cycle of a known circumference L, the finite straight line segment of a known length L, and on the semi-infinite line $[0; +\infty)$. The efficiency of the search in the first two cases is expressed in terms of the time t_f when the search of the cycle is completed or, equivalently, the speed L/t_f of the process. However, in the latter case, the schedule efficiency is better expressed by the speed of the search, represented by

$\inf_\ell \frac{\ell}{t_A(\ell)}$ where $t_A(\ell)$ denotes the time when the search of the segment $[0;\ell]$ is completed. In the *discrete* version of the problem, the infimum $\inf_\ell \frac{\ell}{t_A(\ell)}$ is over $\ell \in \mathbb{N}^*$. On the other hand, in the *continuous* setting, the infimum $\inf_\ell \frac{\ell}{t_A(\ell)}$ is taken over $\ell \geq 1$.

A *schedule* for the robots is defined by a strictly increasing sequence of times $t_0, t_1, \cdots$, as well as, for every robot i and every interval $[t_j, t_{j+1}]$, for $j \geq 0$, a *mode* (walking or searching), a *speed* (respecting the maximum speed of the chosen mode), and a *direction* of movement. A schedule is *correct* if, for every point p of the domain, there exists a time moment at which p is visited by a robot in searching mode. For any fixed robot i, we refer to the individual schedule of robot i as the *trajectory* of robot i. Clearly, this sequence of intervals is finite in the offline case and infinite in the online setting.

Observe that while the model allows to use any speed not exceeding the maximal speed given for the robot's mode, we can restrict consideration only to using its maximal searching and walking speeds. Also notice that any searching schedule may be converted to another one, which has the property that all sub-segments which were being searched have pairwise disjoint interiors. Therefore, when looking for the optimal searching schedule, it is sufficient to restrict consideration to schedules whose searched sub-segments may only intersect at their endpoints.

Previous Work. The Beachcombers' Problem was introduced and studied in [11]. An optimal (offline) algorithm was presented for the problem in which all robots are initially located on one endpoint of a finite segment of known length. Furthermore, a 2-competitive (online) algorithm was presented for the case where all robots are initally collocated on the origin of a semi-infinite line. In [12], the Beachcombers' Problem was studied for the case of more than one starting positions on a finite segment of known length. For a fixed number $t \geq 2$ of starting positions, the t-source Beachcombers' Problem on a finite segment asks to find t starting points on the segment, an assignment of the robots to the starting points, and a search schedule which concludes the search of the finite segment as quickly as possible. It was shown in [12] that this problem is NP-hard for $t = 2$, even when all robots have the same walking speed, that the optimal solution can be computed efficiently when all robots have the same searching speed, and that there exist a deterministic approximation algorithm for $t = 2$ and a randomized approximation algorithm for general t.

Our Contributions. In Sect. 2, we study the online Beachcombers' Problem on the semi-line $[0; +\infty)$. We prove that the LeapFrog algorithm, which was proposed in [11], is in fact optimal in the discrete case. This settles in the affirmative a conjecture from [11]. We also show how to extend this result to the more general continuous online setting.

As regards the offline Beachcombers' Problem, we consider in Sect. 3 the *single-source* Beachcombers' Problem on the cycle. We show that the structure of the optimal solutions to the single-source Beachcombers' Problem on a cycle is identical to the structure of the optimal solutions to the two-source Beach-combers' Problem on a finite segment, as defined in [12]. This implies that the results from [12] for the case of two distinct sources are carried over to the case of the cycle, yielding an NP-hardness result as well as the existence of efficient approximation algorithms for the problem. In particular, the NP-hardness of the single-source Beachcombers' Problem on the cycle seems at first somewhat surprising, in view of the existence of an efficient algorithm generating optimal schedules for the single-source problem on a finite segment.

Furthermore, in Sect. 4, we explain how to modify the arguments from Sect. 3 so as to obtain approximation algorithms for the *multi-source* variant of the Beachcombers' Problem on the cycle and on the finite segment. Our results for the cycle topology provide a partial answer to an open question posed in [11,12], concerning the study of the problem in different domain topologies.

One further important contribution of our work is that, in all variants of the offline Beachcombers' Problem that we discuss, we allow the robots to *change direction of movement* and search points of the domain on both sides of their respective starting positions. This represents a significant generalization compared to the model considered in [12], in which each robot had a fixed direction of movement that was specified as part of the solution to the problem. On an intuitive level, allowing the robots to zigzag should not result in a faster schedule. However, no proof of this intuition had been found until now. We manage to *prove* that changes of direction do not help the robots achieve optimality.

Due to lack of space, proofs are omitted.

Related Work. Searching and exploration have been studied in numerous papers considering graphs or geometric environments (e.g. [1,4,5,7,8,14,16–18,21]). The performance of the searching or exploration is typically expressed by the trajectory length or the time used by the mobile agent.

Many searching and exploration algorithms are studied in the *online* setting, i.e., the target position or sometimes other parameters of the environment are *a priori* unknown (cf. [2,3,9,14,16,19,20]). Efficiency of such algorithms is typically measured by the *competitive ratio*, i.e., the ratio of the time spent by the online algorithm with respect to the time of the optimal offline algorithm.

Most of the papers studying searching and exploration concern single robots. Sets of collaborating mobile robots were studied, e.g., in [10,15,22,23]. Tradeoffs between the number of robots and the time of exploration were derived in [19].

The majority of the research on mobile robots concerns robots having the same mobile speed. Robots with distinct speeds were considered in the context of sensor energy efficiency [25], for designing fast converging population protocols [6], and for patrolling the boundary of an environment [13,24].

2   The Online Beachcombers' Problem on the Semi-line

In this section, we consider two variants of the online beachcombers' problem on the semi-line. The first one corresponds to the online problem presented in [11].

Definition 1 (Discrete Online Beachcombers' Problem). *Given n robots with walking speeds w_i and searching speeds $s_i < w_i$, for $1 \leq i \leq n$, initially collocated at the origin of a semi-line $[0; +\infty)$, the problem consists in finding a correct schedule for this semi-line. The discrete online speed of a schedule A is defined as $\inf_{\ell \in \mathbb{N}^*} \frac{\ell}{t_A(\ell)}$, where $t_A(\ell)$ denotes the time when the search of the segment $[0; \ell]$ is completed.*

Definition 2 (Continuous Online Beachcombers' Problem). *Given n robots with walking speeds w_i and searching speeds $s_i < w_i$, for $1 \leq i \leq n$, initially collocated at the origin of a semi-line $[0; +\infty)$, the problem consists in finding a correct schedule for this semi-line. The continuous online speed of a schedule A is defined as $\inf_{\ell \geq 1} \frac{\ell}{t_A(\ell)}$, where $t_A(\ell)$ denotes the time when the search of the segment $[0; \ell]$ is completed.*

The idea of the LeapFrog algorithm is to make all sufficiently fast robots, forming the so-called swarm of the algorithm, meet at some regular intervals. For this purpose, each robot of the swarm is assigned a specific fraction of such regular interval that it has to search (the robot walks the rest of the interval). For each robot, the assigned searching subinterval is calculated as a function of the walking and searching speeds of all the robots participating in the swarm. The robots repeat the same behavior in each interval, always all of them meeting at its extremities. Although all robots are always used in the optimal offline algorithm presented in [11], some robots whose walking speeds are too slow (informally, not larger than the average speed of the swarm) may not participate in the swarm and thus may be never used in the online LeapFrog algorithm.

The main purpose of this section is to prove the optimality of the Algorithm LeapFrog described in [11]. Our first step toward this goal is to restrict ourselves to particular schedules, which are much simpler to analyze but are nevertheless at least as efficient (in terms of online speeds) as general ones. The following simple lemma holds both for the discrete and the continuous cases.

Lemma 1. *For every correct schedule S, there exists a correct schedule S' whose both online speeds are not smaller than the respective ones of S, and such that every moving agent always moves in the initial direction at the full speed permitted by its current mode. Moreover, the interiors of the segments searched by the different robots do not overlap.*

Let LF be the discrete online speed of Algorithm LeapFrog. Before proving that Algorithm LeapFrog is optimal in terms of *discrete* online speed, we prove the slightly weaker result that no correct schedule can have a *continuous* online speed larger than LF.

Lemma 2. *The continuous online speed of any correct schedule is at most LF.*

Theorem 1. *The discrete online speed of any correct schedule is at most LF.*

Concerning the continuous online speed metrics, it is possible to obtain a slightly more precise result than the one of Lemma 2.

Lemma 3. *If there are at least two robots and Algorithm* LeapFrog *uses all the robots, then any continuous online speed of any correct schedule is less than* LF.

It turns out that simple variations of Algorithm LeapFrog can match the bounds given in Lemmas 2 and 3.

In Algorithm LeapFrog, all agents participating in the swarm are synchronized at every integer point, that is, they all arrive at the same time at every integer point. For any positive integer N, we denote by LeapFrog$_N$ the variant of LeapFrog for which the agents participating in the swarm synchronize every $1/N$ units of distance, instead of every unit as in the original Algorithm LeapFrog. It is easy to check that the continuous online speed of Algorithm LeapFrog$_N$ tends to LF as N tends toward infinity. The family of algorithms $\{$LeapFrog$_N\}_{N\in\mathbb{N}^*}$ is thus optimal in the case when there at least two robots and Algorithm LeapFrog uses all the robots (cf. Lemma 3).

If there is only one robot, then the only reasonable algorithm is the one in which the single robot always searches at its maximal speed. This algorithm is in fact Algorithm LeapFrog, and its continuous online speed is equal, in this special case, to its discrete online speed LF. Lemma 2 thus shows that Algorithm LeapFrog is optimal also in this case.

The remaining case is when there are at least two robots, but the swarm of Algorithm LeapFrog does not use all the robots. In this particular case, we consider the following adaptation LeapFrog′ of Algorithm LeapFrog. Let r, with searching speed s, be some robot not participating in the swarm in Algorithm LeapFrog. In our adaptation LeapFrog′, this robot r searches the semi-line from its beginning at its maximum searching speed s during $1/$LF time units before stopping forever. Let p be the point at which r stops. All the robots of the swarm walk at the walking speed of the slowest walker among them until reaching point p. (Note that this walking speed is larger than LF by construction of the swarm.) At this point, all swarm robots execute Algorithm LeapFrog$_N$ as if p was the origin of the semi-line, with N defined as follows. The integer N is chosen sufficiently large so that, at any time at least $1/$LF, the swarm has always searched one segment of length $1/N$ ahead of the normal Algorithm LeapFrog$_N$. One can prove that the continuous online speed of Algorithm LeapFrog′ is equal to LF, which is optimal by Lemma 2.

3   Single-Source Beachcombers on the Cycle

The purpose of this section is to show that the structure of the optimal solutions to the offline Beachcombers' Problem on the cycle is identical to the structure of the optimal solutions to the two-source Beachcombers' Problem on a finite segment, as defined in [12]. This implies that the results from [12] for the case of two distinct sources are carried over to the case of the cycle, even if the agents are allowed to zigzag. The (offline) single-source Beachcombers' Problem on the cycle is defined as follows:

Definition 3 (BPC – **Beachcombers' Problem on the Cycle**). *Consider a cycle C_L of circumference L and n robots $r_1, r_2, \ldots, r_n$, initially placed at point 0 of the cycle, each robot r_i having searching speed s_i and walking speed w_i, such that $s_i < w_i$. The Beachcombers' Problem consists in finding an efficient correct searching schedule $\mathcal{A}$ of C_L. The speed $S_{\mathcal{A}}$ of the solution to the Beachcombers' Problem equals $S_{\mathcal{A}} = L/t_f$, where t_f is the finishing time of $\mathcal{A}$.*

The (offline) t-source Beachcombers' Problem on the segment was defined in [12] as follows:

Definition 4 (t-SBP – t-**Source Beachcombers' Problem** [12]). *Consider an interval $I_L = [0, L]$ and n robots $r_1, \ldots, r_n$, each robot r_i having searching speed s_i and walking speed w_i, such that $s_i < w_i$. The t-Source Beachcombers' Problem consists in finding an efficient correct searching schedule $\mathcal{A}$ of I_L, in which the robots are divided into at most t groups with each group being initially placed on a particular point of the segment (the source) and having a fixed direction of movement. The speed $S_{\mathcal{A}}$ of the solution to the Beachcombers' Problem equals $S_{\mathcal{A}} = L/t_f$, where t_f is the finishing time of $\mathcal{A}$.*

Note that the model of [12] precludes by definition any change of direction of movement for the robots, since each group of robots has a fixed direction of movement which is specified as part of the solution to the t-SBP problem. On the other hand, in our model for BPC, no such restriction is imposed but we are able to *prove* that changing directions does not help the robots.

In the following propositions and lemmas, we will refer to schedules for BPC, unless it is explicitly stated otherwise.

Proposition 1. *For every correct schedule $\mathcal{S}$, there exists a correct schedule $\mathcal{S}'$ whose completion time is not greater than that of $\mathcal{S}$ and which additionally satisfies the following properties:*

1. *Every pair of arcs searched by the robots under $\mathcal{S}'$ have disjoint interiors.*
2. *During every time interval of $\mathcal{S}'$, every robot i is either stopped or it moves at the maximum speed w_i or s_i, according to its chosen mode during that interval.*

In view of Proposition 1, we will assume in the following that the trajectory of each robot i is characterized by a sequence of arcs $(\mathcal{A}_{i,j})_{0 \le j \le \sigma_i}$ and, for each arc, a mode (searching or walking) and a direction (clockwise or counterclockwise), such that in each arc the robot is moving at the maximum allowed speed. Note that an arc $\mathcal{A}_{i,j}$ may correspond to one or more consecutive time intervals of the schedule.

Lemma 4. *For every correct schedule $\mathcal{S}$, there exists a correct schedule $\mathcal{S}'$ whose completion time is not greater than that of $\mathcal{S}$ and in which the trajectory of every robot in $\mathcal{S}'$ satisfies the following:*

– It either stops at the origin at time 0, or it searches a sequence of arcs in clockwise (resp. counterclockwise) direction, in order of increasing clockwise (resp. counterclockwise) distance from the origin, and then it either stops or it moves counterclockwise (resp. clockwise) to the origin and then searches a sequence of arcs in counterclockwise (resp. clockwise) direction, in order of increasing counterclockwise (resp. clockwise) distance from the origin.
– In between arcs that the robot searches clockwise (resp. counterclockwise), it walks clockwise (resp. counterclockwise) straight from the end of the last searched arc to the beginning of the next one.
– The robot stops at the moment when it searches a non-empty arc for the last time.
– Traversing the circle clockwise from the origin, we first encounter all the arcs that are searched by the robot in the clockwise direction and, subsequently, we encounter all the arcs that are searched by the robot in the counterclockwise direction.

Lemma 5. *For every correct schedule S, there exists a correct schedule S' whose completion time is not greater than that of S and in which, while moving from the origin in a clockwise direction, one first encounters all the arcs that are searched by some robot moving in clockwise direction under S', and then one encounters all the arcs searched by some robot moving in counterclockwise direction under S'.*

We call a schedule that satisfies the properties guaranteed by Proposition 1, Lemmas 4 and 5 *normal*:

Definition 5 (Normal Schedules). *A schedule is called* normal *if every robot's trajectory is either empty (the robot stops at time 0), or it consists of one clockwise or counterclockwise leg, as defined below, or it consists of two legs in opposite directions, such that after the first leg the robot returns to the origin by walking at full speed backwards over the first leg.*

A clockwise (resp. counterclockwise) leg *is a part of a robot's trajectory that starts at the origin and consists of searching at full speed a sequence of arcs in order of increasing clockwise (resp. counterclockwise) distance from the origin. In between searched arcs, the robot walks at full speed in the clockwise (resp. counterclockwise) direction from the end of the last searched arc to the beginning of the next one.*

In addition, a normal schedule satisfies the following properties:

1. *Every pair of searched arcs (not necessarily by the same robot) have disjoint interiors.*
2. *For every robot, each of its legs corresponds to at most one loop around the circle and, if its trajectory has two legs, they do not overlap.*
3. *While moving from the origin in a clockwise direction, one first encounters all the searched arcs that belong to clockwise legs, and then one encounters all the searched arcs that belong to counterclockwise legs.*

It follows from the proofs of Proposition 1, Lemmas 4 and 5 that, for every correct schedule S that is not normal, there exists a correct normal schedule S'

that has smaller or equal completion time. In other words, we can guarantee all of the properties ensured by Proposition 1, Lemmas 4 and 5 simultaneously. In the following, we will assume normal schedules without loss of generality. In fact, a careful examination of the proofs reveals that, in all cases, the modification of S to S' strictly decreases the completion time of at least one robot. This is less obvious in Lemma 5, but it suffices to apply the modification described in the proof for a pair of arcs a, b, such that one of them is the last searched arc in some robot's clockwise leg or the last searched arc in some robot's counterclockwise leg. It is easy to check that if S does not satisfy the property in the statement of Lemma 5, then there exists at least one such pair of searched arcs. We thus have the following:

Lemma 6. *For every non-normal correct schedule S, there exists a normal correct schedule S' whose completion time is not greater than that of S and in which at least one robot requires strictly less time to complete its trajectory.*

With every fixed normal schedule S, we associate the corresponding partition of the circle into pairwise interior-disjoint arcs, each of which is searched by a single robot that is moving in the same direction over a continuous time interval. In view of Lemma 4, we may assume that the origin is not in the interior of any of the arcs.

Definition 6. *Let S be a normal schedule. We denote by $\mathcal{A}_S^+$ (resp. $\mathcal{A}_S^-$) the set of searched arcs that belong to clockwise (resp. counterclockwise) legs of robots. For $a, b \in \mathcal{A}_S^+ \cup \mathcal{A}_S^-$, we write $a \prec b$ if a clockwise traversal starting from the origin encounters arc a before arc b.*

For the purpose of stating the next lemma, given a normal schedule S with completion time T, we will denote by $\mathcal{I}(S)$ the inclusion-maximal set of searched arcs that satisfies the following property: Each arc in $\mathcal{I}(S)$ is searched by a robot that stops strictly earlier than T and $\bigcup_{I \in \mathcal{I}(S)} I$ is a continuous arc that contains the origin. We will denote by $R(S)$ the number of distinct robots that search the arcs in $\mathcal{I}(S)$.

Lemma 7. *Let S be a normal correct schedule with completion time T, such that there exists $\epsilon > 0$ and at least one robot that stops at time $T - \epsilon$. Then, there exists a normal correct schedule S' with completion time at most T and $R(S') > R(S)$.*

Repeated applications of Lemma 7 yield Corollary 1, from which Corollaries 2 and 3 follow immediately:

Corollary 1. *In every optimal and normal schedule, all robots terminate their trajectories simultaneously.*

Corollary 2. *Every optimal schedule is normal.*

Corollary 3. *In every optimal schedule, the trajectory of each robot contains at least one leg.*

We are now ready to further restrict the structure of optimal schedules. We first show that each robot searches only one arc per leg (Lemma 8), then that there are no *crossing robots* (Lemma 9, cf. Definition 7), and then that each robot performs only one leg (Lemma 10).

Lemma 8. *In every optimal schedule, each leg of the trajectory of each robot contains exactly one searched arc.*

Definition 7 (Crossing Robots). *Let S be a normal schedule. We say that a pair of robots i, j cross under S if robot i searches arcs $a_i^+ \in \mathcal{A}_S^+$ and $a_i^- \in \mathcal{A}_S^-$, robot j searches arcs $a_j^+ \in \mathcal{A}_S^+$ and $a_j^- \in \mathcal{A}_S^-$, and $a_i^+ \prec a_j^+ \prec a_i^- \prec a_j^-$.*

Lemma 9. *No optimal schedule contains a pair of crossing robots.*

Lemma 10. *In every optimal schedule, the trajectory of each robot contains exactly one leg.*

Lemma 10 is the main technical tool for connecting the optimal schedules for BPC instances to the optimal schedules for 2-SBP instances.

Lemma 11. *Let $\mathcal{I}$ be an instance of 2-SBP on an interval of length L and let $\mathcal{J}$ be an instance of BPC with the same set of robots on a circle of circumference L. The completion time of the optimal schedule is the same in both instances.*

We obtain now, as immediate corollaries of Lemma 11 and the results in [12], the NP-hardness of BPC, as well as the existence of deterministic and randomized approximation algorithms for BPC.

Theorem 2. BPC *is NP-hard, even when all robots have the same walking speed.*

Theorem 3. BPC *admits a 0.5568-approximation algorithm that runs in $\mathcal{O}(n \log n)$ time.*

Theorem 4. BPC *instances in which all robots have the same search speed can be solved optimally in time $\mathcal{O}(n \log n)$.*

Theorem 5. BPC *admits a randomized algorithm which achieves an expected approximation ratio of $\frac{3}{4}$, needs $\mathcal{O}(n)$ random bits, and runs in $\mathcal{O}(n \log n)$ time.*

4   Multi-Source Beachcombers on the Line and Cycle

We now leverage our techniques from the previous section to obtain results for the multi-source version of the beachcombers' problem on the line and on the cycle, while allowing changes of direction as in BPC (Definition 3). We define the problem t-SBPL$_z$:

Definition 8 (t-SBPL$_z$ – t-Source Beachcombers' Problem on the Line with Zigzags). *Consider a line segment of length L and n robots $r_1, \ldots, r_n$, each robot r_i having searching speed s_i and walking speed $w_i > s_i$. Find an efficient correct searching schedule $\mathcal{A}$ of the segment, in which the robots are divided into at most t groups with each group being initially placed on a particular point of the segment (the source). The speed $S_\mathcal{A}$ of the solution equals $S_\mathcal{A} = L/t_f$, where t_f is the finishing time of $\mathcal{A}$.*

Similarly, we define t-SBPC$_z$ (t-Source Beachcombers' Problem on the Cycle with zigzags), where, instead of a segment of length L, the robots have to search a cycle of circumference L. Note that, in contrast to t-SBP (Definition 4), the robots are allowed to change direction of movement and, in particular, to search segments on both sides of their respective starting points.

By following the arguments for BPC from Sect. 3 and modifying the proofs as necessary, we can prove that, for a fixed choice of starting points and a fixed allocation of the robots to those starting points, the equivalents of Lemma 6 and Corollaries 1, 2 and 3 hold for t-SBPL$_z$ as well, with the only difference that a *normal schedule* for t-SBPL$_z$ is defined as follows:

Definition 9 (Normal Schedules for t-SBPL$_z$). *Given a fixed choice of t starting points and a fixed allocation of the robots to those starting points, a schedule for t-SBPL$_z$ is called* normal *if every robot's trajectory is either empty (the robot stops at time 0), or it consists of one leftward or rightward leg, as defined below, or it consists of two legs in opposite directions, such that after the first leg the robot returns to the origin by walking at full speed backwards over the first leg.*

A rightward (resp. leftward) leg is a part of a robot's trajectory that starts at its assigned source and consists of searching at full speed a sequence of segments toward the right (resp. left) in order of increasing distance from the source. In between searched segments, the robot walks at full speed to the right (resp. left) from the end of the last searched arc to the beginning of the next one.

In addition, a normal schedule satisfies the following properties:

1. *Every pair of searched segments (not necessarily by the same robot) have disjoint interiors.*
2. *The given segment of length L is partitioned into t regions, such that each region is associated with exactly one starting point, and the robots originating from the associated starting point are confined within that region.*

Subsequently, we obtain the following lemma, which corresponds to Lemma 10.

Lemma 12. *Given a fixed choice of t starting points and a fixed allocation of the robots to those starting points, in every optimal t-SBPL$_z$ schedule, the trajectory of each robot contains exactly one leg.*

The following Lemma connects the optimal schedules for t-SBPL$_z$ instances to the optimal schedules for $2t$-SBP instances.

Lemma 13. *Let $\mathcal{I}$ be an instance of $2t$-SBP on a segment of length L and let $\mathcal{J}$ be an instance of t-SBPL$_z$ with the same set of robots on a segment of length L. The completion time of the optimal schedule is the same in both instances.*

We thus obtain, in view of the results for t-SBP [12], the following results for t-SBPL$_z$:

Theorem 6. *t-SBPL$_z$ instances in which all robots have the same search speed can be solved optimally in time $\mathcal{O}(n \log n)$.*

Theorem 7. *t-SBPL$_z$ admits a randomized algorithm which achieves an expected approximation ratio of $1 - \left(1 - \frac{1}{2t}\right)^{2t}$, needs $\mathcal{O}(n \log t)$ random bits, and runs in $\mathcal{O}(n \log n)$ time.*

Finally, we prove that the optimal solution to a t-SBPC$_z$ instance with a given swarm on a cycle of circumference L shares its structure with the optimal solution to a t-SBPL$_z$ instance with the same swarm on a segment of length L. Indeed, the normal schedules for t-SBPC$_z$ are essentially the same as those for t-SBPL$_z$, except that the region associated with each starting point is now an arc of the cycle.

Lemma 14. *Given a fixed choice of t starting points and a fixed allocation of the robots to those starting points, in every optimal t-SBPC$_z$ schedule, the trajectory of each robot contains exactly one leg.*

Lemma 15. *Let $\mathcal{I}$ be an instance of t-SBPL$_z$ on a segment of length L and let $\mathcal{J}$ be an instance of t-SBPC$_z$ with the same set of robots on a cycle of circumference L. The completion time of the optimal schedule is the same in both instances.*

In view of Theorems 6 and 7, we obtain the following results for t-SBPC$_z$:

Theorem 8. *t-SBPC$_z$ instances in which all robots have the same search speed can be solved optimally in time $\mathcal{O}(n \log n)$.*

Theorem 9. *t-SBPC$_z$ admits a randomized algorithm which achieves an expected approximation ratio of $1 - \left(1 - \frac{1}{2t}\right)^{2t}$, needs $\mathcal{O}(n \log t)$ random bits, and runs in $\mathcal{O}(n \log n)$ time.*

5   Concluding Remarks

There are several directions in which the study of the search and exploration using two-speed robots may continue. An obvious one is to improve the approximation ratio for the versions of the problem that are NP-hard. In this respect, we should investigate whether zigzags may help to obtain approximate solutions, at least for particular combinations of searching and walking speeds of the robots (note that we know from the present paper that zigzags never help to obtain *optimal* solutions). Another direction is to study the configurations of robots' speeds and/or environments for which optimal solutions can be computed efficiently. Finally, it is worthwhile to consider different and more general search domains, such as non-simple closed or open curves.

References

1. Albers, S., Henzinger, M.R.: Exploring unknown environments. SIAM J. Comput. **29**(4), 1164–1188 (2000)
2. Albers, S.: Online algorithms: a survey. Math. Program. **97**(1–2), 3–26 (2003)
3. Albers, S., Schmelzer, S.: Online algorithms - what is it worth to know the future? In: Vöcking, B., Alt, H., Dietzfelbinger, M., Reischuk, R., Scheideler, C., Vollmer, H., Wagner, D. (eds.) Algorithms Unplugged, pp. 361–366. Springer, Heidelberg (2011)
4. Alpern, S., Gal, S.: The Theory of Search Games and Rendezvous. Kluwer Academic Publishers, Dordrecht (2002). vol. 55
5. Baeza-Yates, R.A., Culberson, J.C., Rawlins, G.J.E.: Searching in the plane. Inf. Comput. **106**, 234–234 (1993)
6. Beauquier, J., Burman, J., Clement, J., Kutten, S.: On utilizing speed in networks of mobile agents. In: ACM SIGACT-SIGOPS 2010, pp. 305–314. ACM (2010)
7. Beck, A.: On the linear search problem. Isr. J. Math. **2**(4), 221–228 (1964)
8. Bellman, R.: An optimal search problem. Bull. Am. Math. Soc. **62**, 270 (1963)
9. Berman, P.: On-line searching and navigation. In: Fiat, A., Woeginger, G.J. (eds.) Online Algorithms 1996. LNCS, vol. 1442, pp. 232–241. Springer, Heidelberg (1998)
10. Chalopin, J., Flocchini, P., Mans, B., Santoro, N.: Network exploration by silent and oblivious robots. In: Thilikos, D.M. (ed.) WG 2010. LNCS, vol. 6410, pp. 208–219. Springer, Heidelberg (2010)
11. Czyzowicz, J., Gąsieniec, L., Georgiou, K., Kranakis, E., MacQuarrie, F.: The Beachcombers' problem: walking and searching with mobile robots. In: Halldórsson, M.M. (ed.) SIROCCO 2014. LNCS, vol. 8576, pp. 23–36. Springer, Heidelberg (2014)
12. Czyzowicz, J., Gasieniec, L., Georgiou, K., Kranakis, E., MacQuarrie, F.: The multi-source Beachcombers problem. In: Gao, J., Efrat, A., Fekete, S.P., Zhang, Y. (eds.) ALGOSENSORS 2014, LNCS 8847. LNCS, vol. 8847, pp. 3–21. Springer, Heidelberg (2015)
13. Czyzowicz, J., Gąsieniec, L., Kosowski, A., Kranakis, E.: Boundary patrolling by mobile agents with distinct maximal speeds. In: Demetrescu, C., Halldórsson, M.M. (eds.) ESA 2011. LNCS, vol. 6942, pp. 701–712. Springer, Heidelberg (2011)
14. Czyzowicz, J., Ilcinkas, D., Labourel, A., Pelc, A.: Worst-case optimal exploration of terrains with obstacles. Inf. Comput. **225**, 16–28 (2013)
15. Das, S., Flocchini, P., Kutten, S., Nayak, A., Santoro, N.: Map construction of unknown graphs by multiple agents. Theor. Comput. Sci. **385**(1–3), 34–48 (2007)
16. Demaine, E.D., Fekete, S.P., Gal, S.: Online searching with turn cost. Theor. Comput. Sci. **361**(2), 342–355 (2006)
17. Deng, X., Papadimitriou, C.H.: Exploring an unknown graph. In: Foundations of Computer Science, FOCS 1990, pp. 355–361. IEEE (1990)
18. Deng, X., Kameda, T., Papadimitriou, C.H.: How to learn an unknown environment (extended abstract). In: Foundations of Computer Science, FOCS 1991, pp. 298–303. IEEE (1991)
19. Dereniowski, D., Disser, Y., Kosowski, A., Pająk, D., Uznański, P.: Fast collaborative graph exploration. In: Fomin, F.V., Freivalds, R., Kwiatkowska, M., Peleg, D. (eds.) ICALP 2013, Part II. LNCS, vol. 7966, pp. 520–532. Springer, Heidelberg (2013)
20. Fleischer, R., Kamphans, T., Klein, R., Langetepe, E., Trippen, G.: Competitive online approximation of the optimal search ratio. SIAM J. Comput. **38**(3), 881–898 (2008)

21. Fomin, F.V., Thilikos, D.M.: An annotated bibliography on guaranteed graph searching. Theor. Comput. Sci. **399**(3), 236–245 (2008)
22. Fraigniaud, P., Gasieniec, L., Kowalski, D.R., Pelc, A.: Collective tree exploration. Networks **48**(3), 166–177 (2006)
23. Higashikawa, Y., Katoh, N., Langerman, S., Tanigawa, S.: Online graph exploration algorithms for cycles and trees by multiple searchers. J. Comb. Optim. **28**, 480–495 (2012)
24. Kawamura, A., Kobayashi, Y.: Fence patrolling by mobile agents with distinct speeds. In: Chao, K.-M., Hsu, T., Lee, D.-T. (eds.) ISAAC 2012. LNCS, vol. 7676, pp. 598–608. Springer, Heidelberg (2012)
25. Wang, G., Irwin, M.J., Fu, H., Berman, P., Zhang, W., Porta, T.L.: Optimizing sensor movement planning for energy efficiency. ACM Trans. Sens. Netw. **7**(4), 33 (2011)

Radio Aggregation Scheduling

Rajiv Gandhi[1], Magnús M. Halldórsson[2,3], Christian Konrad[2(✉)],
Guy Kortsarz[1], and Hoon Oh[1]

[1] Department of Computer Science, Rutgers University, Camden, NJ, USA
{rajivg,guyk,hoon}@camden.rutgers.edu
[2] ICE-TCS, School of Computer Science, Reykjavik University, Reykjavik, Iceland
{mmh,christiank}@ru.is
[3] RIMS, Kyoto University, Kyoto, Japan

Abstract. We consider the aggregation problem in radio networks: find
a spanning tree in a given graph and a conflict-free schedule of the edges
so as to minimize the latency of the computation. While a large body of
literature exists on this and related problems, we give the first approximation results in graphs that are not induced by unit ranges in the plane.
We give a polynomial-time $\tilde{O}(\sqrt{\bar{d}n})$-approximation algorithm, where $\bar{d}$ is
the average degree and n the number of vertices in the graph, and show
that the problem is $\Omega(n^{1-\epsilon})$-hard (and $\Omega((\bar{d}n)^{1/2-\epsilon})$-hard) to approximate even on bipartite graphs, for any $\epsilon > 0$, rendering our algorithm
essentially optimal. We target geometrically defined graph classes, and
in particular obtain a $O(\log n)$-approximation in interval graphs.

1 Introduction

Wireless sensor networks consist of autonomous sensors that typically monitor physical or environmental conditions. They use wireless communication to
cooperatively aggregate the recorded data and forward it to a central location,
the sink. The information desired is commonly in the form of a *compressible*
function, such as "max" or "average", in which in-network processing can be
used to speed up the processing and greatly reduce transmission energy. At the
same time, interference from simultaneous transmissions must be managed for
successful reception.

In this paper, we consider the data aggregation problem in general graphs, or
radio networks. The objective is to minimize the *latency*, or the longest time it
takes for any message to reach the sink. The task is two-fold: (a) to construct a
directed spanning tree, an *in-arborescence*, and (b) to form a *conflict-free schedule* of the transmissions (the edges) that obeys the ordering of the arborescence.
A schedule is conflict free if whenever a node is to receive a message, none of its

Gandhi and Oh are partly supported by NSF grants CCF 1218620 and CCF
1433220. Halldórsson and Konrad are supported by Icelandic Research Fund grant-
of-excellence no. 120032011. Kortsarz is partly supported by NSF grant CCF
1218620.

P. Bose et al. (Eds.): ALGOSENSORS 2015, LNCS 9536, pp. 169–182, 2015.
DOI: 10.1007/978-3-319-28472-9_13

other neighbors also transmit (causing interference), and a node can transmit to only one of its neighbors at a time.

This problem, which we dub *Radio Aggregation Scheduling* (RAS), has been widely studied under the name *Minimum Latency Aggregation Scheduling* in the wireless networking literature. Most of the existing works consider the setting where nodes are points in the plane with a fixed transmission radius, which corresponds to the case of *unit disc graphs* (UDG). It is, however, well known that wireless environments are *always* much more complicated [1,21] — unless operating in vacuum in outer space. One popular approach in recent years has been to switch to the SINR model of interference, which is known to add more realism. However, its standard form also makes strong assumptions about the geometric nature of communicability and interference and thus ignores the unpredictability seen in practice. To go beyond these assumptions, we initiate here the study of aggregation in more pessimistic models, starting with general graphs. To emphasize the distinction of using graphs rather than planar positions, we refer to the problem as RAS.

By reversing the direction of the aggregation process, we can also view it as a *broadcasting* problem where:

1. [*one-on-one*] a node can only talk to one other node at a time, while
2. [*interference from neighbors*] a node can hear from its neighbor only if none of its other neighbors transmit.

We refer to this communication model as the *radio-unicast* model. It relates closely to two other classic broadcasting problems: *telephone broadcast*, where (1) holds but there are no conflicts from other neighbors (in essence, modeling aggregation in wired networks); and *radio broadcast*, where (2) holds, but a node can transmit to all its neighbors in the same time slot. As we shall see, however, RAS is significantly harder to solve in general than either of these problems.

In the telephone model, in each communication round, the successful transmissions form a (directed) matching. In the radio-unicast model, successful transmissions form what we call a RAS-*legal matching* (see Sect. 2 for precise definitions). For any two edges (s_1, r_1) and (s_2, r_2) in a RAS-legal matching connecting senders s_1, s_2 to receivers r_1, r_2, it is required that neither (s_1, r_2) nor (s_2, r_1) are edges contained in the input graph, thus excluding all potential interference. This is closely related to the notion of an *induced matching*. A matching is induced if the edges of the subgraph induced by the matched vertices are precisely the edges of the matching. A RAS-legal matching hence lies somewhere between a matching and an induced matching, see Fig. 1.

Previous Work on Ras. All previous works on RAS consider the setting where nodes are points located in the plane with unit length transmission radii [3,7,15, 25,26][1]. This corresponds to the study of RAS in unit disc graphs, which has been shown to be NP-complete [7]. All algorithms known for unit disc graphs compute

---

[1] In [15], unit interval graphs as well as grids and tori are considered, which are all subclasses of unit disc graphs.

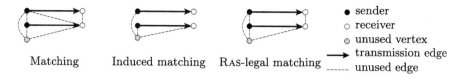

Matching   Induced matching   RAS-legal matching

• sender
○ receiver
◎ unused vertex
⟶ transmission edge
⋯⋯ unused edge

Fig. 1. Left: A matching is a subset of vertex-disjoint edges. Center: The edges of the graph induced by the vertices of an induced matching are precisely the edges of the induced matching. Right: In a RAS-legal matching, every receiver is connected to precisely one sender.

aggregation schedules of lengths $\Theta(Diam + \Delta)$, where $Diam$ is the diameter of the input graph and Δ the maximal degree. Since every aggregation schedule is of length at least $Diam$, these algorithms constitute $O(\Delta)$-approximation algorithms which only give trivial approximation guarantees in graphs with large maximum degree (e.g. if $\Delta = \Theta(n)$). Despite the considerable effort put into the study of RAS on unit disc graphs, no better approximation ratios are known.

One difficulty in obtaining improved approximation ratios in unit disc graphs is to bound the length of an optimal aggregation schedule OPT in terms of properties of the input graph. For instance, in unit interval graphs, it is known that $OPT = \Omega(Diam + \omega(G))$, where $\omega(G)$ is the clique number (size of the largest clique) of the input graph [15]. It is also known how to compute an aggregation schedule of length $O(Diam + \omega(G))$, which hence constitutes an $O(1)$-approximation algorithm (in [15], a 2-approximation is obtained). No interesting bounds on OPT are known for unit disc graphs or any other non-trivial graph class.

Our Contributions. We initiate a systematic study of RAS, starting with general graphs. We prove that it is NP-hard to approximate RAS within a factor of $n^{1-\epsilon}$ (**Theorem 4**) and $(\bar{d}n)^{1/2-\epsilon}$ (**Corollary 1**) even in bipartite graphs, for any $\epsilon > 0$, where n is the number of vertices of the input graph and $\bar{d}$ is the average degree. On the positive side, we present a $\tilde{O}(\sqrt{\bar{d}n})$-approximation[2] algorithm for sparse general graphs (**Theorem 5**), almost matching our lower bound.

Next, we are interested in whether improved algorithms can be obtained for geometrically defined graph classes that contribute to metric-sensitive models of actual wireless environments. We focus here on interval graphs. They can be seen as one-dimensional projections of disc graphs that capture the aspect of different radii, and we present a highly non-trivial $O(\log n)$-approximation algorithm (**Theorem 6**). The key part of our analysis is the identification of subgraphs that provide interesting lower bounds on the length of an optimal aggregation schedule.

Further Related Work. Aggregation problems have been extensively studied in the wireless literature; see the surveys [12,17]. As previously mentioned,

---

[2] We use the notation $\tilde{O}(.)$, which equals the usual $O(.)$ notation where all poly-logarithmic factors are ignored.

RAS has been considered in unit disc graphs [3,7,15,25,26] and O(Δ)-approximation algorithms are known. Furthermore, it has also been shown that, in unit disc graphs, if the interference radius is strictly larger than the transmission radius, then constant factor approximations can be obtained [25]. Optimal algorithms are known for grids and tori [14]. In trees, RAS is equivalent to the telephone broadcast problem, which has a textbook dynamic programming solution [19, Prob. 6.16]. This exhausts the list of previous work known on RAS.

A different setting for aggregation problems is where the nodes are located at points in the plane and can use power control to reach any other node. Kesselman and Kowalski [18] showed that aggregation can then be achieved in O($\log n$) slots. If interference and transmissions follow the geometric SINR model, Moscibroda and Wattenhofer [23] showed that poly-logarithmic slots suffice, which was improved to optimal O($\log n$) [16].

For broadcast in the radio model, Chlamtac and Weinstein [8] proved the first upper bound of O($Diam \cdot \log^2 n$), with $Diam$ being the diameter of the graph, which was improved to O($Diam \cdot \log n + \log^2 n$) soon afterwards by Bar-Yehuda et al. [5]. The best bound known on the number of rounds, O($Diam + \log^2 n$), given by Kowalski and Pelc [22], is optimal in light of results of Alon et al. [2] and Elkin and Kortsarz [10].

The first approximation for telephone broadcast was an additive O($\sqrt{n}$) approximation [20]. This was improved to a multiplicative O($\log^2 n$)-factor by [24], and then to O($\log n$) in [4]. The best approximation known for the problem is O($\log n / \log OPT$) [11], which is O($\log n / \log \log n$), since $OPT \geq \log_2 n$ always holds. The best lower bound known is a factor $3 - \epsilon$, given in [9].

Outline of the Paper. We give formal definitions of our problems in Sect. 2. Then, in Sect. 3, we present our hardness results for general graphs, and in Sect. 4, we present our algorithm for sparse general graphs. Finally, in Sect. 5, interval graphs are discussed.

Due to space restrictions, the proofs of lemmas, theorems, claims and observations marked by (*) are postponed to the full version of this paper.

2 Problem Definition and Notations

Radio Aggregation Scheduling. We are given as input a graph $G = (V, E)$ and a node $s \in V$ which is the *sink* node of the aggregation problem. We view G as a *bidirected graph*, i.e., all edges appear directed in both directions.

We seek a *schedule*, which is a sequence $M_1, M_2, \ldots, M_t$ of directed matchings in G. The union $\cup_i M_i$ of these matchings induces a directed spanning tree (*in-arborescence*) A directed toward s. Each matching M_i corresponds to a set of transmissions that can be successful simultaneously; namely, each matching must be RAS-*legal* in G: if $(u,v), (w,z) \in M_i$ then $(u,z), (w,v) \notin E(G)$. Finally, the edges of A occur in the matchings in order of precedence induced by the arborescence: if $(u,v) \in M_i$ and $(v,w) \in M_j$ then $i < j$. Namely, a node can only forward its message once it has heard from all of its children. Then an optimal solution to the Radio Aggregation Scheduling problem (RAS) is a schedule of minimal length.

Broadcasting in the Radio-Unicast Model. Since reversing the slots of an aggregation schedule gives a broadcast, and vice versa, both viewpoints can be used to tackle RAS. In the broadcast version of the problem, node $s \in V$ is the *source* node and holds a message that is to be sent to all other nodes $V \setminus \{s\}$ in the graph. In each round, we seek a RAS legal matching between the *informed nodes* (those that know the message) and the *uninformed nodes* (those that don't know the message yet). Initially, there is only a single informed node, the source node s. When an uninformed node receives the message, it joins the set of informed nodes and can serve as a sender in upcoming rounds. We denote this communication model where each round induces a RAS-legal matching as the *radio-unicast* model. An optimal solution to the broadcasting problem then is a broadcasting schedule that informs all nodes in the minimal number of rounds.

It turns out that the broadcasting perspective of RAS is more convenient when presenting our algorithms. All our algorithms solve the broadcasting problem in the radio-unicast model.

Notation. Let $G = (V, E)$ be the input graph. Unless stated differently, n denotes the number of vertices of G, $\bar{d}$ the average degree, Δ the maximum degree, and $Diam$ the diameter. Those quantities may also appear as functions, e.g. $\Delta(H)$, $\bar{d}(H)$ and $Diam(H)$ denote the respective quantities of graph H.

We write $dist_G(u, v)$ for the number of hops between nodes u and v in graph G. Let $N_G(u)$ denote the set of neighbors of vertex u in G, and for a set S of vertices, let $N_G(S) = (\cup_{u \in S} N_G(u)) \setminus S$. We write $\deg_G(u)$ the degree of u in G. Furthermore, for a graph G, we denote its vertex set by $V(G)$ and its edge set by $E(G)$. Given a subset of vertices $U \subseteq V$, we denote the subgraph of G induced by the vertices U by $G[U]$.

3 Approximation Hardness of Ras

In this section, we prove that RAS is hard to approximate within factors $n^{1-\epsilon}$ (Theorem 4) and $(\bar{d}n)^{1/2-\epsilon}$ (Corollary 1), for every $\epsilon > 0$. Before giving our lower bound construction, we introduce further required notations and definitions.

Further Definitions. We denote the chromatic number of a graph G with $\chi(G)$, and the independence number (size of a maximum independent set) with $\alpha(G)$. Our lower bound construction relies on *semi-induced matchings* and a specific *graph product* that we discuss first.

A matching is called an *induced matching* if there is no edge from one endpoint of an edge in the matching to an endpoint of another edge in the matching. The *semi-induced matching* has a general definition (see [6]) but we only give the definition for bipartite graphs that is simpler and all we need.

Definition 1 (Semi-induced Matching). *Let $G = (U, V, E)$ be a bipartite graph with a total ordering $u_1, \ldots, u_n$ of U. A semi-induced matching is a matching so that if (u_i, a) and (u_j, b) are in the matching and $i < j$, then there is no edge between u_j and a.*

Let $\text{IM}(G)$ be the size of the largest induced matching of G and $\text{SIM}(G)$ the size of the largest semi-induced matching. Observe that $\text{IM}(G) \leq \text{SIM}(G)$, for any graph G.

Next, we make use of the following graph product:

Definition 2 (Inclusive Graph Product). *The* inclusive graph product *of $G = (V, E)$ and $H = (V', E')$, denoted by $G \vee H$, has vertices $\{(x_G, x_H) \mid x_G \in V, x_H \in V'\}$. A pair of vertices $(x_G, x_H) \in V(G \vee H)$ and $(y_G, y_H) \in V(G \vee H)$ is connected iff $(x_G, y_G) \in E$ or $(x_H, y_H) \in E'$.*

We denote $G^k = G \vee G \vee \ldots \vee G$ when there are k copies of G. This graph has n^k vertices.

The following equalities are folklore for the specific product we chose:

$$\chi(G^k) = \chi(G)^k, \tag{1}$$
$$\alpha(G^k) = \alpha(G)^k. \tag{2}$$

Intermediate Problem: Induced Matching Cover. We shall consider a problem on bipartite graphs that is closely related to RAS. Given a bipartite graph $B = (U, V, E)$, let $\text{IMCOV}(B)$ denote the minimum number of induced matchings that together contain (or cover) all the vertices of V. Suppose that nodes U are informed and nodes V are uninformed. Then, it takes precisely $\text{IMCOV}(B)$ rounds in order to inform V. This is summarized in Observation 1.

Observation 1 (*). *Let $B = (U, V, E)$ be a bipartite graph. Suppose all the vertices in U know the message. Then, the minimum number of rounds it takes to inform V in the radio-unicast model equals $\text{IMCOV}(B)$.*

Lower Bound Construction. In order to prove our hardness result, we will use the construction of Feige and Kilian [13] which shows that it is hard to determine whether a graph G on n vertices has small chromatic number $\chi(G) \leq n^\epsilon$ ("yes instance") or has a small independence number $\alpha(G) \leq n^\epsilon$ ("no instance"), for any $\epsilon > 0$.

Let G be a graph on n vertices as used in the construction of Feige and Kilian. From G, using a construction similar to the one in [6], we will construct a bipartite graph $B_e(G^k)$ on $\Theta(n^k)$ vertices so that:

$$\text{IMCOV}(B_e(G^k)) \leq \chi(G), \text{ and} \tag{3}$$
$$\text{IM}(B_e(G^k)) \leq k \cdot n + \alpha(G)^k. \tag{4}$$

Suppose now that one bipartition of $B_e(H^k)$ is informed and the other one is uninformed. Then, if G is a "yes instance" (i.e. it has small chromatic number), the whole graph can be informed quickly using Inequality 3 and Observation 1.

Suppose now that G is a "no instance" (i.e. it has small independence number). Then, by Inequality 4, $\text{IM}(B_e(G^k))$ is small, too. Using the obvious relationship $\text{IMCOV}(B_e(G^k)) \geq |V(B_e(G^k))|/\text{IM}(B_e(G^k))$, we see that $\text{IMCOV}(B_e(G^k))$ is large which implies that informing the whole graph takes many rounds.

The previous gap-reduction argument is made rigorous in the following. To this end, for a graph G, we first define the graph $B_e(G)$.

Definition 3. *Given a graph $G = (V, E)$, the graph $B(G) = (V, \bar{V}, E_B)$ is a bipartite graph with a copy of V on each side. There is an edge $(v, \bar{u}) \in E_B$ if $(v, u) \in E$. The graph $B_e(G)$ results from $B(G)$ by adding the edges $\{(v, \bar{v}) : v \in V\}$.*

Next, we prove Inequalities 3 and 4 in Claims 2 and 3, respectively.

Claim 2 (*). *Let $G = (V, E)$ be a graph. Then, $B_e(G) = (V, \bar{V}, E')$ can be decomposed into $\chi(G)$ induced matchings that are pairwise disjoint and together contain all of $\bar{V}$, i.e., $\mathrm{IMCov}(B_e(G)) \leq \chi(G)$.*

Claim 3 (*). *Let G be a graph, k an integer. Then, $\mathrm{IM}(B_e(G^k)) \leq k \cdot n + \alpha(G)^k$.*

Finally, we prove our hardness results in Theorem 4 and Corollary 1.

Theorem 4. *The RAS problem is hard to approximate on bipartite graphs within a factor of $N^{1-\delta}$, for any $\delta > 0$, where N is the number of vertices.*

Proof. We use the gap reduction of Feige and Kilian [13]: for any $\epsilon > 0$, it is hard to distinguish between the case ("yes" instance) when a graph G is n^ϵ-colorable, i.e., when $\chi(G) \leq n^\epsilon$, and the case ("no" instance) when there is no independent set of size at least n^ϵ, i.e., $\alpha(G) < n^\epsilon$.

Let ϵ be such that $1/\epsilon = 2\lceil 1/\delta \rceil$, and let $k = 1/\epsilon$. Consider $B_e(G^k) = (V_k, \bar{V}_k, E_k)$ and let H_k be the graph obtained by adding to $B_e(G^k)$ a complete binary tree of depth $O(\log |V_k|)$ whose set of leaves contains V_k. H_k is clearly bipartite, too. We show that it is hard to approximate the number of rounds in a RAS schedule of H_k.

Suppose that the root of the binary tree is the source node of the broadcast problem. Let OPT denote the length of a shortest broadcast schedule. Observe that informing the nodes of the complete binary tree, and thus also the nodes in V_k, requires only $O(\log n)$ slots. Informing $\bar{V}_k$ after V_k has been informed takes $\mathrm{IMCov}(B_e(G^k))$ rounds, by Observation 1. Thus, $OPT = \mathrm{IMCov}(B_e(G^k)) + O(\log n)$.

If G is a yes-instance, $\chi(G) \leq n^\epsilon$, so by Claim 2 and Inequality 1,

$$\mathrm{IMCov}(B_e(G^k)) \leq \chi(G^k) = \chi(G)^k \leq n^{k\epsilon} = n.$$

and hence

$$OPT = \mathrm{IMCov}(B_e(G^k)) + O(\log n) = O(n).$$

If G is a no-instance, $\alpha(G)^k \leq n^{k\epsilon} = n$, so by Claim 3, $\mathrm{IM}(B_e(G^k)) = O(n)$, and

$$OPT \geq \mathrm{IMCov}(B_e(G^k)) \geq \frac{|V_k|}{\mathrm{IM}(B_e(G^k))} = \Omega(n^{k-1}).$$

The ratio between the bounds for the two cases is $\Omega(n^{k-2})$. Recalling that the size of H_k is given by $N = |H_k| = \Theta(n^k)$, we get that the approximation hardness is $\Omega(n^{k-2}) = \Omega(N/n^2) = \Omega(N^{1-\frac{2}{k}}) = \Omega(N^{1-\delta})$. $\square$

Corollary 1 (*). *The* RAS *problem is hard to approximate on bipartite graphs within a factor of* $(\overline{d}N)^{\frac{1}{2}-\delta}$, *for any* $\delta > 0$, *where* N *is the number of vertices.*

Corollary 1 renders our $\tilde{O}(\sqrt{\overline{d}n})$-approximation algorithm that we present in the next section essentially best possible.

The graphs used in the proofs of Theorem 4 and Corollary 1 have a diameter of $O(\log n)$. By adding additional edges, their diameters can be reduced to 2. This shows that unlike in the radio model, broadcasting in the radio-unicast model is no easier in graphs of low diameter.

4 $\tilde{O}(\sqrt{\overline{d}n})$-approximation Algorithm

We now present a $\tilde{O}(\sqrt{\overline{d}n})$-approximation algorithm for RAS in general graphs $G = (V, E)$ with average degree $\overline{d}$. We consider the broadcasting perspective in the radio-unicast model. Before presenting our algorithm, we discuss simulation results that allow us to reuse existing algorithms designed for the telephone and the radio models.

Simulation Between Models. We derive now (rather straightforward) bounds on RAS schedules, utilizing its relationship to better studied broadcast problems.

Recall that in the *telephone model*, there are no conflicts if two neighbors of a node both transmit. However, a node can only transmit to one of its neighbors in a given round. In the *radio model*, when a node transmits, its message goes to all of its neighbors. However, an uninformed neighbor receives the message only if exactly one of its neighbors is transmitting in that round.

Our problem shares the unicast transmission rule with the telephone model and the reception conflicts with the radio model. Algorithms for these models can be simulated in our models.

Lemma 1. *A round in the radio (telephone) model can be simulated in* Δ ($2\Delta - 1$) *rounds in the unicast-radio model, respectively.*

Proof. Suppose a set S of nodes transmits in a given round in the radio model. Assume without loss of generality that the neighbors of each node are ordered in an arbitrary order. We can then simulate it with Δ rounds, where in round i, each node in S forwards the message to its i-th neighbor.

Suppose a directed matching M that corresponds to the transmissions of a round in the telephone model. For each edge $e \in M$, there are at most $2(\Delta - 1)$ edges in M within distance 2 in G. We can color the edges in M "first-fit" using $2\Delta - 1$ colors so that each color class induces a RAS-legal matching.   □

Simulating the algorithm of Kowalski and Pelc for radio broadcast [22], and using Lemma 1, we obtain the following corollary.

Corollary 2. *There is a polynomial-time algorithm for* RAS *that computes an aggregation schedule of length* $O(\Delta(Diam + \log^2 n))$ *and thus constitutes a* $O(\Delta + \Delta \log^2(n)/Diam)$-*approximation algorithm.*

In the previous corollary, we used the fact that $Diam$ is a trivial lower bound on the length of an optimal schedule. In light of the hardness results in Sect. 3, this bound is close to best possible.

Complete binary trees with degrees at least $\log^2 n$ provide examples showing that the $O(\Delta(Diam + \log^2 n))$ bound of Corollary 2 generally cannot be improved.

Center Selection. Our algorithm uses as a subroutine solutions to a classic facility location problem. In CENTER SELECTION, we are given a graph $G = (V, E)$, a set $X \subseteq V$ of possible sites for centers, a set $C \subseteq V$ of clients, and a parameter k. We wish to find a set $S \subseteq X$ of k centers, such that the maximum distance from a client to the nearest center is minimized. For a set of centers $S \subseteq X$, let $\rho(G, S, C) := \max_{v \in C} dist_G(v, S)$ be the *covering radius* of S in G. The objective of CENTER SELECTION is to find an $S \subseteq X$ of cardinality k which minimizes $\rho(G, S, C)$.

A greedy algorithm, which we denote by GREEDY-CS(G, X, C, k), gives a 3-approximation to this problem.[3]

Lemma 2 (*). GREEDY-CS *is a 3-appr. algorithm for* CENTER SELECTION.

Ras Scheme. In Algorithm 1, we present an algorithm for the broadcast problem in the radio-unicast model. We assume that the optimal value OPT (length of a shortest broadcast scheme) is known by the algorithm. This can be ensured e.g. by running the algorithm multiple times trying the different values $\{\log n, \ldots, n\}$ for OPT and returning the best solution ($\log n$ is an obvious lower bound).

Let $s \in V$ be the source node. To keep the presentation simple, we assume that $\deg_G(s) \geq \sqrt{\bar{d}n}$. If this is not the case, then we first inform an arbitrary node s' of degree at least $\sqrt{\bar{d}n}$ in at most OPT rounds which then takes the role of s. Clearly, the length of a minimum length schedule of the modified instance with source s' is at most by OPT longer than the length of a minimum length schedule with source node s. Hence, by solving the instance with source node s', we may lose an additive $2 \cdot OPT$ term. However, since our obtained approximation factor is polynomial, this factor is negligible. Last, if no node of degree at least $\sqrt{\bar{d}n}$ exists, then we simply apply the simulation result of Corollary 2, and we immediately obtain an $\tilde{O}(\sqrt{\bar{d}n})$-approximation algorithm.

First, our algorithm informs the large-degree nodes, i.e., nodes L of degree at least $K = \sqrt{\bar{d}n}$. The number of large degree nodes is bounded by $|L| \leq K$, as otherwise the degree sum of the graph would be greater than $K^2 = \bar{d}n = 2|E(G)|$. Thus, by transmitting serially on shortest paths (with no transmissions occurring simultaneously), the nodes in L can be informed in time $O(K \cdot OPT)$. In order to inform the small-degree nodes $V \setminus L$, we simulate the radio-broadcast algorithm of [22] on the subgraph $G[C]$, where $C = V \setminus L$. To make this work in

---

[3] While the result is surely well known, we were not aware of a reference for this particular version, and thus include the algorithm and a proof in the full version of this paper for completeness.

Algorithm 1. Broadcast in the radio-unicast model for sparse general graphs

---

Require: $G = (V, E)$ input graph, let $K = \sqrt{dn}$; s source node of degree at least K
1: Let $L \leftarrow \{v : \deg_G(v) \geq K\}$, $C = V \setminus L$, and $X = N(L) \cap C$
2: Inform the nodes in L sequentially along shortest paths from s
3: Let $S \leftarrow \text{GREEDY-CS}(G[C], X, C, K \cdot OPT)$
4: Inform all nodes in S using single hops from L
5: Simulate the radio broadcast algorithm of [22] on $G[C]$ until all nodes are informed

---

the desired number of rounds, we have to ensure that for each node in C, there is an informed node within distance $O(OPT)$ in $G[C]$. In the following lemma, we show that the set S found by the greedy center selection algorithm guarantees this property.

Lemma 3. *Each node in C is within distance at most $3 \cdot OPT$ from a node in S in the induced subgraph $G[C]$, i.e., $\rho(G[C], S, C) \leq 3 \cdot OPT$.*

Proof. Let Q be the set of nodes in C that are informed (directly) by nodes in L in the optimal broadcasting scheme. At most $|L|$ of them can be informed in a single round, so $|Q| \leq |L| \cdot OPT \leq K \cdot OPT$. The nodes $v \in C \setminus Q$ must then all satisfy $dist_{G[C]}(v, Q) \leq OPT$ and thus $\rho(G[C], Q, C) \leq OPT$. The center selection algorithm GREEDY-SC positions $K \cdot OPT \geq |Q|$ nodes, that by Lemma 3 yields a 3-approximation of the covering radius, giving $\rho(G[C], S, C) \leq 3 \cdot \rho(G[C], Q, C) \leq 3 \cdot OPT$.   □

The previous lemma is the main ingredient of the analysis of our main result:

Theorem 5 (*). *There is a polynomial time randomized approximation algorithm for RAS with approximation factor $\tilde{O}(\sqrt{dn})$.*

5   Interval Graphs

Let $G = (V, E)$ be an interval graph. For an interval $v \in V$, denote by $l(v)$ and $r(v)$ its left and right boundaries. For $x, y \in \mathbb{R}$, let $G[x, y]$ denote the subgraph of G induced by the intervals that are entirely contained in $[x, y]$, that is, $V(G[x, y]) = \{v \in V : l(v) \geq x$ and $r(v) \leq y\}$. Furthermore, denote by $len(v)$ the length of interval v. We write l_{max} for the length of a longest interval in G. W.l.o.g., we assume that all interval boundaries are integers in $\{1, \ldots, 2n\}$, and all interval boundaries are distinct (every interval graph has such a representation).

Before presenting our algorithm, we show that the clique number of an interval graph G (the size of a largest clique in G) provides a lower bound for the length of an optimal schedule. This lemma is similar to Lemmas 2 and 3 of [14].

Lemma 4 (*). $OPT \geq \omega(G)/2$.

Next, our algorithm relies on the subroutine DIAM-PATH(G) that, given a connected interval graph G, returns a shortest-distance path that dominates all vertices of G.

DIAM-PATH(G). Let $u_1 \in V(G)$ be the interval with smallest left boundary, and let $u_2 \in V(G)$ be the interval with largest right boundary. Let $V_p \subseteq V(G)$ be the subset of *proper intervals*, that is, the set of intervals $v \in V(G)$ that are not contained in another interval. In other words, $v \in V_p$ if, and only if, there is no $v' \in V(G)$ with $l(v') < l(v) < r(v) < r(v')$. Since all interval boundaries are distinct, both u_1 and u_2 are proper intervals and hence in V_p. DIAM-PATH(G) returns a shortest path from u_1 to u_2 in the graph $G[V_p]$. This "diameter path" has length at most $Diam(G)$.

Algorithm. Similar to our algorithm for sparse general graphs, we assume that the value of OPT is known. Furthermore, assume that the input graph G is connected. We will decompose G hierarchically as follows. Let $G_1 = G$ and let $P_1 = $ DIAM-PATH(G_1). Furthermore, for integers $i \geq 1$, let $U_i \subseteq V$ be the subset of intervals whose lengths are contained in $((\frac{1}{2})^i l_{max}, (\frac{1}{2})^{i-1} l_{max}]$. Then, we define the subgraph $H_1 = G[V(P_1) \cup U_1]$ consisting of intervals of the largest length class plus a diameter path, where $V(P_1)$ denotes the intervals contained in path P_1. As P_1 is a diameter path, $V(P_1)$ can be informed in $Diam(G)$ time. In Lemma 5, we will argue that the subgraph H_1 is 4-claw-free[4], and, using this property, we will show in Lemma 6 that U_1 can be informed in $O(OPT)$ rounds. Thus, overall in $O(OPT)$ rounds, the nodes $V(H_1)$ are informed.

Next, given the subgraph G_i, we define inductively $G_{i+1} \subseteq G_i$ to be the subgraph induced by the set of yet uninformed intervals, that is, $G_{i+1} = G[V(G_i) \setminus V(H_i)]$. Let P_{i+1} be a collection of diameter paths of the connected components of G_{i+1} as computed by DIAM-PATH, and let $H_{i+1} = G_{i+1}[V(P_{i+1}) \cup (U_{i+1} \cap V(G_{i+1}))]$ consisting of yet uninformed intervals of length class $i+1$ and a collection of diameter paths, where $V(P_{i+1})$ denotes the intervals contained in the diameter paths P_{i+1}. Similar as before, once $V(P_{i+1})$ has been informed, by Lemma 6, we can inform $V(H_{i+1})$ in $O(OPT)$ time. The key part of our argument is that $V(P_{i+1})$ can be informed by $V(P_i)$ in $O(OPT)$ time, which is proved in Lemma 7. Our argument shows that given an interval $v \in V(P_i)$, there are at most $O(OPT^2)$ intervals in $V(P_{i+1})$ that intersect with v, and we prove that they can be informed in $O(OPT)$ time. Thus, for every i, the nodes $V(H_i)$ can be informed in $O(OPT)$ rounds.

As $l_{max} \leq 2n$ and every interval is of length at least 1, there are $O(\log n)$ length classes. Hence, in $O(\log(n) \cdot OPT)$ rounds, all nodes $V(G)$ can be informed.

Analysis. We are going to prove the following theorem:

Theorem 6. *There is a polynomial-time algorithm for* RAS *in interval graphs with approximation factor* $O(\log n)$.

The theorem follows from the previous description of the algorithm together with the main Lemmas, Lemmas 6 and 7. In Lemma 6, we show that nodes $V(H_i)$ can be informed in $O(OPT)$ rounds if nodes $V(P_i)$ are informed, and in Lemma 7, we show that nodes $V(P_i)$ can be informed in $O(OPT)$ rounds if $V(P_{i-1})$ are informed.

---

[4] A graph is 4−claw-free, if it doesn't contain the complete bipartite graph $K_{1,4}$ as an induced subgraph.

We first state simple observations about the employed quantities in our algorithm.

Observation 7. *All intervals in subgraph G_i are of length at most $(\frac{1}{2})^{i-1} l_{max}$.*

Observation 8. *No interval in $V(H_i) \setminus V(P_i)$ contains an interval of P_i, that is, for every $v \in V(H_i) \setminus V(P_i)$ there is no $u \in V(P_i)$ such that $l(v) < l(u) < r(u) < r(v)$.*

Observation 8 follows by construction of P_i. The path P_i is constructed via algorithm DIAM-PATH which only chooses proper intervals.

Next, we show that the graphs H_i do not contain $K_{1,4}$ as an induced subgraph.

Lemma 5 (*). *For any i, the subgraph H_i is 4-claw-free.*

Last, we prove the main lemmas, Lemmas 6 and 7, that show that the subtasks of our algorithm can all be performed in O(OPT) rounds.

Lemma 6. *Suppose that the vertices of P_i have been informed. Then, $V(H_i)$ can be informed in O(OPT) rounds.*

Proof. We color the vertices of P_i alternately with four colors, where each color is used on every fourth vertex. Since P_i is a diameter path, nodes with the same color must have disjoint neighborhoods. Processing the colors in sequence, the nodes of each color inform their neighbors in H_i in parallel. Since H_i is 4-claw-free, each node has at most $3\omega(H_i) \leq 3\omega(G) \leq 6 \cdot OPT$ neighbors, and using Lemma 4, the lemma follows. □

Lemma 7. *Nodes D_{i+1} can be informed by nodes D_i in O(OPT) rounds.*

Proof. Let $\phi_{i+1} : D_{i+1} \to D_i$ be a mapping so that $\phi_{i+1}(v) = u \Rightarrow u \in N(v)$. Next, produce a 4-coloring of D_i with color classes $D_i^1, \ldots, D_i^4$, as in the proof of Lemma 6. Iterate now through the color classes D_i^j. In each iteration, all nodes $u \in D_i^j$ inform the nodes $\phi_{i+1}^{-1}(u)$ simultaneously as follows: Let $C_1 \ldots C_k$ denote the connected components of $G[\phi_{i+1}^{-1}(u)]$. Node u informs every OPT-th interval of every connected component C_j. If $|C_j| < OPT$ then an arbitrary interval of C_j is informed. Thus, u requires O($k + |\phi_{i+1}^{-1}(u)|/OPT$) rounds. In Claim 9, we will prove that $k = $ O(OPT) and $|\phi_{i+1}^{-1}(u)| = $ O(OPT^2).

Claim 9 (*). *$|\phi_{i+1}^{-1}(u)| = $ O(OPT^2) and the number of components of $G[\phi^{-1}(u)]$ is O(OPT).*

Thus, the previous step requires O(OPT) rounds. Next, the informed nodes of $\phi_{i+1}^{-1}(u)$ inform the uninformed nodes of $\phi_{i+1}^{-1}(u)$. Since $\phi_{i+1}^{-1}(u)$ is a collection of paths, and since for every uninformed node of $\phi_{i+1}^{-1}(u)$ there is an informed node within distance OPT, this step clearly can be done in O(OPT) rounds. This completes the proof. □

References

1. Aguayo, D., Bicket, J., Biswas, S., Judd, G., Morris, R.: Link-level measurements from an 802.11 b mesh network. ACM SIGCOMM Comput. Commun. Rev. **34**(4), 121–132 (2004)
2. Alon, N., Bar-Noy, A., Linial, N., Peleg, D.: A lower bound for radio broadcast. J. Comput. Syst. Sci. **43**, 290–298 (1991)
3. An, M.K., Lam, N.X., Huynh, D.T., Nguyen, T.N.: Minimum data aggregation schedule in wireless sensor networks. I. J. Comput. Appl. **18**(4), 254–262 (2011)
4. Bar-Noy, A., Guha, S., Naor, J., Schieber, B.: Message multicasting in heterogeneous networks. SIAM J. Comput. **30**(2), 347–358 (2000)
5. Bar-Yehuda, R., Goldreich, O., Itai, A.: On the time-complexity of broadcast in multi-hop radio networks: an exponential gap between determinism and randomization. J. Comput. Syst. Sci. **45**(1), 104–126 (1992)
6. Chalermsook, P., Laekhanukit, B., Nanongkai, D.: Graph products revisited: tight approximation hardness of induced matching, poset dimension and more. In: SODA, pp. 1557–1576. SIAM (2013)
7. Chen, X., Hu, X., Zhu, J.: Minimum data aggregation time problem in wireless sensor networks. In: Jia, X., Wu, J., He, Y. (eds.) MSN 2005. LNCS, vol. 3794, pp. 133–142. Springer, Heidelberg (2005)
8. Chlamtac, I., Weinstein, O.: Distributed "wave" broadcasting in mobil multi-hop radio networks. In: ICDCS, pp. 82–89 (1987)
9. Elkin, M., Kortsarz, G.: A combinatorial logarithmic approximation algorithm for the directed telephone broadcast problem. SIAM J. Comput. **35**(3), 672–689 (2005)
10. Elkin, M., Kortsarz, G.: Polylogarithmic additive inapproximability of the radio broadcast problem. SIAM J. Discrete Math. **19**(4), 881–899 (2005)
11. Elkin, M., Kortsarz, G.: Sublogarithmic approximation for telephone multicast. J. Comput. Syst. Sci. **72**(4), 648–659 (2006)
12. Fasolo, E., Rossi, M., Widmer, J., Zorzi, M.: In-network aggregation techniques for wireless sensor networks: a survey. IEEE Wirel. Commun. **14**(2), 70–87 (2007)
13. Feige, U., Kilian, J.: Zero knowledge and the chromatic number. J. Comput. Syst. Sci. **57**(2), 187–199 (1998)
14. Gagnon, J., Narayanan, L.: Minimum latency aggregation scheduling in wireless sensor networks. In: Gao, J., Efrat, A., Fekete, S.P., Zhang, Y. (eds.) ALGO-SENSORS 2014, LNCS 8847. LNCS, vol. 8847, pp. 152–168. Springer, Heidelberg (2015)
15. Guo, L., Li, Y., Cai, Z.: Minimum-latency aggregation scheduling in wireless sensor network. J. Comb. Optim., 1–32 (2014)
16. Halldórsson, M.M., Mitra, P.: Wireless connectivity and capacity. In: SODA (2012)
17. Incel, O.D., Ghosh, A., Krishnamachari, B.: Scheduling algorithms for tree-based data collection in wireless sensor networks. In: Nikoletseas, S., Rolim, J.D.P. (eds.) Theoretical Aspects of Distributed Computing in Sensor Networks, pp. 407–445. Springer, Heidelberg (2011)
18. Kesselman, A., Kowalski, D.: Fast distributed algorithm for convergecast in ad hoc geometric radio networks. In: WONS, pp. 119–124. IEEE (2005)
19. Kleinberg, J., Tardos, É.: Algorithm Design. Pearson Education, Boston (2006)
20. Kortsarz, G., Peleg, D.: Approximation algorithms for minimum-time broadcast. SIAM J. Discrete Math. **8**(3), 401–427 (1995)
21. Kotz, D., Newport, C., Gray, R.S., Liu, J., Yuan, Y., Elliott, C.: Experimental evaluation of wireless simulation assumptions. In: MSWiM, pp. 78–82 (2004)

22. Kowalski, D.R., Pelc, A.: Optimal deterministic broadcasting in known topology radio networks. Distrib. Comput. **19**(3), 185–195 (2007)
23. Moscibroda, T., Wattenhofer, R.: The complexity of connectivity in wireless networks. In: INFOCOM (2006)
24. Ravi, R.: Rapid rumor ramification: approximating the minimum broadcast time (extended abstract). In: FOCS, pp. 202–213 (1994)
25. Wan, P.J., Huang, S.C.H., Wang, L., Wan, Z., Jia, X.: Minimum-latency aggregation scheduling in multihop wireless networks. In: MOBIHOC, pp. 185–194. ACM (2009)
26. Xu, X., Wang, S., Mao, X., Tang, S., Li, X.Y.: An improved approximation algorithm for data aggregation in multi-hop wireless sensor networks. In: FOWANC, pp. 47–56. ACM (2009)

Gathering of Robots on Meeting-Points

Serafino Cicerone[1], Gabriele Di Stefano[1], and Alfredo Navarra[2]([✉])

[1] Dipartimento di Ingegneria e Scienze dell'Informazione e Matematica,
Università degli Studi dell'Aquila, Via Vetoio, Coppito, 67100 L'Aquila, Italy
{serafino.cicerone, gabriele.distefano}@univaq.it
[2] Dipartimento di Matematica e Informatica, Università degli Studi di Perugia,
Via Vanvitelli 1, 06123 Perugia, Italy
alfredo.navarra@unipg.it

Abstract. We consider the gathering problem of oblivious and asynchronous robots moving in the plane. When $n > 2$ robots are free to gather anywhere in the plane, the problem has been solved in [Cieliebak et al., *SIAM J. on Comput., 41(4), 2012*]. We propose a new natural and challenging model that requires robots to gather only at some predetermined points in the plane, herein referred to as *meeting-points*.

Robots operate in standard Look-Compute-Move cycles. In one cycle, a robot perceives the robots' positions and the meeting-points (Look) according to its own coordinate system, decides whether to move toward some direction (Compute), and in the positive case it moves (Move). Cycles are performed asynchronously for each robot. Robots are anonymous and execute the same distributed and deterministic algorithm.

In the new proposed model, we fully characterize when gathering can be accomplished. We design an algorithm that solves the problem for all configurations with $n > 0$ robots but those identified as ungatherable.

1 Introduction

The gathering task is a basic primitive in robot-based computing systems. It has been extensively studied in the literature under different assumptions. The problem asks to design a distributed algorithm that allows a team of robots to meet at some common place. Varying on the capabilities of the robots as well as on the environment where they move, very different and challenging aspects must be faced (see, e.g. [2,7,9–11,14,15], and references therein).

In this paper we consider a very minimal setting. We are interested in robots placed in $\mathbb{R}^2$ where they can freely move but they must meet at some predetermined points, herein called *meeting-points*. We call this new problem the *Gathering on Meeting-Points* problem, shortly GMP.

The work has been partially supported by the Italian Ministry of Education, University, and Research (MIUR) under national research projects: PRIN 2010N5K7EB "ARS TechnoMedia – Algoritmica per le Reti Sociali Tecno-Mediate" and PRIN 2012C4E3KT "AMANDA – Algorithmics for MAssive and Networked DAta", and by the National Group for Scientific Computation (GNCS-INdAM).

P. Bose et al. (Eds.): ALGOSENSORS 2015, LNCS 9536, pp. 183–195, 2015.
DOI: 10.1007/978-3-319-28472-9_14

Initially, no robots occupy the same location, and they are assumed to be: *Dimensionless*: modeled as geometric points in the plane; *Anonymous*: no unique identifiers; *Autonomous*: no centralized control; *Oblivious*: no memory of past events; *Homogeneous*: they all execute the same deterministic algorithm; *Asynchronous*: there is no global clock that synchronize their actions; *Silent*: no direct way of communicating; *Unoriented*: no common coordinate system, no compass, no chirality. Robots are equipped with sensors and motion actuators, and operate in *Look-Compute-Move* cycles (see, e.g. [11]). In one cycle a robot takes a snapshot of the current global configuration (Look) in terms of relative robots and meeting-points positions, according to its own coordinate system. Successively, in the Compute phase it decides whether to move toward a specific direction or not, and in the positive case it moves (Move).

During the Look phase, robots are assumed to perceive *multiplicities*, that is, whether a same point is occupied by one or more robots, but not the exact number. In the literature, this is called *global-weak multiplicity detection* [7, 11, 12]. Herein we simply call it *multiplicity detection*. Note that robots always detect whether a meeting-point and one or more robots occupy the same location.

Cycles are performed asynchronously, i.e., the time between Look, Compute, and Move phases is finite but unbounded, and it is decided by an adversary for each robot. Moreover, during the Look phase, a robot does not perceive whether other robots are moving or not. Hence, robots may move based on outdated perceptions. In fact, due to asynchrony, by the time a robot takes a snapshot of the configuration, this might have drastically changed when it starts moving. The scheduler determining the Look-Compute-Move cycles timing is assumed to be fair, that is, each robot performs its cycle within finite time and infinitely often. In the literature, this kind of scheduler is called Asynchronous (ASYNCH). Different options for the scheduler are: Fully-synchronous (FSYNCH), where all robots are awake and run their Look-Compute-Move cycle concurrently and each phase of the cycle has exactly the same duration for all robots; Semi-synchronous (SSYNCH), that coincides with the FSYNCH model with the only difference that not all robots are necessarily activated during a cycle.

The distance traveled within a move is neither infinite nor infinitesimally small. More precisely, the adversary has also the power to stop a moving robot before it reaches its destination, but there exists an unknown constant $\delta > 0$ such that if the destination point is closer than δ, the robot will reach it, otherwise the robot will be closer to it of at least δ. Note that, without this assumption, an adversary would make it impossible for any robot to ever reach its destination.

Considering the model without meeting-points, the problem has been solved in [5] for any number of robots $n > 2$. Adding meeting-points can sometimes help in designing a gathering algorithm while sometimes can play for the adversary. In fact, meeting-points are like anchors in the plane that never move, and hence if they are "favorably" placed, they may suggest the final gathering point. Contrary, when the placement of the meeting-points induces nasty symmetries, then they can be completely useless in terms of orientation, and it might be a real trouble for the robots to agree on a common meeting-point where to gather.

The rationale behind the choice of introducing meeting-points is twofold. From the one hand, we believe the model is theoretically interesting, as it is a hybrid scenario in between the classical environment where robots freely move in the plane (see, e.g., [1,5]), and the more structured one where robots must move on the vertices of a graphs (see, e.g., [8,13]), implemented here by the set of meeting-points. On the other hand, meeting-points for gathering purposes might be a practical choice when robots move in specific environments where not all places can be candidate to serve as gathering points.

Optimization issues have been addressed in [3,4]. The same strategies cannot be applied here since a wider set of configurations must be now considered.

Our Results. The first contribution is that of introducing the so called meeting-points for the well-know gathering problem under the Look-Compute-Move model. Although the new formulation of the gathering problem seems to be rather close to the original one [5], it turns out to require challenging strategies. In fact, there exist ungatherable configurations, characterized by some symmetries, regardless the number of robots, and the most of configurations have been approached with new stigmergy methodologies since the previous techniques cannot be applied, even those proposed in [3,4].

We fully characterize when GMP can be solved. We exploit the ungatherability results from [3], holding also in the stronger FSYNCH setting. Then, for all other configurations, we design a distributed algorithm that solves the problem for any number $n > 0$ robots. The new algorithm works in the weakest ASYNCH setting.

2   Definitions and General Ungatherability Results

In this section we formally define the GMP problem, and then we recall from [3]: the view of a configuration, relations between symmetries and the view, ungatherability results, and the concept of Weber-points of a configuration.

Problem Definition. The system is composed of n mobile *robots*. At any time, the multiset $R = \{r_1, r_2, \ldots, r_n\}$, with $r_i \in \mathbb{R}^2$, contains the *positions* of all the robots. The set $U(R) = \{x \mid x \in R\}$ contains the *unique* robots' positions. M is a finite set of fixed *meeting-points* in the plane representing the only locations in which robots can be gathered. The pair $\mathcal{C} = (R, M)$ represents a system *configuration*. A configuration $\mathcal{C}$ is *initial* at time t if at that time all robots have distinct positions (i.e., $|U(R)| = n$). A configuration $\mathcal{C}$ is *final* at time t if (*i*) at that time each robot computes or performs a null movement and (*ii*) there exists a point $m \in M$ such that $r_i = m$ for each $r_i \in R$; in this case we say that the robots have gathered on point m at time t.

We study the GATHERING ON MEETING-POINTS problem (shortly, GMP), that asks to transform an initial configuration into a final one. A *gathering algorithm* for GMP is a deterministic distributed algorithm that brings the robots in the system to a final configuration in a finite number of cycles from any given initial configuration, regardless the adversary. We say that an initial configuration $\mathcal{C}$ is *ungatherable* if there are no gathering algorithms for GMP with respect to $\mathcal{C}$.

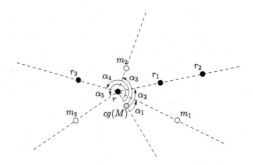

Fig. 1. The counter-clockwise order in which a robot perceives the configuration from r is $(r, m_1, r_1, r_2, m_2, r_3, m_3)$ and $\mathcal{V}^-(r) = (0°, d(r, cg(M)), \mathbf{r}, \alpha_1, d(r, m_1), \mathbf{m}, \alpha_2, d(r, r_1), \mathbf{r}, \alpha_2, d(r, r_2), \mathbf{r}, \alpha_3, d(r, m_2), \mathbf{m}, \alpha_4, (r, r_3), \mathbf{r}, \alpha_5, d(r, m_3), \mathbf{m})$.

Configuration View and Symmetries. Given two distinct points u and v in the plane, let $d(u, v)$ denote their distance, $line(u, v)$ denote the straight line passing through these points, and (u, v) (resp. $[u, v]$) denote the open (resp. closed) segment containing all points in this line that lie between u and v. The half-line starting at point u (but excluding the point u) and passing through v is denoted by $hline(u, v)$. Given two lines $line(c, u)$ and $line(c, v)$, we denote by $\sphericalangle(u, c, v)$ the angle (ranging from zero to less than 360°) centered in c and with sides $hline(c, u)$ and $hline(c, v)$.

Given a configuration $\mathcal{C} = (R, M)$, $cg(M)$ is the *center of gravity* of points in M, that is the point whose coordinates are the mean values of the coordinates of the points of the set. In [3] it has been defined a data structure called *view* and computable by each robot r (according to its local coordinate system) for any point $p \in R \cup M$. Essentially, a robot r that needs to evaluate the view of a point p, first computes $cg(M)$ and then, starting from the direction given by $hline(p, cg(M))$ and looking around from p (in clockwise and counter-clockwise manner), it determines the order $(p = p_0, p_1, \ldots, p_{|U(R)|+|M|})$, $p_i \in R \cup M$, in which all robots and meeting-points appear. From such order of points, the configuration's view is produced by replacing each point p_i with a triple α_i, d_i, x_i formed by, in order and for $i > 0$, $\alpha_i = \sphericalangle(p_0, p, p_i)$, $d_i = d(p, p_i)$, and $x_i \in \{\mathbf{r}, \mathbf{m}, \mathbf{x}\}$ according whether p_i is a robot position, a meeting-point, or a position where a multiplicity occurs (cf. Fig. 1). The triple associated to p_0 represents the point p where d_0 is equal to $d(p, cg(M))$. Finally, by considering $\mathbf{r} < \mathbf{m} < \mathbf{x}$, it is possible to order the two strings $\mathcal{V}^-(p)$ and $\mathcal{V}^+(p)$ associated to the view of each point p according to clockwise or counter-clockwise look. So, the view of p is $\mathcal{V}(p) = \min\{\mathcal{V}^+(p), \mathcal{V}^-(p)\}$, and then $\mathcal{V}(\mathcal{C}) = \bigcup_{p \in U(R) \cup M}\{\mathcal{V}(p)\}$. Notice that, even if robots do not have a common understanding of the handedness (chirality), by computing $\mathcal{V}(\mathcal{C})$ they all get the same information.

Robots can use $\mathcal{V}(\mathcal{C})$ not only to share a common view about $\mathcal{C}$ but also to determine whether a configuration is "symmetric" or not. Let $\varphi : \mathbb{R}^2 \to \mathbb{R}^2$ a map from points to points in the plane. It is called an *isometry* or distance preserving if for any $a, b \in \mathbb{R}^2$ one has $d(\varphi(a), \varphi(b)) = d(a, b)$. Examples of isometries in the plane are *translations, rotations* and *reflections*. An isometry φ is a translation if there exists no point x such that $\varphi(x) = x$; it is a rotation if there exists a unique point x such that $\varphi(x) = x$ (and x is called *center of rotation*); it is a reflection if there exists a line ℓ such that $\varphi(x) = x$ for each point $x \in \ell$ (and ℓ is called *axis of reflection*).

An *isometry of an initial configuration* $\mathcal{C} = (R, M)$ is an isometry in the plane that maps robots to robots (i.e., points of R into R) and meeting-points to meeting-points (i.e., points of M into M). Isometries for $\mathcal{C}$ do not include translations as the sets R and M are finite.

If $\mathcal{C}$ admits only the identity isometry, then $\mathcal{C}$ is said *asymmetric*, otherwise it is said *symmetric* (i.e., $\mathcal{C}$ admits rotations or reflections). If $\mathcal{C}$ is symmetric due to an isometry φ, a robot cannot distinguish its position at $r \in R$ from $r' = \varphi(r)$. As a consequence, two robots (e.g., one on r and one on $\varphi(r)$) can decide to move simultaneously, as any algorithm is unable to distinguish between them. In such a case, there might be a so called *pending move*, that is, wlog r performs its entire Look-Compute-Move cycle while r' does not terminate the Move phase, i.e. its move is pending. Clearly, all the other robots performing their cycles are not aware whether there is a pending move, that is they cannot deduce the global status from their view. This fact greatly increases the difficulty to devise a gathering algorithm for symmetric configurations.

The following results states that each robot can use the view $\mathcal{V}(\mathcal{C})$ to determine whether $\mathcal{C}$ is symmetric or not.

Lemma 1 [3]. *An initial configuration* $\mathcal{C} = (R, M)$, $|M| > 1$, *admits a reflection (rotation, resp.) if and only if there exist* $p, q \in R \cup M$, *such that* $\mathcal{V}^+(p) = \mathcal{V}^-(q)$ *with* p *and* q *not necessarily distinct* $(\mathcal{V}^+(p) = \mathcal{V}^+(q))$, *with* $p \neq q$, *resp.).*

From this results we get that, for an asymmetric configuration $\mathcal{C}$, it is unique the point (robot or meeting-point) having the minimum view.

Lemma 2. *Let* $\mathcal{C} = (R, M)$ *be a non-rotational initial configuration, and* ℓ *be a line passing through* $cg(M)$. *If the line perpendicular to* ℓ *at* $cg(M)$ *is not a reflection axis for* $\mathcal{C}$, *then* ℓ *admits a North-South orientation.*

The above lemma implies that, under certain conditions, all robots can agree about the North of a line ℓ passing through $cg(M)$, and in case about the "northernmost" robot or meeting-point on ℓ.

Ungatherability Results. In this section we recall a sufficient condition for a configuration to be ungatherable: if this applies then GMP is not solvable. Note that the results hold also for the case of the synchronous environments FSYNCH.

Corollary 1 [3]. *An initial configuration* $\mathcal{C} = (R, M)$ *is ungatherable even in* FSYNCH *if it admits a rotation with center* c *and* $c \notin R \cup M$ *or it admits a reflection with axis* ℓ *and* $\ell \cap (R \cup M) = \emptyset$.

So, if a configuration admits a reflection (rotation, resp.) then the gathering is possible only if on the axis (center, resp.) there are robots or meeting-points.

Weber-Points. Let $\mathcal{C} \stackrel{\text{\tiny def}}{=} (R, M)$ be an initial configuration. We define the *Weber-distance* of $\mathcal{C}$ as the value $\Delta(\mathcal{C}) = \min_{m \in M} \sum_{r \in R} d(r, m)$. The name of *Weber-distance* is due to the following remark: given a set of points $T \subseteq \mathbb{R}^2$, the *Weber-point* of T is a well known concept and corresponds to a point p such that $p = \operatorname{argmin}_{p' \in \mathbb{R}^2} \sum_{t \in T} d(t, p')$. It is well known that (i) if the points in T are not on a line, then the Weber-point of T is unique [16], and (ii) the Weber-point of T is not computable in general [6]. The *Weber-distance* of a point $m \in M$ in $\mathcal{C}$ is denoted as $wd(\mathcal{C}, m)$ and is defined as $wd(\mathcal{C}, m) = \sum_{r \in R} d(r, m)$. Hence, a point $m \in M$ is called *Weber-point* of $\mathcal{C}$ if $wd(\mathcal{C}, m)$ is minimum, that is $wd(\mathcal{C}, m) = \Delta(\mathcal{C})$. Symbol $wp(\mathcal{C})$ is used to denote the set containing all the Weber-points of $\mathcal{C}$ (notice that the $wp(\mathcal{C})$ may contain more that one point and that such points can be easily computed since M is finite).

We recall now a characterization about the set of Weber-points after the move of a robot toward a Weber-point. From now on, we use the sentence "robot r moves toward a meeting-point m" to mean that r *performs a straight move toward m and the final position of r lies on the interval $(r, m]$*.

Lemma 3 [3]. *Let $\mathcal{C} = (R, M)$ be a configuration with $m \in wp(\mathcal{C})$ and $r \in R$. If $\mathcal{C}' = (R', M)$ is the configuration obtained after r moved toward m, then all the Weber-points in $wp(\mathcal{C}')$ lie on $hline(r, m)$ and $m \in wp(\mathcal{C}') \subseteq wp(\mathcal{C})$.*

3   Gathering for GMP

In this section we provide a solution for the GMP problem. We start by providing a partition of the set $\mathcal{I}$ containing all the possible initial configurations for GMP. According to Corollary 1 there are configurations in $\mathcal{I}$ that are ungatherable. The class of such configurations is denoted by $\mathcal{U}$ and contains any initial configuration $\mathcal{C}$ fulfilling one of the following conditions:

- $\mathcal{C}$ admits a rotation with center c and $c \notin R \cup M$;
- $\mathcal{C}$ admits a reflection with axis ℓ and $\ell \cap (R \cup M) = \emptyset$.

In the remaining of this section we provide a gathering algorithm for the GMP problem in the most general ASYNCH setting when the input configuration (R, M) is restricted to $\mathcal{I} \setminus \mathcal{U}$. Moreover we assume $|R| > 1$ and $|M| > 1$, as otherwise the solution is straightforward: in fact, if $|R| = 1$ it is sufficient that the only robot reaches a meeting-point and if $|M| = 1$ all the robots can move toward the only meeting-point.

Before starting the description of the algorithm we introduce some additional concepts and notation. Given a configuration $\mathcal{C}$, let $O_1, O_2, \ldots, O_t$, $t \geq 1$, be all the circles centered in $cg(M)$ such that $O_i \cap R \neq \emptyset$ for each $1 \leq i \leq t$. Moreover, ρ_i represents the radius of O_i, and we assume $\rho_i < \rho_j$ if $i < j$. If a robot is on $cg(M)$, then $\rho_1 = 0$. Let O_M be the smallest circle that is centered in $cg(M)$ and contains all points in M, and let ρ_M be its radius.

All the initial configurations processed by the algorithm, along with those with one multiplicity created during the execution, are partitioned as follows:

Procedure: COMPUTE
Input: Configuration $\mathcal{C} = (R, M)$

1 Let $c = cg(M)$, $d = \max\{\rho_{t-1}, \rho_M\}$, $d' = \max\{\rho_t, \rho_M\}$;
2 if $\mathcal{C} \in \mathcal{S}_1$ then move toward the meeting-point with unique multiplicity;
3 if $\mathcal{C} \in \mathcal{S}_2$ then move toward $cg(M)$;
4 if $\mathcal{C} \in \mathcal{S}_3$ then
5   if r *on* $cg(M)$ then move at distance $\rho_2/2$ from $cg(M)$ in any direction;
6 if $\mathcal{C} \in \mathcal{S}_4$ then
7   if r *on* O_1 then move at distance $\rho_2/2$ from $cg(M)$ on $hline(cg(M), r)$;
8 if $\mathcal{C} \in \mathcal{S}_5$ then
9   NUMGUARDS $= 0$;
10   Call GUARDS()   /* GUARDS modifies NUMGUARDS */ ;
11   if NUMGUARDS $\neq 0$ then
12   Call MAKEMULTIPLICITY();
13 if $\mathcal{C} \in \mathcal{S}_6$ then Call ATMOST3BOTS();

$\mathcal{S}_0$: any final configuration $\mathcal{C}$ where all the robots form one multiplicity on some meeting-point;

$\mathcal{S}_1$: any configuration $\mathcal{C} \notin \mathcal{S}_0$ with only one multiplicity on some meeting-point;

$\mathcal{S}_2$: any $\mathcal{C} = (R, M) \notin \bigcup_{i=0}^{1} \mathcal{S}_i$, with $cg(M) \in M$;

$\mathcal{S}_3$: any $\mathcal{C} \notin \bigcup_{i=0}^{2} \mathcal{S}_i$ admitting a rotation;

$\mathcal{S}_4$: any $\mathcal{C} = (R, M) \notin \bigcup_{i=0}^{2} \mathcal{S}_i$ with one robot r on O_1 such that $0 < \rho_1 < \rho_2/2$ and $(R \setminus \{r\}, M) \in \mathcal{S}_3$.

$\mathcal{S}_5$: any $\mathcal{C} \notin \bigcup_{i=0}^{4} \mathcal{S}_i$, with more than 3 robots.

$\mathcal{S}_6$: any $\mathcal{C} \notin \bigcup_{i=0}^{4} \mathcal{S}_i$, with at most 3 robots.

It easily follows that $\{\mathcal{U}, \mathcal{S}_2, \mathcal{S}_3, \ldots, \mathcal{S}_6\}$ is a partition of $\mathcal{I}$. Note that configurations in classes $\mathcal{S}_0$ and $\mathcal{S}_1$ are the only non-initial ones handled by the algorithm.

The general algorithm, executed by each robot, is represented by Procedure COMPUTE. It is divided into six parts, according to the subdivision of the non-final configurations. The procedure and the sub-procedures represent what a generic robot r executes during the Compute phase once it has detected the class which the perceived configuration belongs to.

The general strategy of the algorithm is to transform each initial configuration $\mathcal{C} \in \bigcup_{i=2}^{6} \mathcal{S}_i$ into a configuration $\mathcal{C}' \in \mathcal{S}_1$ by moving robots to create a multiplicity on some meeting-point m. Once this occurs, all robots can always detect m and the gathering can be easily finalized. This approach is easy to obtain when there is a meeting-point m on $cg(M)$ (i.e., configurations in $\mathcal{S}_2$) since m is always recognizable by all robots. Whereas, it is not applied in rotational (or quasi-rotational) configurations with a robot on $cg(M)$ (i.e., configurations in $\mathcal{S}_3$ or $\mathcal{S}_4$) that are transformed in configurations in $\mathcal{S}_5$ or $\mathcal{S}_6$.

Then, the core of the algorithm is given for cases $\mathcal{S}_5$ and $\mathcal{S}_6$ where some sub-procedures later defined are invoked. For handling configurations in such classes, the strategy of our algorithm is composed of four phases:

1. select one or two robots, denoted as guard(s);
2. place suitably the guard(s), if required;

3. crate a multiplicity by means of robots not designated as guard(s);
4. finalize the gathering on the created multiplicity.

The selection of the guard(s) is done among the robots furthest from $cg(M)$. Due to the limited number of symmetries that a configuration can admit, it is always possible to select one or two robots that are moved away from $cg(M)$ so that they are always recognizable. The algorithm selects two guards, which are equivalent, only for reflexive configurations where a single guard cannot be pointed out. As final positioning for the guards, we chose a sufficiently large distance from $cg(M)$ that depends on the radius of the current O_{t-1}. Once guards are correctly placed, all other robots cooperate in order to create a multiplicity on a meeting-point. In practice, stigmergy paradigms are applied, that is guards are used in place of a compass so that all robots get oriented by observing their positions. From this orientation they deduce what will be the final gathering point and hence move there. Such a point is a meeting-point m that maintains its peculiarity while robots move toward it. For instance, if we chose among the meeting-points the northernmost of minimum Weber distance, then as soon as robots start moving toward it, its Weber distance decreases. Eventually, the meeting-point m will remain the only one of minimum Weber distance, according to Lemma 3.

According to the assumed multiplicity detection, once a multiplicity is created, robots are no longer able to compute the Weber distance accurately. Hence, our strategy assures to create the first multiplicity over m, and once this happens all robots move toward it without creating other multiplicities. Moves are always computed without creating undesired multiplicities. Clearly, the movement of the guards (during the above phase 2) cannot create any multiplicity since they are moved from O_t to the outer space. Afterward, since robots move straightly toward m, then two robots meet only at the final destination point, unless they move along the same direction. In such a case, we make robots move without overtaking each other. In particular if a robot r is moving toward a point p and there is another robot r' in the open segment (r, p), then r moves toward a point p' on (r, p) such that $d(r, p') = \frac{d(r,r')}{2}$. In this way, undesired multiplicities are never created. Once a multiplicity is created on m, it is then easy to move all other robots (including the guards) toward it, by exploiting the multiplicity detection. Hence the gathering is easily finalized.

The next theorem provides the correctness of our algorithm.

Theorem 1. *There exists a gathering algorithm that solves the* GMP *problem for an initial configuration* C *if and only if* $C \in \mathcal{I} \setminus \mathcal{U}$.

3.1   Classes $\mathcal{S}_1, \mathcal{S}_2, \mathcal{S}_3,$ and $\mathcal{S}_4$

In this section we describe how configurations in the first four classes are handled by our algorithm. Concerning classes $\mathcal{S}_1$ and $\mathcal{S}_2$, robots can move concurrently toward the unique multiplicity, or $cg(M)$, respectively.

Lemma 4. *Given a configuration* C *in class* $\mathcal{S}_1$ *or* $\mathcal{S}_2$, *Procedure* COMPUTE *leads to a configuration* C' *in class* $\mathcal{S}_0$, *eventually.*

Consider now the case where the initial configuration $\mathcal{C}$ admits a rotation $(\mathcal{C} \in \mathcal{S}_3)$ or it "almost" admits a rotation $(\mathcal{C} \in \mathcal{S}_4)$. Since in case of rotations it is generally impossible to identify specific robots suitable for the role of guards, when $\mathcal{C}$ is in $\mathcal{S}_3$ the algorithm first breaks this symmetry by moving the robot in the center of the rotation c. Notice that a robot must be on c as otherwise either $\mathcal{C} \in \mathcal{S}_2$ or $\mathcal{C} \in \mathcal{U}$. Class $\mathcal{S}_4$ has been introduced to assure that the robot moving from c reaches a target (a distance $\rho_2/2$ from c) and stops there before the positioning of the guard(s) starts.

Lemma 5. *Given an initial configuration $\mathcal{C} = (R, M)$ in class $\mathcal{S}_3$ or $\mathcal{S}_4$, Procedure* COMPUTE *leads to a configuration $\mathcal{C}'$ in class $\mathcal{S}_5$ or $\mathcal{S}_6$, eventually.*

3.2   Class $\mathcal{S}_5$

In this section, configurations of class $\mathcal{S}_5$ are addressed. As described at the beginning of the section, our algorithm makes use of a stigmergy paradigm, that is some robots (namely, the "guards") are used in place of a compass so that all robots get oriented by observing their positions. We now formally define the guards of a configuration.

Definition 1. *Let $\mathcal{C} = (R, M)$ be an initial configuration with $\rho_t \geq d = 3 \cdot \max\{\rho_{t-1}, \rho_M\}$. If $O_t \cap R = \{r\}$ then r is a guard. Assume $O_t \cap R = \{r_1, r_2\}$ and r_1, r_2 are symmetric with respect to an axis ℓ of M. If $\mathcal{C}$ is reflexive according to ℓ or there exists a meeting-point which is unique according to some property, then r_1 and r_2 are two guards.*

According to the notion of guards, in order to finalize the gathering, the configuration evolves in four different stages: the first two stages are devoted to (1) select one or two robots that are guards or that can be moved until they become guards, and (2) suitably move the guards, if necessary. These first two stages are realized by means of Procedure GUARDS. The third stage concerns (3) making a multiplicity by means of robots that are not guards. Procedure MAKEMULTIPLICITY realizes it. The last stage requires (4) finalizing the gathering on the created multiplicity according to Lemma 4.

We now shortly introduce Procedure GUARDS applied by robots in order to detect whether the guards exist or if they have to be created. Procedure GUARDS is invoked when it is recognized that the initial configuration $\mathcal{C}$ taken as input belongs to $\mathcal{S}_5$. Procedure GUARDS checks how many robots reside on O_t and then calls the corresponding subroutine. Note that, according to Definition 1, guards are at distance at least $3d$ from the center $cg(M)$, while all the remaining robots and all the meeting-points are at distance at most d from $cg(M)$. In this way, once guards are recognized, the other robots start moving toward a specific meeting-point and they will never exceed distance d from $cg(M)$.

Conversely, when there are no guards, the configuration must be transformed so that a new configuration with one or two guards is created. During the transformation, one or two robots must be selected and moved far from $cg(M)$ until

---

Procedure: GUARDS
Input: Configuration $\mathcal{C} = (R, M)$ with circles O_i, $i = 1, 2, \ldots, t$

1 **if** $O_t \cap R = \{r_1\}$ **then** Call CHECKONE($\mathcal{C}, r_1$);
2 **if** $O_t \cap R = \{r_1, r_2\}$ **then** Call CHECKTWO($\mathcal{C}, r_1, r_2$);
3 **if** $|O_t \cap R| > 2$ **then** Call CHECKMANY($\mathcal{C}$);

Procedure: CHECKONE
Input: Configuration $\mathcal{C} = (R, M)$ and robot r_1 on O_t

1 **if** $\rho_t \geq 3d$ **then** NUMGUARDS = 1; **exit**;
2 Let $x = hline(c, r_1) \cap O_{t-1}$;
3 Let $\mathcal{C}'$ be the configuration obtainable from $\mathcal{C}$ by assuming r_1 on x;
4 **if** $\rho_t - \rho_{t-1} \leq d$ and $\mathcal{C}'$ is reflexive with axis $\ell \neq line(c, x)$ and $\mathcal{C}' \in \mathcal{S}_5$ **then**
5 | Let r_2 be the robot symmetric to r_1 in $\mathcal{C}'$ with respect to ℓ;
6 | **if** $r = r_2$ **then**
7 | | move r on $hline(c, r)$ at distance ρ_t from c;   /* Possible pending move */
8 **else**
9 | **if** $r = r_1$ **then** move r on $hline(c, x)$ at distance $3d$ from c;

---

reaching a position compatible with the definition of guards. In particular, if two guards must be created, then the designed robots must move at distance $3d$ from $cg(M)$ in two steps: first they are moved both at distance $2d$, and afterward $3d$. This double step is done so as the difference between the distances of the two guards (that might move asynchronously) from $cg(M)$ is always kept below d in order to distinguish the case where only one guard must be created. During the phase in which robots are moved to create guards, d is initially defined as the maximum between ρ_{t-1} and ρ_M. Sometimes we make use of $d' = \max\{\rho_t, \rho_M\}$ instead of d; this happens when the guard(s) have not started moving yet, and O_t is occupied also by robots that will not become guards.

Once GUARDS completed its task (and, due to the adversary, it may require several but finite many executions made by the guard robots), the variable NUMGUARDS is set to one or two and hence MAKEMULTIPLICITY starts. Of course, also this procedure may require several executions to complete its task. Then, according to Lines 8–12 of Procedure COMPUTE, Procedure GUARDS is called again each time MAKEMULTIPLICITY restarts. In such executions, GUARDS has only to recognize that guard(s) are settled.

So, the next lemmata can be stated.

Lemma 6. *Given a configuration $\mathcal{C} = (R, M)$ in class $\mathcal{S}_5$, Procedure GUARDS eventually leads to a configuration $\mathcal{C}' \in \mathcal{S}_5$ with 1 or 2 guards. Moreover,*

(i) *if $\mathcal{C}'$ has one guard, then either $\mathcal{C}'$ is asymmetric or it admits a reflection axis with the guard on the axis;*
(ii) *if $\mathcal{C}'$ has two guards, then either $\mathcal{C}'$ is reflexive and the two guards are symmetric, or $\mathcal{C}'$ is asymmetric, ℓ is a reflection axis for M, and the two guards are symmetric with respect to ℓ.*

Lemma 7. *If Procedure GUARDS returns a configuration $\mathcal{C}$ with 1 or 2 guards, then Procedure MAKEMULTIPLICITY leads to a configuration $\mathcal{C}' \in \mathcal{S}_1$, eventually.*

Procedure: CHECKTWO
Input: Configuration $\mathcal{C} = (R, M)$ and robots r_1 and r_2 on O_t

1 Let $c = cg(M)$;
2 **if** $\mathcal{C}$ *is reflexive* **then**
3   Let ℓ be the axis of symmetry;
4   **if** $r1$ *and* r_2 *are on* ℓ **then**
5   Let r' be the northernmost robot between r_1 and r_2;
6   **if** $r = r'$ **then** move r on $hline(c, r)$ at distance $3d'$ from c;
7   **else**
8   **if** $\rho_t \geq 3d$ **then** NUMGUARDS = 2; **exit**;
9   **if** $\rho_t \geq 2d$ **then**
10   **if** $r \in \{r_1, r_2\}$ **then** move r on $hline(c, r)$ at distance $3d$ from c;
11   **else**
12   **if** $r \in \{r_1, r_2\}$ **then** move r on $hline(c, r)$ at distance $2d$ from c;

13 **else**
14   **if** $\rho_t \geq 3d$ **then**
15   Let ℓ be the bisector of $\alpha = \sphericalangle(r_1, c, r_2)$;
16   **if** ℓ *is an axis of reflection for* M **then**
17   **if** $M \cap \ell \neq \emptyset$ **then** NUMGUARDS = 2; **exit**;
18   **else**
19   Let M' be the set of meeting-points in $M \cap O_M$, closest to ℓ, and with minimum Weber-distance;
20   **if** $|M'| = 1$ **then** NUMGUARDS = 2; **exit**;

21   **if** r *is the robot on* O_t *with minimum view* **then**
22   move r on $hline(c, r)$ at distance $3d'$ from c

Procedure: CHECKMANY
Input: Configuration $\mathcal{C} = (R, M)$

1 **if** $\mathcal{C}$ *is reflexive with axis* ℓ **then**
2   Let r_1 and r_2 be the robots on O_t that are not on ℓ but closest to it (and the northernmost in case of ties);
3   **if** $r \in \{r_1, r_2\}$ **then** move r on $hline(c, r)$ at distance $2d'$ from c;
4 **else**
5   **if** r *is the robot on* O_t *with minimum view* **then**
6   move r on $hline(c, r)$ at distance $3d'$ from c

3.3   Class $\mathcal{S}_6$

Since each configuration $\mathcal{C} \in \mathcal{S}_6$ has two or three robots only, then the approach of Sect. 3.2 that makes use of guards cannot be always applied since there might be not enough remaining robots to create a multiplicity. Here we briefly summarize the strategy implemented by Procedure ATMOST3BOTS().

If $\mathcal{C}$ has only two robots r_1 and r_2 and is reflexive without robots on the axis ℓ, then the approach is to move both the symmetric robots by small steps toward a meeting-point m on the axis. Small steps are required in order to maintain ℓ as recognizable. During the movements, the following invariant is used: the triangles (r_1, m, h_1) and (r_2, m, h_2), where h_1 and h_2 are the projections of r_1 and r_2 on ℓ, respectively, remain similar after the movements. The small movements are repeated until the two robots reach m.

Procedure: MAKEMULTIPLICITY
Input: Configuration $\mathcal{C} = (R, M)$

1 **if** NUMGUARDS $= 1$ **then**
2   Let g be the guard and let $\ell = hline(c, g)$;
3   **if** $M \cap \ell \neq \emptyset$ **then**
4   Let $m \in M \cap \ell$ be the meeting-point closest to g;
5   **if** $r \neq g$ **then** move r toward m;

6   **else**
7   Let $X = \bigcup_{m \in M} hline(c, m) \cap O_t$;
8   **if** $r = g$ **then** move r on O_t toward any closest point in X;

9 **if** NUMGUARDS $= 2$ **then**
10   Let g_1, g_2 be the guards and ℓ be the bisector of $\sphericalangle(g_1, c, g_2)$;
11   Let $M' = M \cap \ell$;
12   **if** $|M'| \neq \emptyset$ **then**
13   Let $M'_W \subseteq M'$ be the set of meeting-points with minimum Weber-distance;
14   **if** $|M'_W| = 2$ **then**
15   Let m be the northernmost meeting-point in M'_W;
16   **if** $r \notin \{g_1, g_2\}$ *and* r *is not on* ℓ **then** move r toward m;

17   **else**
18   **if** $r \notin \{g_1, g_2\}$ **then** move r toward m, being $M'_W = \{m\}$;

19   **else**
20   Let M'' be the set of meeting-points on O_M, closest to ℓ, and with minimum Weber-distance;
21   **if** $M'' = \{m\}$ **then**
22   **if** $r \notin \{g_1, g_2\}$ **then** move r toward m;

23   **else**
24   **if** r *is the robot on* ℓ *with minimum view* **then**
25   move r toward any $m \in M''$;

In all the other cases one guard is created. In particular: if the there are three robots, once the guard is created the remaining robots can create a multiplicity (and hence a configuration in $\mathcal{S}_1$ is created); if there are two robots, once the guard is created then (1) the other robot can move toward the meeting-point closest to the guard, and (2) the guard can move toward the occupied meeting-point until finalizing the gathering.

4   Conclusion

We have studied a new version of the gathering problem of anonymous and oblivious robots in the plane. Robots are required to gather at some predetermined meeting-points. Robots operate in the Look-Compute-Move cycle model empowered with the multiplicity detection. We provide a new deterministic distributed gathering algorithm that solves the problem for all initial configurations but those proved to be ungatherable. Introducing meeting-points is a natural and challenging choice, and the resolution of the gathering problem within this model is of main interest in robot-based computing systems.

Revisiting other existing models for the gathering or even other problems with respect to the meeting-points represents intriguing research directions.

References

1. Bouzid, Z., Das, S., Tixeuil, S.: Gathering of mobile robots tolerating multiple crash faults. In: IEEE 33rd International Conference on Distributed Computing Systems (ICDCS), pp. 337–346 (2013)
2. Chalopin, J., Dieudonné, Y., Labourel, A., Pelc, A.: Fault-tolerant rendezvous in networks. In: Esparza, J., Fraigniaud, P., Husfeldt, T., Koutsoupias, E. (eds.) ICALP 2014, Part II. LNCS, vol. 8573, pp. 411–422. Springer, Heidelberg (2014)
3. Cicerone, S., Di Stefano, G., Navarra, A.: Minimum-traveled-distance gathering of oblivious robots over given meeting points. In: Gao, J., Efrat, A., Fekete, S.P., Zhang, Y. (eds.) ALGOSENSORS 2014, LNCS 8847. LNCS, vol. 8847, pp. 57–72. Springer, Heidelberg (2015)
4. Cicerone, S., Di Stefano, G., Navarra, A.: MinMax-distance gathering on given meeting points. In: Paschos, V.T., Widmayer, P. (eds.) CIAC 2015. LNCS, vol. 9079, pp. 127–139. Springer, Heidelberg (2015)
5. Cieliebak, M., Flocchini, P., Prencipe, G., Santoro, N.: Distributed computing by mobile robots: gathering. SIAM J. Comput. **41**(4), 829–879 (2012)
6. Cockayne, E.J., Melzak, Z.A.: Euclidean constructibility in graph-minimization problems. Math. Mag. **42**(4), 206–208 (1969)
7. D'Angelo, G., Di Stefano, G., Navarra, A.: Gathering asynchronous and oblivious robots on basic graph topologies under the look-compute-move model. In: Alpern, S., Fokkink, R., Gąsieniec, L., Lindelauf, R., Subrahmanian, V.S. (eds.) Search Theory: A Game Theoretic Perspective, pp. 197–222. Springer, New York (2013)
8. D'Angelo, G., Di Stefano, G., Navarra, A.: Gathering on rings under the look-compute-move model. Distrib. Comput. **27**(4), 255–285 (2014)
9. Degener, B., Kempkes, B., Langner, T., Meyer auf der Heide, F., Pietrzyk, P., Wattenhofer, R.: A tight runtime bound for synchronous gathering of autonomous robots with limited visibility. In: 23rd ACM Symposium on Parallelism in Algorithms and Architectures (SPAA), pp. 139–148 (2011)
10. Farrugia, A., Gąsieniec, L., Kuszner, Ł., Pacheco, E.: Deterministic rendezvous in restricted graphs. In: Italiano, G.F., Margaria-Steffen, T., Pokorný, J., Quisquater, J.-J., Wattenhofer, R. (eds.) SOFSEM 2015-Testing. LNCS, vol. 8939, pp. 189–200. Springer, Heidelberg (2015)
11. Flocchini, P., Prencipe, G., Santoro, N.: Distributed Computing by Oblivious Mobile Robots. Synthesis Lectures on Distributed Computing Theory. Morgan & Claypool Publishers, San Rafael (2012)
12. Izumi, T., Izumi, T., Kamei, S., Ooshita, F.: Randomized gathering of mobile robots with local-multiplicity detection. In: Guerraoui, R., Petit, F. (eds.) SSS 2009. LNCS, vol. 5873, pp. 384–398. Springer, Heidelberg (2009)
13. Klasing, R., Markou, E., Pelc, A.: Gathering asynchronous oblivious mobile robots in a ring. Theoret. Comput. Sci. **390**, 27–39 (2008)
14. Kranakis, E., Krizanc, D., Markou, E.: The Mobile Agent Rendezvous Problem in the Ring. Morgan & Claypool, San Rafael (2010)
15. Pelc, A.: Deterministic rendezvous in networks: a comprehensive survey. Networks **59**(3), 331–347 (2012)
16. Weiszfeld, E.: Sur le point pour lequel la somme des distances de n points donnés est minimum. Tohoku Math. **43**, 355–386 (1936)

Mutual Visibility with an Optimal Number of Colors

Gokarna Sharma[1]([⊠]), Costas Busch[2], and Supratik Mukhopadhyay[2]

[1] Department of Computer Science, Kent State University, Kent, OH 44242, USA
gsharma2@kent.edu
[2] School of Electrical Engineering and Computer Science, Louisiana State University,
Baton Rouge, LA 70803, USA
{busch,supratik}@csc.lsu.edu

Abstract. We consider the following fundamental MUTUAL VISIBILITY problem: Given a set of n identical autonomous point robots in arbitrary distinct positions in the Euclidean plane, find a schedule to move them such that within finite time they reach, without collisions, a configuration in which they all see each other. The robots operate following Look-Compute-Move cycles and a robot r_i can not see other robot r_j if there lies a third robot r_l in the line segment connecting the positions of r_i and r_j. Moreover, n is not assumed to be known to the robots. We study this problem in the robots with lights model, where each robot has an externally visible persistent light which can assume colors from a fixed set of colors and the color set is identical to all the robots. This model corresponds to the classical model of oblivious robots when the number of colors $c = 1$ in the color set. Therefore, we focus here on the objective of minimizing the number of colors required to successfully solve MUTUAL VISIBILITY. Di Luna *et al.* [16] presented two algorithms and showed that MUTUAL VISIBILITY can always be solved without collisions with $c = 3$ colors for both semi-synchronous and asynchronous robots under both rigid and non-rigid moves. In this paper, we present and analyze an improved algorithm which requires only $c = 2$ colors and works for both semi-synchronous and asynchronous robots under both rigid and non-rigid moves; this is optimal since any algorithm for MUTUAL VISIBILITY needs at least 2 colors in the robots with lights model, when n is not known. We employ a non-trivial technique which moves all the interior robots first to the boundary of the initial convex hull and then the robots in the boundary of the hull (except the corners) to outside of the hull until all the robots eventually become corners, without the need of any third color. Our result is interesting in the sense that asynchronicity and non-rigidity of robot movements have no effect on MUTUAL VISIBILITY with respect to the number of colors needed to successfully solve it. We also provide an improved solution to the CIRCLE FORMATION problem.

1 Introduction

Consider a set of n autonomous point robots (n is not known) operating in the Euclidean plane $\mathbb{R}^2$ which are anonymous, indistinguishable, and without any

© Springer International Publishing Switzerland 2015
P. Bose et al. (Eds.): ALGOSENSORS 2015, LNCS 9536, pp. 196–210, 2015.
DOI: 10.1007/978-3-319-28472-9_15

direct means of communication. Initially, each robot is in the distinct positions of $\mathbb{R}^2$ and equipped with a local coordinate system and sensor capabilities (i.e., vision) to determine the positions of other robots. The local coordinate system of a robot may be different with that of other robots. The robots execute the same algorithm. They operate in *Look-Compute-Move* cycles, i.e., when a robot becomes active, it uses its vision to get a snapshot of its surroundings (*Look*), computes a destination point based on the snapshot (*Compute*), and finally moves towards the destination (*Move*). There has been an extensive research in distributed computing literature under the assumption that the robots are *oblivious* – each robot has no memory of its past Look-Compute-Move cycles – and visibility is *unobstructed* - three collinear robots are mutually visible to each other [2,8,9,12,20,23].

In this paper, we consider *obstructed visibility* [1,3–6,10] – a robot r_i can see robot r_j iff there is no other robot in the line segment connecting their positions – and study this fundamental MUTUAL VISIBILITY problem: Starting from the arbitrary distinct positions in $\mathbb{R}^2$, determine a schedule to reposition the robots without collisions such that, within finite time, they reach a configuration where they all see each other. Although obstructed visibility is considered before in the classical oblivious robots model for the SPREADING problem [4] and in the so-called *fat robots* model [1,6,12,18], the technique of [4] cannot be generalized for MUTUAL VISIBILITY, since it works only in the one-dimensional space $\mathbb{R}^1$, and the techniques of [1,6,12,18] are also not suitable since collisions are allowed.

We study MUTUAL VISIBILITY in the *robots with lights* model initially suggested by Peleg [19], where each robot has an externally visible light that can assume colors from a fixed set and communicate with other robots using these lights [7,11,12,14,19,22]; the reason for considering robots with lights model is that there is no MUTUAL VISIBILITY algorithm in the oblivious robots model when n is not known. The lights are not erased at the end of each cycle in this model and the robots are oblivious, except the direct communication capability provided by lights. Moreover, this model generalizes the classical oblivious robots model since a light with only one possible color acts as no light. Therefore, the objective in the robots with lights model is to minimize the number of colors. We have the following theorem for the optimality of any MUTUAL VISIBILITY algorithm in the robots with lights model [16].

Theorem 1 (Optimality). *Any algorithm for the* MUTUAL VISIBILITY *problem needs at least 2 colors under the robots with lights model, if n is not known to the robots.*

We consider both semi-synchronous and asynchronous robots under both rigid and non-rigid moves. In the *semi-synchronous* (SSYNCH) setting, the robots operate in rounds but all robots may not be activated in a round and the robots that are activated perform their cycles in a perfectly synchronous setting, whereas in the *asynchronous* (ASYNCH) setting, there is no common notion of time and no assumption is made on the duration of each operations and the delay between them except that it is finite. We say that the moves of robots are RIGID if an adversary does not have the power to stop a moving robot before reaching its destination, otherwise the moves of robots are NON-RIGID. The only constraint on NON-RIGID moves is that the robot moves at least a minimum distance $\delta > 0$ (otherwise MUTUAL VISIBILITY might never be solved).

In the literature, Di Luna et al. [17] were the first to study this problem. They gave an algorithm that solves it with $c = 6$ colors in SSYNCH and with $c = 10$ colors in ASYNCH under both RIGID and NON-RIGID moves. Later, Vaidyanathan et al. [21] gave an algorithm in the *fully synchronous* (FSYNCH) setting that optimizes time complexity but not the number of colors (i.e., their algorithm needs $c = 12$ colors). Recently, Di Luna et al. [16] gave two algorithms, one for RIGID SSYNCH robots and the other for the rest of the combinations, and proved the following theorem.

Theorem 2 (Di Luna et al. *[16]*)**.** MUTUAL VISIBILITY *is solvable without collisions for (a)* SSYNCH *robots under* RIGID *moves with 2 colors, always; (b)* SSYNCH *robots under* NON-RIGID *moves with 3 colors, always; (c)* ASYNCH *robots under* RIGID *moves with 3 colors, always; and (d)* ASYNCH *robots under* NON-RIGID *moves with 3 colors, if the robots agree on the direction of one axis.*

In this paper, we give an algorithm which guarantees the following theorem.

Theorem 3 (Our Result). MUTUAL VISIBILITY *is solvable without collisions for (a)* SSYNCH *robots under both* RIGID *and* NON-RIGID *moves, and* ASYNCH *robots under* RIGID *moves with 2 colors, always; and (b)* ASYNCH *robots under* NON-RIGID *moves with 2 colors, if the robots agree on the direction of one axis.*

Our algorithm asks robots to position themselves in the vertices of the convex polygon so as to solve MUTUAL VISIBILITY. It first differentiates the robots in the boundary of the convex polygon from the robots that are in the interior of that polygon, and asks the internal robots to move to the boundary of that polygon. After all internal robots reached the boundary, it differentiates the robots that are in the vertices (corners) with those in the edges of that polygon. The robots in the vertices never move and the algorithm asks the robots in the edges to move in the area outside of the hull without making the corner robots internal. Non-trivial technique of repeated moving of internal robots to the hull boundary and the side robots to outside of the hull is used until all side robots eventually become the vertexes of the polygon, without the need of any additional color besides 2 colors needed to differentiate internal robots from the robots in the convex hull boundary. Our technique is interesting because the same technique works for both SSYNCH and ASYNCH robots under both RIGID and NON-RIGID moves; Di Luna et al. [16] used two different techniques, one for RIGID SSYNCH robots and the other for the rest of the combinations. Moreover, our technique shows that asynchronicity and non-rigidity of robot moves have no impact on the solvability of MUTUAL VISIBILITY with respect to the number of colors (except for NON-RIGID ASYNCH robots, where we give a 2-color solution under the assumption that the robots agree on one axis).

As a byproduct, our algorithm solves the CIRCLE FORMATION problem – position robots on the perimeter of a circle – under obstructed visibility through a simple modification. Our algorithm uses 2 colors for SSYNCH robots and 3 colors for ASYNCH robots under both RIGID and NON-RIGID moves. The best previous solution to this problem uses 3 colors for SSYNCH robots under NON-RIGID moves and 4 colors for ASYNCH robots under both RIGID and NON-RIGID moves [16].

We proceed as follows. We discuss model and some preliminaries in Sect. 2. We then present our algorithm and analyze it for RIGID SSYNCH robots in Sect. 3, and analyze it for NON-RIGID SSYNCH, RIGID ASYNCH, and NON-RIGID ASYNCH robots in Sects. 4, 5, and 6, respectively. In Sect. 7, we discuss how to modify our algorithm to solve CIRCLE FORMATION problem. Many proofs and details are omitted due to space constraints.

2 Model and Preliminaries

Consider a set of n anonymous robots $\mathcal{R} := \{r_1, r_2, \ldots, r_n\}$ operating in the Euclidean plane $\mathbb{R}^2$; n is not assumed to be known. Each robot r_i has its own coordinate system centered in itself and it knows its position with respect to its coordinate system. We denote by $p_i(t) \in \mathbb{R}^2$ the position occupied by robot $r_i \in \mathcal{R}$ at time t. A robot r_i sees robot r_j at time t if and only if the line segment $\overline{p_i(t)p_j(t)}$ does not contain any other robot at time t. Two robots r_i and r_j are said to *collide* at time t if $p_i(t) = p_j(t)$. If no ambiguity arises, we omit t from $r_i(t)$ and $r_j(t)$, and use r_i to denote both the robot r_i and its position p_i. Moreover, robots have their own unit of distance which may not agree on the unit of measure of other robots. Robots may not agree on the orientation of their coordinate system, i.e., there is no common notion of clockwise direction.

Each robot r_i is equipped with an externally visible persistent light which can assume any color from a fixed finite set of colors $\mathcal{C}$. The colors in $\mathcal{C}$ are the same for all robots in $\mathcal{R}$. We use variable $r_i.light$ to denote the light of a robot r_i. The color of the light of a robot r at time t can be seen by all robots that are visible to r at time t. Each robot executes the same algorithm locally every time it is activated.

A *configuration* $\mathbb{C}$ is a set of n tuples in $\mathcal{C} \times \mathbb{R}^2$ which defines the position and color of a robot. Let $\mathbb{C}_t$ denote the configuration at time t. Let $\mathbb{C}_t(r_i)$ denote the configuration $\mathbb{C}_t$ for robot r_i. A configuration $\mathbb{C}$ is *obstruction-free* if $\forall r_i \in \mathcal{R}$, we have that $|\mathbb{C}(r_i)| = n$ (all robots can see each other). Let $\mathbb{H}_t$ denote the convex hull formed by $\mathbb{C}_t$ which can be easily computed using Graham's convex hull algorithm [15]. Let $\partial \mathbb{H}_t = \mathcal{V}_t \cup \mathcal{S}_t$ denote the robots in the boundary of $\mathbb{H}_t$, where $\mathcal{V}_t \subseteq \mathcal{R}$ are the set of robots lying at the corners of $\mathbb{H}_t$ and $\mathcal{S}_t \subseteq \mathcal{R}$ are the set of robots lying at the sides (or edges) of $\mathbb{H}_t$. The robots in the set $\mathcal{V}_t$ are called *corner robots* and in the set $\mathcal{S}_t$ are called *side robots*. The robots in $\mathcal{V}_t \cup \mathcal{S}_t$ are called *external robots*. The robots in the set $\mathcal{I}_t = \mathbb{H}_t \setminus \partial \mathbb{H}_t$ are called *internal robots*. Given a robot $r_i \in \mathcal{R}$, we denote by $\mathbb{H}_t(r_i)$ the convex hull of $\mathbb{C}_t(r_i)$. Given two points $a, b \in \mathbb{R}^2$, we denote by $\overleftrightarrow{ab}$ the line that contains them. We denote by $|\overline{ab}|$ the length of the line segment $\overline{ab}$ connecting points a and b.

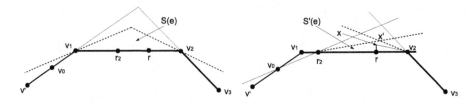

Fig. 1. An illustration of safe zone for edge $e = \overline{v_1 v_2}$ and also for the robots in e.

Moreover, given $a, b, d \in \mathbb{R}^2$, we use $\angle abd$ to denote the angle with vertex b and sides ab and bd.

At any time t, a robot $r_i \in \mathcal{R}$ is either active or inactive. When active, r_i performs a sequence of *Look-Compute-Move* (LCM) operations: a robot takes the snapshot of the positions of the robots visible to it in its own coordinate system (*Look*); executes its algorithm using the snapshot which returns a destination point $x \in \mathbb{R}^2$ and a color $c \in \mathcal{C}$ (*Compute*); and sets its own light to color c and moves towards the computed destination $x \in \mathbb{R}^2$ (if x is different than its current position), if any (*Move*). We consider two schedulers for the activation of the robots in $\mathcal{R}$: SSYNCH and ASYNCH. In SSYNCH, the time is discrete and at each time instant t a subset of the robots (from empty set to all of $\mathcal{R}$) are activated and perform their LCM operations instantaneously, ending at time $t + 1$. Therefore, we use round t in SSYNCH instead of time t. In ASYNCH, each robot acts independently from the others and the duration of the LCM operations of each robot is finite but unpredictable, i.e., there is no common notion of time. More precisely, we assume that the Look operation is instantaneous, but the Compute and Move operations can take an unpredictable but bounded amount of time, unknown to the robot. Moreover, there may be unpredictable but bounded delay between Look-Compute and Compute-Move operations.

We assume that the execution starts at time 0. Therefore, at time $t = 0$, the robots start in an arbitrary configuration $\mathbb{C}_0$ occupying distinct positions in $\mathbb{R}^2$ and the color of the light of each robot is set to *Off*. The MUTUAL VISIBILITY problem is defined as follows: Given any $\mathbb{C}_0$, reach in finite time an obstruction-free configuration without collisions. An algorithm is said to solve MUTUAL VISIBILITY if it always achieves an obstruction-free configuration regardless of the choices of the adversary and from any arbitrary initial configuration. We measure the quality of the algorithm by counting the number of distinct colors in the set $\mathcal{C}$ needed to solve MUTUAL VISIBILITY.

Let $e = \overline{v_1 v_2}$ be a line segment connecting two corner robots v_1 and v_2 of $\mathbb{H}_t$. A safe zone is the portion of plane outside $\mathbb{H}_t$ such that the corner robots v_1 and v_2 of $\mathbb{H}_t$ remain as the corner robots despite the movements of the side robots in that area. Following Di Luna *et al.* [17], the *safe zone* of e, denoted as $S(e)$, consists of the portion of plane outside $\mathbb{H}_t$, such that for all points $x \in S(e)$, we have that $\angle x v_1 v_2 \leq \frac{180 - \angle v_0 v_1 v_2}{4}$ and $\angle v_1 v_2 x \leq \frac{180 - \angle v_1 v_2 v_3}{4}$ (the left of Fig. 1). Side robots may not always be able to compute $S(e)$ exactly due to the mutual obstructions of visibility which lead to different local views. However, if there is only one side robot in e, then it can compute $S(e)$ exactly. When there are more than one robot in e, $S'(e)$ computed by a robot based on its local view is such that $S'(e) \subseteq S(e)$ (the right of Fig. 1 for robot r).

3   MUTUAL VISIBILITY for RIGID SSYNCH Robots

In this section, we consider the MUTUAL VISIBILITY problem for RIGID SSYNCH robots. We present and analyze an algorithm (Algorithm 1) and prove that it solves MUTUAL VISIBILITY for RIGID SSYNCH robots with only 2 colors. We will prove later that this algorithm works also for NON-RIGID SSYNCH, RIGID ASYNCH, and NON-RIGID ASYNCH robots with only 2 colors, whereas two different algorithms were designed in [16] and needed 3 colors except RIGID SSYNCH robots. Algorithm 1 uses Algorithms 2–4 as subroutines.

Our algorithm works in two phases similar to the algorithm of Di Luna et al. [17]: (i) interior depletion (ID) and (ii) side depletion (SD). However, our algorithm uses 2 colors to perform these phases in contrast to at least 6 colors needed in [17]. The goal of the ID phase is to reach a configuration $\mathbb{C}_{ID}$ in which there are no robots in the interior of $\mathbb{H}_{ID}$, i.e., all the robots in $\mathcal{R}$ are in the set $\partial\mathbb{H}_{ID}$. The goal of the SD phase is to move all the robots in the set $\mathcal{S}_{ID}$ to outside of $\mathbb{H}_{ID}$ to become corner robots. After all robots in $\mathcal{S}_{ID}$ become corner robots, our algorithm terminates solving MUTUAL VISIBILITY. Note that the robots in $\mathcal{V}_{ID}$ never move and only the robots in $\mathcal{S}_{ID}$ move to become corner robots in the SD phase. Two colors are used: $\mathcal{C} = \{Off, External\}$ and $\mathcal{C}$ is same for all the robots in $\mathcal{R}$. The restriction of only 2 colors in the set $\mathcal{C}$ makes this problem challenging. Therefore, our algorithm executes the SD phase multiple times and it consists of the execution of the ID and SD phases repeatedly until all robots in $\mathcal{R}$ become corners. We prove later that this is indeed possible without any additional color besides 2 colors in $\mathcal{C}$. We also prove that the SD phase starts only after the ID phase finishes, i.e., they are executed in sequence one after another without overlap.

Interior Depletion: The objective of this phase to have all robots that are already in $\partial\mathbb{H}_{ID}$ and the newly become side robots colored *External*. This phase works as follows. In $\mathbb{C}_0$, the lights of all robots in $\mathcal{R}$ are set to *Off*. If a robot r_i with light *Off* is activated at some round k, and it sees that $\mathbb{C}_k(r_i)$ contains a region of plane that is free of robots and wider than 180° (wide exactly 180°), then r_i knows it is a corner (a side) and sets its light to *External* (Lines 17, 19 of Algorithm 1, Line 1 of Algorithm 2, Line 1 of Algorithm 4). The side robot with light *External* does not move as long as it can see robots whose light is still *Off* (Line 2 of Algorithm 4). This guarantees that the SD phase starts only after the ID phase finishes. If a robot r_i with light *Off* is activated at some round k, and its sees that it is in the interior of $\mathbb{H}_k(r_i)$, it moves to position itself on one of its nearest visible edges of $\partial\mathbb{H}_k(r_i)$ using the technique described in Algorithm 3; we omit detailed description of the technique due to space constraints. The edge of $\partial\mathbb{H}_k(r_i)$ is visible to r_i when it is occupied by only robots with lights *External*.

Algorithm 1. MUTUAL VISIBILITY algorithm for robot r_i for any time $k > 0$

---

1 // Look-Compute-Move cycle for robot r_i
2 $\mathbb{C}_k(r_i) \leftarrow$ configuration $\mathbb{C}_k$ for robot r_i (including r_i);
3 $\mathbb{H}_k(r_i) \leftarrow$ convex hull of the positions of the robots in $\mathbb{C}_k(r_i)$;
4 **if** $|\mathbb{C}_k(r_i)| = 1$ **then** Terminate;
5 **else if** $\mathbb{H}_k(r_i)$ *is a line segment* **then**
6   **if** $|\mathbb{C}_k(r_i)| = 2$ **then**
7   Let $r_j \in \mathbb{C}_k(r_i)$;
8   **if** $r_i.light = Off$ **then**
9             $r_i.light \leftarrow External$;
10   Move orthogonal to line $\overleftrightarrow{r_i r_j}$ by any non-zero distance;
11   **else if** $r_j.light = External$ **then** Terminate;
12   **else if** $|\mathbb{C}_k(r_i)| = 3$ **then**
13   Let $r_j, r_l \in \mathbb{C}_k(r_i)$;
14   **if** $r_i.light = Off \wedge r_j.light = External \wedge r_l.light = External$ **then**
15             $r_i.light \leftarrow External$;
16   Move orthogonal to line $\overleftrightarrow{r_j r_l}$ by any non-zero distance;
17 **else if** r_i is a vertex robot of $\mathbb{H}_k(r_i)$ **then** $Corner(r_i, \mathbb{C}_k(r_i), \mathbb{H}_k(r_i))$;
18 **else if** r_i is an interior robot of $\mathbb{H}_k(r_i)$ **then** $Internal(r_i, \mathbb{C}_k(r_i), \mathbb{H}_k(r_i))$;
19 **else if** r_i is a side robot of $\mathbb{H}_k(r_i)$ **then** $Side(r_i, \mathbb{C}_k(r_i), \mathbb{H}_k(r_i))$;

Algorithm 2. $Corner(r_i, \mathbb{C}_k(r_i), \mathbb{H}_k(r_i))$

---

1 **if** $r_i.light = Off$ **then** $r_i.light \leftarrow External$;
2 **else if** $\forall r \in \mathbb{C}_k(r_i)$, $r.light = External \wedge$ *no three robots in* $\mathbb{C}_k(r_i)$ *are in a line* $\wedge$ *no robot* $r \in \mathbb{C}_k(r_i) \backslash \{r_i\}$ *is in the interior of* $\mathbb{H}_k(r_i)$ **then** Terminate;

We will prove later that all the robots will be positioned on the boundary of $\mathbb{H}_0$ without collisions at the end of this phase. Note that only 2 colors are used.

Side Depletion: This phase is similar to the edge depletion phase of [17]; however the edge depletion phase of [17] requires at least $c = 6$ colors whereas our algorithm needs only $c = 2$ colors. Therefore, this phase is crucial is minimizing the number of colors to 2 in solving MUTUAL VISIBILITY and a non-trivial technique is needed to make it possible; note that only 2 colors are used by our algorithm and [17] in the ID phase.

In this phase, a side robot, as soon as it is activated, moves from its edge $e = \overline{v_1 v_2}$ to a point in the safe zone $S(e)$ as defined in Sect. 2 without changing its color. Specifically, if a side robot r_i with light $External$ is activated at some round k, it checks whether all the robots in its local view $\mathbb{C}_k(r_i)$ have light $External$ and there is no robot in the interior of $\mathbb{H}_k(r_i)$ (Line 2 of Algorithm 4). If this is the case, it computes a point x in $S(e)$ based on its local view and moves to x without changing its color (Lines 3–7 of Algorithm 4). Due to the assumptions of the SSYNCH model, not all side robots in e may reach to their safe zones $S(e)$. When robots in e are activated at round $k + 1$ and they find that they are

---

Algorithm 3. $Internal(r_i, \mathbb{C}_k(r_i), \mathbb{H}_k(r_i))$

---

1   **if** $r_i.light = Off \wedge$ *there is no robot* $r \in \mathbb{C}_k(r_i)\backslash\{r_i\}$ *with* light *External in the interior of* $\mathbb{H}_k(r_i)$ **then**

2   Order the robots in $\mathbb{H}_k(r_i)$ starting from any arbitrary robot v_1 in clockwise order so that $\mathcal{T} = \{v_1, \ldots, v_{last}\}$, where v_1 is the first robot and v_{last} is the last robot;

3   Let c, d be any pair of two consecutive robots in $\mathcal{T}$ with $c.light = External$ and $d.light = External$;

4   Let HP_{cd} be the half-plane divided by line parallel to $\overleftrightarrow{cd}$ that passes through r_i such that $r_i \notin HP_{cd}$ (but the robots c, d are in HP_{cd});

5       $Q \leftarrow$ set of line segments $\overline{cd}$ such that there is no robot in HP_{cd} (including $\overline{cd}$) with light *Off*;

6   **if** Q *is empty* **then return**;

7   **else**

8           $\overline{ef} \leftarrow$ the line segment in Q between two robots e, f that is closest to r_i;

9           $m \leftarrow$ midpoint of $\overline{ef}$;

10         $\alpha \leftarrow$ angle $\angle emr_i$;

11   Move to the point $x = \frac{\alpha \cdot |\overline{ef}|}{180°}$ from endpoint e in $\overline{ef}$;

12 **else if** $r_i.light = External$ **then** $r_i.light \leftarrow Off$;

Algorithm 4. $Side(r_i, \mathbb{C}_k(r_i), \mathbb{H}_k(r_i))$

---

1   **if** $r_i.light = Off$ **then** $r_i.light \leftarrow External$;

2   **else if** $\forall r \in \mathbb{C}_k(r_i), r.light = External \wedge$ *no robot* $r \in \mathbb{C}_k(r_i)\backslash\{r_i\}$ *is in the interior of* $\mathbb{H}_k(r_i)$ **then**

3   Order the robots in counterclockwise order of r_i such that the order is $\mathcal{T}_i = \{v_3, v_2, r_i, r, v_0\}$, where v_3 is the first robot non-collinear in clockwise direction of r_i with $v_3.light = External$, v_2 is the robot that is collinear with r_i in clockwise direction with $v_2.light = External$, and r is collinear in counterclockwise direction with $r.light = External$, and v_0 is the first non-collinear robot in counterclockwise direction with $v_0.light = External$;

4   Compute angles $\alpha = 180 - \angle v_0 r r_i$ and $\beta = 180 - \angle r_i v_2 v_3$, and set $\gamma = \min\{\alpha/4, \beta/4\}$;

5   Compute a point x' such that $\angle x' v_2 r_i = \gamma$ and a point x'' such that $\angle x'' r r_i = \gamma$;

6       $x \leftarrow x'$ or x'' whichever is nearest to e;

7   Move perpendicular to e with destination x;

---

internal robots in their local convex hulls at round $k + 1$, they change their color of *Off* indicating that they now become internal robots (Line 12 of Algorithm 3). Among the robots that moved to $S(e)$ at least one becomes corner robot and it does not move in any future rounds. The newly become internal robots deplete again to $\partial\mathbb{H}$ before any other robot in the boundary moves to $S(e)$.

A robot r_i terminates at round k and does not move in future rounds as soon as it satisfies the condition that all the robots in its local view $\mathbb{C}_k(r_i)$ have light

External, no three robots in $\mathbb{C}_k(r_i)$ are in a line, and no robot in $\mathbb{C}_k(r_i)$ is in the interior of $\mathbb{H}$ (Line 2 of Algorithm 2). Due to vision, we can easily see that robots can decide on whether they satisfy all the conditions for the termination. When all robots in $\mathcal{R}$ terminate, the configuration is strictly convex, and therefore MUTUAL VISIBILITY is solved.

There are two special cases in our algorithm. The first special case is when $n = 1$ (Line 4 of Algorithm 1). The second special case is when $\mathbb{H}_0$ is a line segment. These cases are handled using the technique described in Lines 4–16 of Algorithm 1.

3.1   Analysis of the Algorithm

We now analyze our algorithm for RIGID SSYNCH robots. We first provide the analysis of our algorithm for the ID phase. Specifically, we show that, starting from any initial configuration $\mathbb{C}_0$, the ID phase finishes in finite time and no collision of robots occur during the execution. Particularly, we prove the following lemmas.

Lemma 1. *Given any initial configuration* $\mathbb{C}_0$, $\exists k \in \mathbb{N}^+$ *such that* $\mathcal{I}_k = \emptyset$ *in* $\mathbb{C}_k$.

Lemma 2. *Given any initial configuration* $\mathbb{C}_0$, *no collisions of robots occur until* $\mathcal{I}_k = \emptyset$ *is reached at some round* $k \in \mathbb{N}^+$ *in* $\mathbb{C}_k$.

We now provide the main result from the analysis of the ID phase.

Theorem 4. *Given any initial configuration* $\mathbb{C}_0$, *there is some round* $k \in \mathbb{N}^+$ *in which the robots in* $\mathcal{R}$ *occupy different positions of* $\mathbb{H}_k$, *and both the corner robots and side robots have light External.*

We now show that the detection of the absence of internal robots in the local convex hull $\mathbb{H}(r_i)$ of each robot r_i implies the global absence of internal robots from $\mathbb{H}$. In particular, we have the following lemma which is similar to [17, Lemma 2].

Lemma 3. *Given a robot* $r_i \in \mathcal{R}$ *with light External and a round* $k \in \mathbb{N}^+$, *if all the robots in* $\mathbb{C}_k(r_i)$ *have light External and no robot is in the interior of* $\mathbb{H}_k(r_i)$, *then* $\mathbb{C}_k$ *does not contain internal robots with respect to* $\mathbb{H}_0$.

Denote by $\mathbb{C}_{ID}$ the configuration $\mathbb{C}_k$ at round $k \in \mathbb{N}^+$ after the ID phase of our algorithm such that $\mathcal{I}_k = \emptyset$ and by $\mathbb{H}_{ID}$ the convex hull created by $\mathbb{C}_{ID}$. We have the following lemma that shows how many other robots a side robot r_i sees in $\mathbb{C}_{ID}$.

Lemma 4. *Given a configuration* $\mathbb{C}_{ID}$ *and a side robot* r_i *in some edge* $e = \overline{v_1 v_2}$ *of* $\mathbb{H}_{ID}$ *with neighbors* r_l *and* r_r. *Let* $V(\mathbb{H}_{ID \setminus e})$ *denote the robots in* $\mathbb{H}_{ID}$ *except the robots in* e *(including the endpoint robots). Robot* r_i *sees all the robots in the set* $V(\mathbb{H}_{ID \setminus e}) \cup \{r_l, r_r\}$.

According to Lemma 4, suppose no robot is moved to $S(e)$ in Fig. 2, i.e., the configuration is $\mathbb{C}_{ID}$. Then, the side robot r in $e = \overline{v_1 v_2}$ sees all the robots in all the edges of $\mathbb{H}_{ID}$ except edge e (i.e., the robots in $V(\mathbb{H}_{ID \setminus e})$) and the robots r_2 and v_2 in e (its neighbors in e). We now analyze our algorithm for the SD phase. We start with the following observation which shows that SD phase starts only after the ID phase finishes.

Observation 1. *Given any initial configuration $\mathbb{C}_0$, if a configuration $\mathbb{C}_k$ is reached at some round $k \in \mathbb{N}^+$ such that $\mathcal{I}_k = \emptyset$, then the side depletion phase starts at round k', where $k' = k + 1$.*

Lemma 5. *Given a configuration $\mathbb{C}_{ID}$ and an edge $e = \overline{v_1 v_2}$ of $\mathbb{H}_{ID}$, if a robot $r_i \in e$ moves from e, it moves inside the safe zone $S(e)$.*

Lemma 6. *Consider a corner robot r_i of $\mathbb{H}_{k'}$ with $r_i.light = External$, $\forall k \in \mathbb{N}^+$ such that $k > k'$, even if the robots in any edge e of $\mathbb{H}_{k'}$ moved to $S(e)$, r_i is a corner robot of $\mathbb{H}_k$ and $r_i.light = External$.*

Lemma 7. *Let $X(e) > 0$ be the number of side robots in $e = \overline{v_1 v_2}$ that moved to $S(e)$ at some round k, then at least a robot in $X(e)$ becomes corner and does not move in any future round $k' > k$.*

We showed in Lemma 4 the number of robots that are visible to a side robot r_i in any edge e of $\mathbb{H}_{ID}$. We now prove the following lemma for (i) the number of robots that are visible to a robot that is still in e when $X(e)$ robots are moved to $S(e)$ and (ii) the number of robots that are visible to the robots in $X(e)$. This visibility proof is crucial since we ought to guarantee that before any other robot in any edge e move to $S(e)$, all the robots that became internal due to the movement of the side robots at some previous round become side robots again.

Lemma 8. *Given two configurations $\mathbb{C}_k$ and $\mathbb{C}_{k'}$, $k' = k + 1$, and an edge $e = \overline{v_1 v_2}$ of $\mathbb{H}_k$ such that $r_i \in e$ in $\mathbb{C}_k$ and $r_i \in S(e)$ in $\mathbb{C}_{k'}$. Let $X(e) > 0$ denote the side robots in e that moved to $S(e)$ at round k. Moreover, let $V(\mathbb{H}_{k \setminus e})$ denote the robots in $\mathbb{H}_k$ except the robots in e (including the endpoint robots). Furthermore, let e_l and e_r are the edges of $\mathbb{H}_k$ in the left and right of e. Then,*

- *If there is no side robot in both e_l and e_r, each robot $r \in X(e)$ sees robots in the set $V(\mathbb{H}_{k \setminus e})$ and at least two other robots in e or in $S(e)$.*
- *If there are side robots in both e_l and e_r, each robot $r \in X(e)$ sees at least robots in the set $V(\mathbb{H}_{k \setminus e}) \setminus \{v_0, v_3\}$ and at least two other robots in e or in $S(e)$, where v_0 and v_3 are the nearest robots from e in e_l and e_r.*
- *If there are side robots in either e_l or e_r, each robot $r \in X(e)$ sees at least robots in the set $V(\mathbb{H}_{k \setminus e}) \setminus \{v_0\}$ (if no side robots in e_r) or $V(\mathbb{H}_{k \setminus e}) \setminus \{v_3\}$ (if no side robots in e_l) and at least two other robots in e or in $S(e)$.*
- *If there is no side robot in e at round k', then each robot $r \in X(e)$ sees the robots in the set $V(\mathbb{H}_{k \setminus e})$ and at least two other robots in the set $X(e) \cup \{v_1, v_2\}$.*
- *If there is a robot $r_m \in e$ at round k' and $X(e) = 1$, then r_m sees the robots in the set $V(\mathbb{H}_{k \setminus e}) \cup \{r_l, r_r\} \cup X(e)$, where r_l and r_r are the neighbors of r_m in e in $\mathbb{C}_{k'}$.*

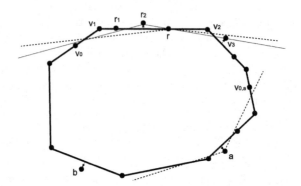

Fig. 2. An illustration of the robots in the set $V(\mathbb{H}_{k\backslash e})$ that a robot in $S(e)$ or in e sees

- If there is a robot $r_m \in e$ at round k' and $X(e) \geq 2$, then r_m sees the robots in the set $V(\mathbb{H}_{k\backslash e}) \cup \{r_l, r_r\}$ and at least two robots in $X(e)$.

The claims above also hold at round k' when the side robots in $V(\mathbb{H}_{k\backslash e})$ move to their respective $S(e)$'s at round k.

An example is given in Fig. 2 for the illustration of Lemma 8. Robot r_2 that moved to $S(e), e = \overline{v_1 v_2}$, sees all the robots in $V(\mathbb{H}_{k\backslash e})$ except v_3 in some cases, since r may fall in the line segment $\overline{r_2 v_3}$ when v_3 moved to its $S(e)$. The robots a and b see all the robots even if they moved to their respective $S(e)$. The side robot r sees all the robots of $V(\mathbb{H}_{k\backslash e})$ even if a and b moved to their $S(e)$ and the robots r_1, r_2 and v_2 of e.

Lemma 9. *Given a configuration $\mathbb{C}_k$ and an edge $e = \overline{v_1 v_2}$ of $\mathbb{H}_k$, no robot in any edge e of $\mathbb{H}_k$ moves to $S(e)$ if there is an internal robot with light $\in \{$Off, External$\}$.*

Lemma 10. *Given a configuration $\mathbb{C}_{ID}$, let $e = \overline{v_1 v_2}$ be an edge with $q \geq 1$ side robots. Eventually all these robots will become corner robots and set their lights to External.*

We now provide the main result from the analysis of our algorithm.

Theorem 5. *Our algorithm solves* MUTUAL VISIBILITY *for* RIGID SSYNCH *robots with 2 colors in the robots with lights model.*

Proof. We have from Theorem 4 that from any initial non-collinear configuration $\mathbb{C}_0$, we reach a configuration $\mathbb{C}_{ID}$ where $\mathcal{I}_{ID} = \emptyset$ in finite time. We have from Lemma 3 that robots can locally detect whether the configuration $\mathbb{C}_{ID}$ is reached and start executing the SD phase (Observation 1). We also have from Lemma 6 that the corner robots can only increase during the execution of the algorithm. We have from Lemma 10 that side robots in each edge e of $\mathbb{H}_{ID}$ eventually become corners. Therefore, starting from any non-collinear configuration $\mathbb{C}_0$, all robots eventually become corners of a convex hull and can not obstruct each other, solving MUTUAL VISIBILITY.

Therefore, it only remains to show that, starting from any initial collinear configuration, the robots correctly evolve into some non-collinear configuration from which applying the analysis above, the robots become corners of a convex hull and terminate. If $n \leq 3$, we can immediately prove that robots become corners and terminate through a case analysis. For $n = 1$, when the only robot becomes active it sees no other robot, that is, it changes its color to *External* and immediately terminates. For $n = 2$, the robot changes its color to *External* when it becomes active for the very first time and moves orthogonal to line $\overline{r_i r_j}$ that connects it to the only other robot r_j it sees in $\mathbb{C}(r_i)$. When r_i realizes later that $|\mathbb{C}(r_i)|$ is still 2 and $r_j.light =$ *External*, it simply terminates. For $n = 3$, if r_i realizes that both of its neighbors in $\mathbb{C}(r_i)$ have light set to *External* and are collinear with it, it sets its light to *External* and moves orthogonal to that line. When it becomes active next time, it finds itself at one of the corner and simply terminates as it sees all the other robots in the corners of the hull with light set to *External*. For $n \geq 4$, let a and b be two robots that occupy the vertices of the line segment $\mathbb{H}_0$ (i.e., the endpoint robots of $\mathbb{H}_0$). Nothing happens until a or b is activated, setting its light to *External*, and moving orthogonal to $\mathbb{H}_0$. After a or b moved, the other robots in $\mathbb{H}_0$, when become active, realize that they are not in a line anymore and enter the normal execution of our algorithm. It is also easy to see that after the line segment $\mathbb{H}_0$ evolves in a polygonal shape, it does not become line segment anymore. Note that in the whole process only 2 colors are used in the color set $\mathbb{C}$. Therefore, the theorem follows.   □

4   Mutual Visibility for Non-Rigid SSynch Robots

In this section, we consider Mutual Visibility for Non-Rigid SSynch robots. We analyze the same algorithm (Algorithm 1) and show that it also solves Mutual Visibility for Non-Rigid SSynch robots with only 2 colors.

For the ID phase, Lemmas 1 and 2 hold also in this setting. This is because the internal robots that are activated in a round perform their LCM operations in perfect synchrony and move at least a minimum distance of $\delta > 0$ in every round. Moreover, the side robots in $\partial \mathbb{H}_0$ do not move to $S(e)$ until they see robots with light *Off*, and until the ID phase is not finished, the robots in $\partial \mathbb{H}_0$ see at least a robot with light *Off*. Therefore, Theorem 4 also holds for the ID phase for Non-Rigid SSynch robots.

For the SD phase, Lemmas 3–8, 10, and Observation 1 hold without changes. We can guarantee that Lemma 9 also holds for Non-Rigid SSynch robots but proof argument needs to change. Therefore, we prove the following lemma similar to Lemma 9 for Non-Rigid SSynch robots.

Lemma 11. *For Non-Rigid SSynch robots, given a configuration $\mathbb{C}_k$ and an edge $e = \overline{v_1 v_2}$ of $\mathbb{H}_k$, no internal robot r_i in any edge e of $\mathbb{H}_k$ moves to $S(e)$ if there is an internal robot with light $\in \{$Off, External$\}$.*

Finally, we have the following main theorem for Non-Rigid SSynch robots.

Theorem 6. *Our algorithm solves Mutual Visibility for Non-Rigid SSynch robots with 2 colors in the robots with lights model.*

5   MUTUAL VISIBILITY for RIGID ASYNCH Robots

In this section, we consider the MUTUAL VISIBILITY problem for RIGID ASYNCH robots. We analyze the same algorithm (Algorithm 1) and show that it also solves MUTUAL VISIBILITY for RIGID ASYNCH robots with only 2 colors.

For the ID phase, only the collision avoidance proof gets slightly more complex, otherwise the algorithm works similar to RIGID SSYNCH robots. Therefore, we obtain Theorem 4 for RIGID ASYNCH robots after we prove the following lemma for collision avoidance. Note that due to ASYNCH robots, we use a specific time $t \in \mathbb{R}$ instead of a round $k \in \mathbb{N}$ in this and next section.

Lemma 12. *Given any initial configuration $\mathbb{C}_0$, no collision of robots occur until $\mathcal{I}_k = 0$ is reached at some round $k \in \mathbb{N}^+$ in $\mathbb{C}_k$, even if the RIGID ASYNCH robots execute our algorithm.*

Therefore, we prove the following theorem for RIGID ASYNCH robots.

Theorem 7. *Our algorithm solves MUTUAL VISIBILITY for RIGID ASYNCH robots with 2 colors in the robots with lights model.*

6   MUTUAL VISIBILITY for NON-RIGID ASYNCH Robots

In this section, we discuss how our algorithm (Algorithm 1) can be modified to solve MUTUAL VISIBILITY for NON-RIGID ASYNCH robots with only $c = 2$ colors. Similar to the algorithm of Di Luna *et al.* [16], we need the assumption of the robots to agree on one axis, say North (i.e. y-axis). The idea here is similar to the algorithm of Di Luna *et al.* [16]: If an internal robot sees that the light of every robot that lies to the North of it is set to *External*, it moves. If there are more than one robot that are eligible to move, then only two endpoint robots are allowed to move. The termination is guaranteed by making sure that each robot always moves straight to the North.

Theorem 8. *Our algorithm solves MUTUAL VISIBILITY for NON-RIGID ASYNCH robots with only 2 colors in the robots with lights model, provided that the robots agree on one axis.*

We obtain our main result, Theorem 3, combining the results of Theorems 5–8.

7   Circle Formation Problem

Our algorithm can be modified to solve the CIRCLE FORMATION problem where the goal is to place the robots in the perimeter of a circle [8–10,13]. The best previous solution [16] used 3 colors for NON-RIGID SSYNCH robots and 4 colors for ASYNCH robots under both RIGID and NON-RIGID motions. We have the following theorem.

Theorem 9. CIRCLE FORMATION *problem is solvable without collisions by (a) both* RIGID *and* NON-RIGID *robots with 2 colors in* SSYNCH, *always; (b)* RIGID *robots with 3 colors in* ASYNCH, *always; and (c)* NON-RIGID *robots with 3 colors in* ASYNCH, *if the robots agree on the direction of one axis.*

References

1. Agathangelou, C., Georgiou, C., Mavronicolas, M.: A distributed algorithm for gathering many fat mobile robots in the plane. In: PODC, pp. 250–259 (2013)
2. Agmon, N., Peleg, D.: Fault-tolerant gathering algorithms for autonomous mobile robots. In: SODA, pp. 1070–1078 (2004)
3. Bolla, K., Kovacs, T., Fazekas, G.: Gathering of fat robots with limited visibility and without global navigation. In: Rutkowski, L., Korytkowski, M., Scherer, R., Tadeusiewicz, R., Zadeh, L.A., Zurada, J.M. (eds.) EC 2012 and SIDE 2012. LNCS, vol. 7269, pp. 30–38. Springer, Heidelberg (2012)
4. Cohen, R., Peleg, D.: Local spreading algorithms for autonomous robot systems. Theor. Comput. Sci. **399**(1–2), 71–82 (2008)
5. Cord-Landwehr, A., Degener, B., Fischer, M., Hüllmann, M., Kempkes, B., Klaas, A., Kling, P., Kurras, S., Märtens, M., auf der Heide, F.M., Raupach, C., Swierkot, K., Warner, D., Weddemann, C., Wonisch, D.: Collisionless gathering of robots with an extent. In: Černá, I., Gyimóthy, T., Hromkovič, J., Jefferey, K., Králović, R., Vukolić, M., Wolf, S. (eds.) SOFSEM 2011. LNCS, vol. 6543, pp. 178–189. Springer, Heidelberg (2011)
6. Czyzowicz, J., Gasieniec, L., Pelc, A.: Gathering few fat mobile robots in the plane. Theor. Comput. Sci. **410**(6–7), 481–499 (2009)
7. Das, S., Flocchini, P., Prencipe, G., Santoro, N., Yamashita, M.: The power of lights: synchronizing asynchronous robots using visible bits. In: ICDCS, pp. 506–515 (2012)
8. Défago, X., Souissi, S.: Non-uniform circle formation algorithm for oblivious mobile robots with convergence toward uniformity. Theor. Comput. Sci. **396**(1–3), 97–112 (2008)
9. Dieudonné, Y., Labbani-Igbida, O., Petit, F.: Circle formation of weak mobile robots. TAAS **3**(4), 16:1–16:20 (2008)
10. Dutta, A., Gan Chaudhuri, S., Datta, S., Mukhopadhyaya, K.: Circle formation by asynchronous fat robots with limited visibility. In: Ramanujam, R., Ramaswamy, S. (eds.) ICDCIT 2012. LNCS, vol. 7154, pp. 83–93. Springer, Heidelberg (2012)
11. Efrima, A., Peleg, D.: Distributed models and algorithms for mobile robot systems. In: van Leeuwen, J., Italiano, G.F., van der Hoek, W., Meinel, C., Sack, H., Plášil, F. (eds.) SOFSEM 2007. LNCS, vol. 4362, pp. 70–87. Springer, Heidelberg (2007)
12. Flocchini, P., Prencipe, G., Santoro, N.: Distributed computing by oblivious mobile robots. Synth. Lect. Distrib. Comput. Theory **3**(2), 1–185 (2012)
13. Flocchini, P., Prencipe, G., Santoro, N., Viglietta, G.: Distributed computing by mobile robots: solving the uniform circle formation problem. In: Aguilera, M.K., Querzoni, L., Shapiro, M. (eds.) OPODIS 2014. LNCS, vol. 8878, pp. 217–232. Springer, Heidelberg (2014)
14. Flocchini, P., Santoro, N., Viglietta, G., Yamashita, M.: Rendezvous of two robots with constant memory. In: Moscibroda, T., Rescigno, A.A. (eds.) SIROCCO 2013. LNCS, vol. 8179, pp. 189–200. Springer, Heidelberg (2013)
15. Graham, R.L.: An efficient algorithm for determining the convex hull of a finite planar set. Inf. Process. Lett. **1**(4), 132–133 (1972)
16. Luna, G.A.D., Flocchini, P., Chaudhuri, S.G., Poloni, F., Santoro, N., Viglietta, G.: Mutual visibility by luminous robots without collisions. To appear in Information and Computation (2015). arxiv.org/abs/1503.04347
17. Di Luna, G.A., Flocchini, P., Gan Chaudhuri, S., Santoro, N., Viglietta, G.: Robots with lights: overcoming obstructed visibility without colliding. In: Felber, P., Garg, V. (eds.) SSS 2014. LNCS, vol. 8756, pp. 150–164. Springer, Heidelberg (2014)

18. Luna, G.A.D., Flocchini, P., Poloni, F., Santoro, N., Viglietta, G.: The mutual visibility problem for oblivious robots. In: CCCG (2014)
19. Peleg, D.: Distributed coordination algorithms for mobile robot swarms: new directions and challenges. In: Pal, A., Kshemkalyani, A.D., Kumar, R., Gupta, A. (eds.) IWDC 2005. LNCS, vol. 3741, pp. 1–12. Springer, Heidelberg (2005)
20. Suzuki, I., Yamashita, M.: Distributed anonymous mobile robots: formation of geometric patterns. SIAM J. Comput. **28**(4), 1347–1363 (1999)
21. Vaidyanathan, R., Busch, C., Trahan, J.L., Sharma, G., Rai, S.: Logarithmic-time complete visibility for robots with lights. In: IPDPS, pp. 375–384 (2015)
22. Viglietta, G.: Rendezvous of two robots with visible bits. In: Flocchini, P., Gao, J., Kranakis, E., der Heide, F.M. (eds.) ALGOSENSORS 2013. LNCS, vol. 8243, pp. 286–301. Springer, Heidelberg (2014)
23. Yamashita, M., Suzuki, I.: Characterizing geometric patterns formable by oblivious anonymous mobile robots. Theor. Comput. Sci. **411**(26–28), 2433–2453 (2010)

Mobile Agents Rendezvous in Spite of a Malicious Agent

Shantanu Das[1], Flaminia L. Luccio[2]([✉]), and Euripides Markou[3]

[1] LIF, Aix-Marseille University, Marseille, France
[2] DAIS, Università Ca' Foscari Venezia, Venezia, Italy
luccio@unive.it
[3] DIB, University of Thessaly, Lamia, Greece

Abstract. We examine the problem of rendezvous, i.e., having multiple mobile agents gather in a single node of the network. Unlike previous studies, we need to achieve rendezvous in presence of a very powerful adversary, a malicious agent that moves through the network and tries to block the *honest* agents and prevents them from gathering. The malicious agent can be thought of as a *mobile fault* in the network. The malicious agent is assumed to be arbitrarily fast, has full knowledge of the network and it cannot be exterminated by the honest agents. On the other hand, the honest agents are assumed to be quite weak: They are asynchronous and anonymous, they have only finite memory, they have no prior knowledge of the network and they can communicate with the other agents only when they meet at a node. Can the honest agents achieve rendezvous starting from an arbitrary configuration in spite of the malicious agent? We present some necessary conditions for solving rendezvous in spite of the malicious agent in arbitrary networks. We then focus on the ring and mesh topologies and provide algorithms to solve rendezvous. For ring networks, our algorithms solve rendezvous in all feasible instances of the problem, while we show that rendezvous is impossible for an even number of agents in unoriented rings. For the oriented mesh networks, we prove that the problem can be solved when the honest agents initially form a connected configuration without holes if and only if they can see which are the occupied nodes within a two-hops distance. To the best of our knowledge, this is the first attempt to study such a powerful and mobile fault model, in the context of mobile agents. Our model lies between the more powerful but static fault model of *black holes* (which can even destroy the agents), and the less powerful but mobile fault model of *Byzantine agents* (which can only imitate the honest agents but can neither harm nor stop them).

S. Das—This work has been partially supported by the ANR - MACARON project (anr-13-js02-0002).

F.L. Luccio—This work has been partially supported by the PRIN 2010 Project *Security Horizons*.

E. Markou—Part of this work has been done while this author was visiting Università Ca' Foscari Venezia. This research has been co-financed by the European Union (European Social Fund – ESF) and Greek national funds through the Operational Program "Education and Lifelong Learning" of the National Strategic Reference Framework (NSRF) — Research Funding Program: THALIS-NTUA (MIS 379414).

P. Bose et al. (Eds.): ALGOSENSORS 2015, LNCS 9536, pp. 211–224, 2015.
DOI: 10.1007/978-3-319-28472-9_16

Keywords: Asynchronous · Mobile agents · Rendezvous problem · Malicious agent

1   Introduction

One of the fundamental problems in distributed computing with mobile robots or agents is the problem of gathering all agents at a single location, known as the *rendezvous* problem. Rendezvous is important for example, for coordination between the agents or for sharing information or for planning a collaborative task. This problem has been well studied for the fault-free environment but there are very few results on solving rendezvous in the presence of faults, in particular, in the presence of a hostile entity that could prevent the agents from achieving their task. As in most previous works, we model the environment as a connected graph with multiple mobile agents moving along the edges of the graph; the objective is to gather them at a single node of the graph. In this context, the hostile entity may be either stationary (e.g. a harmful node in the graph) or mobile (e.g. a virus propagating on a network). Methods for protecting mobile agents from malicious host nodes have been proposed, e.g. based on the identification of the malicious host [14]. However, the issue of protecting a network (hosts and mobile agents) from a malicious and mobile entity is still wide open (see [18] and references therein).

A model for a particularly harmful node which has been extensively studied is the *black hole*, where a node which contains a stationary process destroys all mobile agents upon visiting it, without leaving any trace. In this case, although the hostile entity is very powerful, it is stationary; the mobile agents can simply avoid the black hole once its location is known. Thus, the main issue is locating the black hole [14,17,19]. Locating and avoiding a malicious entity that is also mobile and moves from node to node of the graph, seems to be a more difficult problem. A recent result considers the problem of rendezvous in the presence of *Byzantine agents* [12]. A Byzantine agent is indistinguishable from the legitimate or *honest* agents, except that it may behave in an arbitrary manner and may provide false information to the honest agents in order to induce them to make mistakes, thus preventing the rendezvous of the honest agents. Thus, the issue here is identifying the Byzantine agents and distinguishing them from the honest agents. Note that the Byzantine agent cannot actively harm the agent or physically prevent the agents from gathering. In this paper, we consider a more powerful adversary called a *malicious agent* which can actively block the movement of an honest agent to the node occupied by the malicious agent. For example, when two honest agents are close to each other, the malicious agent can enter the path between the two agents and prevent them from meeting. We investigate the feasibility of rendezvous in the presence of such a powerful adversary. In particular, the malicious agent is more powerful than the honest agents; it can move arbitrarily fast through the graph, has full information about the current configuration (i.e. the graph and location of the agents), and has knowledge of the next action to be taken by each honest agent. On the other hand, the honest agents are relatively weak; they are anonymous finite automata, they move asynchronously without any prior knowledge of the graph and they can communicate only locally on meeting another agent at the same node. We remark here that the malicious agent is distinguishable from the honest agents, so the question of identifying the malicious agent (as in Dieudonne et al. [12]), does not arise here.

We believe this is an interesting model for studying mobile faults in a graph, that has never been considered before. In some sense this model can be seen as an extension of the model of networks with delay faults. For example, Chalopin et al. [8] consider the problem of rendezvous in the presence of an adversary that can prevent an agent from moving for an arbitrary but finite time. In their case, the agent cannot be blocked forever as in our scenario. Our model can also be contrasted with the model for network decontamination or, cops and robbers search games on graphs, where a team of good agents (called cops) tries to capture a fast fugitive (robber). The fugitive or hostile entity is exterminated as soon as one of the cops reaches it. Thus the behavior of the hostile entity, in this case, is opposite to that of the malicious agent in our model – instead of blocking the honest agents, the hostile entity tries to get away from the good agents.

In terms of practical motivation for this research, we can think of the malicious agent as representing a virus that may spread around the network. While in the classical decontamination problem the aim is to extinguish the virus, in our setting the virus cannot be extinguished and has to be contained in one part of the network, thus dividing the network into *unstrusted* and *trusted* subnetworks. This scenario can be compared to the problem of *botnets*, i.e. a subnet of compromised computers (bots), typically used for denial-of-service attacks on the internet. The untrusted subnetwork in our model can be seen as a botnet, and the *botmaster* who controls the bots represents the malicious agent. An honest agent that resides on a node protects the trusted network from the untrusted one by running some protection mechanism (e.g. a firewall, an intrusion detection mechanism, etc.). Thus the malicious agent cannot enter a node already occupied by an honest agent. On the other hand the botnet is dynamic, and it may reduce its dimension (i.e., when the botmaster leaves the host) or it may increase it only on hosts not occupied, i.e., not protected by an agent. Honest agents may expand towards the untrusted hosts which are not controlled by the botmaster anymore by running botnet detection mechanisms (see, e.g., [25]). We are then interested in solving the rendezvous problem in the trusted subnetwork, and we want to study how this malicious behaviour affects the solvability of the *Rendezvous* problem.

Related Work: The rendezvous problem has been studied for agents moving on graphs [2] or for robots moving on the plane [9], using either deterministic or randomized algorithms. In the fault-free scenario, the rendezvous problem can be solved relatively easily, even in asynchronous networks, when the network has an asymmetry (e.g., a distinguished node), and can be explored by the agents, since the mobile agents can simply be instructed to meet at such a distinguished node. However, this is not the case for symmetric networks, or when the agents is incapable of visiting all nodes of the network, and the rendezvous problem in such settings is non-trivial and not always solvable even in simple topologies such as the ring network [21]. Symmetry-breaking for the rendezvous problem can be achieved by attaching unique identifiers to the agents (see, e.g., [10,24]), or in the anonymous case using tokens as in e.g., [6,11]. With respect to hostile environments, the *Rendezvous* problem has been studied when there is a black hole or other stationary faults in the network [7,13,23]. Another model for hostile nodes has been presented in [3,20], where the authors have studied how a more severe (than a black hole) behaviour of a malicious host affects the solvability of the *Periodic Data Retrieval* problem in asynchronous networks. A well studied problem in the context of a mobile adversary is the problem of graph searching where a team of mobile agents has to decontaminate the infected sites and prevent any reinfection of cleaned areas. This problem is equivalent to the one of capturing a fast and invisible fugitive moving in the network. For results on this and related problems see, e.g., [4,15,16,22].

Gathering of mobile agents has been also studied in the plane when there are faulty agents which may crash [1,5] and in networks with delay faults [8] or in the presence of Byzantine agents [12], as mentioned before. However, to the best of our knowledge, the rendezvous problem has never been studied under the presence of hostile agents that may block other agents from having access to parts of the network.

Our Results: In this paper we consider a network modelled as a connected undirected graph with multiple honest agents located at distinct nodes of the graph. There is also a hostile entity which is mobile, called the malicious agent. It cannot harm the honest agents but can prevent them from visiting a node: an honest mobile agent cannot visit a node which is occupied by a malicious agent and vice versa. We are interested in solving the rendezvous of all honest agents in this hostile environment. Our objective is to study the feasibility of rendezvous with minimal assumptions. Thus, we consider the weakest possible model for the honest agents. The honest agents are finite state automata with local communication capability and having no prior knowledge of the network. In Sect. 2 we show some configurations in which the problem is unsolvable and we discuss properties that must be respected by any correct algorithm for the problem. For the rest of the paper, we consider ring and mesh networks – two topologies that can be explored even by a finite automaton. In Sect. 3 we present a rendezvous algorithm for ring networks. For oriented rings, we have a universal algorithm that achieves rendezvous starting from any initial configuration, despite the existence of a malicious agent. We prove that the problem is unsolvable for any even number of agents in unoriented rings. Finally, we present an algorithm for rendezvous of any odd number of agents in unoriented rings, thus solving the problem in all solvable instances. In Sect. 4 we consider oriented mesh topologies and we prove that the problem can be solved when the agents initially form a connected configuration without holes if and only if they can detect which are the occupied nodes within a distance of two hops. We show that this latter capability is necessary to achieve rendezvous even for connected configurations without holes. We conclude in Sect. 5 with a discussion about future research directions for this new model. For space limitation, proofs of some lemmas and theorems have been omitted; these can be found in the full version of the paper.

2   Preliminaries

2.1   Our Model

We represent the network by a graph $G = (V, E)$ composed by $|V| = n$ anonymous nodes or *hosts* and $|E|$ edges or connections between nodes. Each host is connected to other hosts by bidirectional asynchronous FIFO links (i.e., an agent cannot overtake another agent moving in the same edge), and it is capable of serving agents by a mutual exclusive mechanism (i.e., an agent at a node u must finish its computation and move or decide to stay, before any other agent at u starts its computation or another agent visits u). The links incident to a host are distinctly labelled but this port labelling (unless explicitly mentioned), is not globally consistent. In the network there are some *mobile agents* which are independent computational processes with some constant internal memory. The agents are initially scattered in the network (i.e., at most one agent at a node), and can move along its edges. An agent arriving at a node u, learns the label of the incoming port, the degree of u and the labels of the outgoing ports. We assume there are k *honest* anonymous identical agents $A_1, A_2, \ldots A_k$, and one *malicious* agent M which may deviate from the proper operations. The initial locations of the honest and malicious agents are decided by an adversary. We describe below the capabilities and behaviour of honest and malicious agents.

Honest Agents: An honest agent located at a node u can see all other agents at u (if any), and can also read their states. It can also read the degree of u and the labels of the outgoing ports. The agents are anonymous, cannot exchange messages and cannot leave messages at nodes. They are identical finite state automata, hence they have some constant memory. The agents do not know n and k. Two agents travelling on the same edge in different directions do not notice each other, and cannot meet on the edge. Their goal is to rendezvous at a node.

Malicious Agent: We consider a worst case scenario in which the malicious agent M is a very powerful entity compared to honest agents: It can move arbitrarily fast inside the network (since the model is asynchronous and the adversary is combined with the malicious agent) and it can permanently 'see' the positions of all the other agents. It has unlimited memory and knows the transition function of the honest agents. When it resides at a node u it prevents any honest agent A from visiting u (i.e., it "blocks" A): if an agent A attempts to visit u it receives a signal that M is in u (botnet detection) and in that case we say that A *bumps* into M. The malicious agent can neither visit a node which is already occupied by some honest agent, nor cross some honest agent in a link. It also obeys the FIFO property of the links (i.e., it cannot overpass an honest agent which is moving on a link).

We call a node u *occupied* (respectively, *free* or *unoccupied*) when one or more (no) *honest* agents are in u. We notice here that some of our impossibility results hold even for stronger models, e.g., when honest agents have unlimited memory, distinct identities, knowledge about the size of the network, visibility, etc. Our algorithm for the ring topology only requires the capabilities of the honest agents mentioned above while for the mesh topology we assume that the honest agents also have the ability to scan whether a node within a two-hops distance, is occupied or not.

2.2 Basic Properties

In this section we show a special class of configurations for which the problem is unsolvable. Intuitively, those are configurations in which the malicious agent can keep separated at least two agents forever.

Definition 1. *Let C be a configuration of a number of agents in a graph G with a malicious agent. The configuration C is called* **separable** *if there is a connected vertex cut-set F composed of free nodes which, when removed, disconnects G so that not all occupied nodes are in the same connected component.*

Lemma 1. *Rendezvous is impossible for any initial configuration in a graph G which is separable, even if the agents have unlimited memory, distinct identities and can always see their current configuration.*

Proof. Let C be an initial configuration which is separable, and let F be a connected vertex cut-set, whose removal disconnects G so that not all occupied nodes are in the same connected component. Let u, v be two occupied nodes which are in different connected components of G and let A, B be the honest agents located at u, v respectively. Due to asynchronicity an adversary can introduce delays to A's and B's movements while at the same time the malicious agent, which has been initially placed at a node in F, can move everywhere in F (since F has only free nodes and it is connected) preventing agents A, B, from visiting any node in F. Since all paths between u and v include at least one node of F, agents A, B can never meet, no matter how powerful they are.   □

Hence for every initial separable configuration the problem is unsolvable. A natural question is whether there are non-separable initial configurations for which the problem is unsolvable. The answer is yes and one can easily find such configurations. We now state sufficient conditions under which the problem is unsolvable for a separable (initial or not) configuration of agents.

Definition 2. *Let C_t be a configuration at time $t \geq 0$ (i.e., initial or not) of a number of agents in a graph G with a malicious agent. The configuration C_t is called* **separating** *if C_t is separable and either C_t is an initial configuration or the following conditions hold:*

- *there is a node $x_m \in F_t$ (F_t is any vertex cut-set of C_t as defined in Definition 1) and a path of nodes $(x_0, x_1, \ldots, x_m)$ so that x_0 is free at time 0 and,*

- the sequence of nodes $(x_0, x_1, \ldots, x_m)$ can be partitioned in $k \leq t+1$ contiguous subsequences $(x_0^0, \ldots, x_i^0), (x_{i+1}^1, \ldots, x_j^1), \ldots, (x_{l+1}^k, \ldots, x_m^k)$, where $0 \leq i < j < l < m$ and,
- the nodes $(x_w^s, \ldots, x_r^s)$ belonging to the same subsequence s are free at time s, where $0 \leq s \leq k$ and nodes $(x_w^k, \ldots, x_r^k)$ are free at time t.

Lemma 2. *Rendezvous is impossible for any separating configuration in a graph G, even if the agents have unlimited memory, distinct identities and can always see their current configuration.*

Intuitively, Lemma 2 states that if $\mathcal{C}_t$ is a separable configuration, and in $\mathcal{C}_t$ there is a free node x so that either: (i) x has been always free or, (ii) there are paths of nodes which eventually become free and they form a connection between a free node at $\mathcal{C}_0$ and x, then there are at least two agents in $\mathcal{C}_t$ which will never meet. Hence, any correct algorithm for the solution of the problem should avoid creating a separating configuration.

3 Rendezvous in a Ring Network

In this section we will study the rendezvous problem in bidirectional rings with a malicious agent M. Notice that in a ring topology there are no separable (and hence no separating either) configurations, since there cannot exist a connected cut-set composed of free nodes whose removal would disconnect the ring. However, since the ring is highly symmetric, rendezvous is impossible even if the agents have unlimited memory and have full knowledge of the configuration, since an adversary can keep synchronized the agents so that they always take the same actions at the same time and therefore they maintain their initial distances (the malicious agent can keep on moving synchronized with the honest agents). Thus, in order to solve the problem we need to add some constraints to the model. A natural step is to assume that there is a special node labeled o^* in the ring which can be recognized by the agents. Note that the malicious agent is so powerful that it could place itself on o^* and never move from there. Our strategies also work under this scenario. We now present algorithms for solving the problem in oriented and unoriented rings.

3.1 Oriented Ring

In an oriented ring, the two incident edges at each node are labelled as clockwise or counter-clockwise in a consistent manner; so, all agents agree on the ring orientation.

The idea of the algorithm is the following. Each agent moves in the clockwise direction until it meets o^* or bumps into M. For the first three times that the agent bumps into M without meeting o^*, it reverses its direction and continues moving in the opposite direction. Due to the FIFO property and the fact that the agent cannot pass over M, we can show that if an agent bump into M after reversing directions at least three times, then the other agents should

---

Algorithm 1. RV-OR : Rendezvous of $k \geq 2$ agents in oriented rings

---

Let $i := 0$;
DIR := Clockwise;
while *not Stopped* **do**
  Move DIR until you bump into M or meet o^* or a stopped agent;
  i=i+1;
  if *you met a Stopped agent* **then**
    Become Stopped and Exit loop;
  if $i = 1$ *or* $i = 2$ **then**
    if *Current node is* o^* **then**
      Become Stopped and Exit loop;
    else if *Bumped into* M **then**
      Reverse direction (DIR := inverse(DIR));
  if $i = 3$ **then**
    if *Current node is* o^* *or Bumped into* M **then**
      Reverse direction (DIR := inverse(DIR));
  if $i = 4$ **then**
    if *Current node is* o^* *or Bumped into* M **then**
      Become Stopped and Exit Loop ;

---

have bumped into M at least twice, without meeting the special node o^* (see Lemma 3). After an agent has already bumped into M three times, the next time it bumps into M or meets o^* it stops. On the other hand, if the agent meets o^* before it bumps into M twice, then the agent stops at o^*, and all the other agents will arrive at o^* after bumping into M at most once. The algorithm called RV-OR is presented below.

Lemma 3. *During the execution of the algorithm, if an agent bumps into M in the fourth iteration of the* while *loop, then any other agent must have bumped into M at least two times.*

Lemma 4. *Algorithm RV-OR solves rendezvous of $k \geq 2$ agents in spite of one malicious agent, in any oriented ring containing one special node o^*.*

3.2 Unoriented Rings

In unoriented rings, each agent has its own notion of clockwise and the agents may not agree on the clockwise direction. In this case rendezvous is not always feasible.

Lemma 5. *For any even number $k \geq 2$, the rendezvous problem for k honest agents and one malicious agent cannot be solved in any bidirectional unoriented anonymous ring with a special node o^*, even if the agents know k.*

We now present an algorithm for solving rendezvous of k agents, for any *odd* integer k, in an unoriented asynchronous ring network. Notice that in an unoriented ring, if we follow an algorithm similar to Algorithm RV-OR it is possible that the agents form two distinct groups that gather at two distinct nodes. However, since the total number of agents is odd, exactly one of the two groups would have even number of agents, thus one of the agents of this group could move to collect all the other agents. The algorithm must ensure that there are at most two groups of agents, i.e. there are at most two distinct nodes where the agents stop in the initial phase. In our algorithm, an agent stops at o^* only if it has seen it at least three times, while moving in the same direction. This implies that this agent has traversed the complete ring two times and while M has moved at least once around the ring. So, there could be no agents moving in the opposite direction. On the other hand if some agent stops while bumping into M, then any agent moving in the same direction would reach this node with the stopped agent before reaching M or o^*. In all cases, there will be at most two nodes where the agents stop. When two or more agents have gathered at a node v, one of the agents called the *searcher*[1] reverses direction and moves to search for the other agents. The searcher only stops when it reaches the other node w containing stopped agents. If the number of agents gathered at node w is even then the searcher becomes a *Collector* and it collects all agents and returns to node v. Note that the agent does not need to count the number of other agents as the algorithm depends only on the parity of the size of the group of agents. The complete algorithm, called RV-UR is presented in a following table.

Lemma 6. *Consider an anonymous ring consisting of n nodes, including a special node o^* and one malicious agent. If $k \geq 2$ honest agents execute Algorithm RV-UR, then, after a total number of $O(kn)$ edge traversals the honest agents correctly rendezvous, if k is odd.*

The following result summarizes the results of this section:

Theorem 1. *In any anonymous and asynchronous ring with a special node o^* and one malicious agent, k honest agents having constant memory and no knowledge about their number, can solve the rendezvous problem if and only if either the ring is oriented or k is odd.*

We briefly consider the case when there could be multiple malicious agents in the network. In this case, rendezvous is feasible only if all the malicious agents are located in a continuous segment of the ring with no honest agent in between. This scenario is equivalent to the one with a single malicious agent and thus the same algorithm would work in this case.

4 Rendezvous in an Oriented Mesh Network

We now study the problem in an oriented mesh network. In view of Lemma 1, rendezvous is impossible for separable initial configurations. Hence, in this section

---

[1] We select as searcher the second agent that arrives at node v.

Algorithm 2. RV-UR : Rendezvous in unoriented rings

Case 0. *Initial* state
  Move clockwise until:
    Case 0.1. You meet node o^* **unoccupied for the third time:**
      Change state to *stopper*;
    Case 0.2. You bump into M **trying to move from a node that hosts only you:**
      Change state to *stopper*;
    Case 0.3. You meet an agent not at node o^***:**
        Case 0.3.1. The agent you meet is alone and is a *stopper***:**
          Change state to *transformer-1*;
        Case 0.3.2. Every other agent at the node is at state *final***:**
          Change state to *stopper*;
        Case 0.3.3. You meet a *stopper* **and at least one agent at state** *final***:**
          Change state to *transformer-2*;

Case 1. State *transformer-1*
  Wait until all other agents change to state *final*;
  Change state to *searcher*;

Case 2. State *searcher*
  Move counter-clockwise until you bump into M while you try to move from a node u:
    Case 2.1. You see one or more agents at u **and all of them are at state** *final***:**
      Change state to *stopper*;
    Case 2.2. You see no agent at u **or an agent not at state** *final***:**
      Change state to *collector*;

Case 3. State *stopper*
  Wait until:
    Case 3.1. You see a *transformer-1* **or** *transformer-2***:** Change state to *final*;
    Case 3.2. You see a *collector***:** Follow *collector*;
    Case 3.3. You see a *terminator***:** Change state to *terminator*;

Case 4. State *collector*
  Wait until every other agent at the node changes its state to *stopper*;
  Collect everyone;
  Move clockwise collecting every agent you meet, until you meet an agent at state *final*;
  Change state to *terminator*;

Case 5. State *final*
  Wait until:
    Case 5.1. You see a *collector***:** Change state to *stopper*;
    Case 5.2. You see a *terminator***:** Change state to *terminator*;

Case 6. State *transformer-2*
  Wait until every other agent at the node changes its state to *final*;
  Change state to *final*;

Case 7. State *terminator*
  Wait until every other agent at the node changes its state to *terminator*;
  Exit;

---

we study the problem for a special class of non separable initial configurations and we give an algorithm that solves the problem for this type of configurations. In particular, we focus on initial configurations where the induced subgraph of the occupied nodes is connected without holes, i.e., there is no connected set of unoccupied nodes surrounded by occupied nodes. At the end of the section we discuss the solvability of the problem in other classes of initial non separable configurations.

First observe that even in configurations that consist of a simple path of occupied nodes, the problem is unsolvable in the considered model due to network asynchronicity: Initially all agents have the same input and thus (following any potentially correct algorithm), they should all try to move; however, an adversary may slowdown all agents, except for one not located at the endpoints of the path, hence creating a separating configuration. Thus, by Lemma 2 the problem is unsolvable. Therefore, the agents need to be able to gain some knowledge about their current configuration before they move in order to avoid creating a separating configuration. We enhance our model by giving the agents, the capability to discover all occupied nodes within a distance of d-hops.

Definition 3. *We say that an agent A located at a node x can see (or scan) at a distance d or it has d-visibility if A can decide for any node u within a distance of d hops from x, whether u is occupied or not by an honest agent.*

We emphasize that, if a node u scanned by agent A is occupied, A cannot tell how many agents are in u, or read their states. When the agents have a $d-$visibility capability we assume that moves are instantaneous, i.e., an agent cannot be traveling along an edge while another agent is scanning its neighbourhood. Unfortunately, as we show below, even when the agents have $1-$visibility (i.e., they can only scan their neighbours), the problem remains unsolvable for some connected without holes configurations.

Lemma 7. *The rendezvous problem is unsolvable in an oriented mesh with a malicious agent for initial connected without holes configurations, even when the agents are capable of scanning their adjacent nodes.*

Hence we further equip the agents with the capability of discovering the occupied nodes within a two-hops distance. In that case, as we show below, the problem can be solved for any connected without holes initial configuration.

We present an algorithm which instructs the agents to move only to occupied nodes in a way that they maintain the connectivity and they do not create holes. In order to describe the algorithm we define eleven local configurations as shown in Fig. 1. In these configurations, empty circles represent free nodes, while circles containing black dots represent occupied nodes. The remaining vertices on the figures represent nodes which may be either occupied or free. The agent (let us call it A) which is located below a horizontal arrow in cases (a-g), moves horizontally as depicted by the arrow. The agent (let us call it B) which is located left of a vertical arrow in cases (h-m), moves vertically as depicted by the arrow. Hence the algorithm can be described as follows:

Algorithm RV-MESH: *If an agent has a view (within two hops) like the one of agent A or B described before, then this agent moves towards the direction shown by the corresponding arrow; otherwise the agent does not move.*

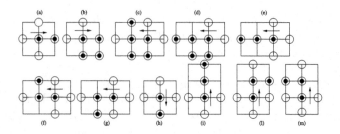

Fig. 1. View of the scanning agent located below (cases a-g) or left (cases h-m) of the depicted arrow. Occupied nodes are depicted as cycles containing black dots, while free nodes are depicted as empty cycles. Nodes which are within two hops from the scanning agent but not shown, can be either occupied or free. The scanning agent will move East in cases (a, b), West in cases (c, d, e, f, g), South in case (h), and North in cases (i, l, m).

Nodes which are within two hops from the scanning agent and are not shown in those configurations can be either occupied or free. If the location of the scanning agent is close to the border of the mesh and some of the nodes in those eleven configurations do not exist, then the agent acts as it would act if those nodes existed in its view and were free. Moreover, while an agent A located at a node u is executing its scan or compute phase then no other operation can take place at u before A moves or decides to stay (i.e., no other agent at u can start scanning and no other agent can arrive at u). That is, operations at a node u are executed in mutual exclusion. Notice that if two adjacent agents want to swap positions they can only do it at the same time.

Lemma 8. *Given an $n \times m$ oriented mesh, for any connected configuration without holes of at least three occupied nodes, there is at least one agent whose view is in one of the configurations depicted in Fig. 1.*

Lemma 9. *Given an $n \times m$ oriented mesh, consider a connected configuration of k agents in two occupied nodes. According to Algorithm RV-MESH, after a total number of at most $k+1$ edge traversals there will be only one occupied node.*

Lemma 10. *Given an $n \times m$ oriented mesh, consider any connected configuration without holes of k agents occupying at least 3 nodes. After any number of moves according to Algorithm RV-MESH, the resulting configuration is also connected without holes. Furthermore, the number of occupied nodes will strictly decrease after at most k edge traversals, reaching the value of only one occupied node after at most $O(k^2)$ edge traversals.*

In view of Lemmas 7, 8, 9 and 10 we have:

Theorem 2. *The rendezvous problem for $k \geq 2$ agents can be solved for any initial connected without holes configuration of agents in an $n \times m$ oriented mesh if and only if the agents are able to discover the occupied nodes within a distance of two-hops.*

If the initial non separable configuration is different from the one considered above, then even the 2-visibility capability is not sufficient anymore to solve rendezvous. In fact the problem remains unsolvable for connected configurations with holes even when the agents are able to discover the occupied nodes within any constant distance. The problem is also unsolvable for some disconnected non separable configurations. Hence it appears that for many initial non separable configurations in an oriented mesh, the combination of the asynchronicity and the limited view (to any constant fraction of the complete view) makes the problem unsolvable.

5   Conclusion

In this paper we studied deterministic protocols for the rendezvous of $k \geq 2$ honest agents in asynchronous networks with a malicious agent which can prevent the agents from reaching any node it occupies. We have presented algorithms for oriented and unoriented ring networks which gathers the honest agents within $O(kn)$ edge traversals for all feasible instances of the problem. We have also presented a deterministic protocol for oriented $n \times m$ meshes which leads the agents to rendezvous within $O(k^2)$ edge traversals for any initial connected without holes configuration when the agents can discover the occupied nodes within a distance of two-hops (which is a necessary condition). Given the novelty of the model there are many interesting open questions. The first is whether the problem can be solved in unoriented meshes for connected configurations without holes when the agents are capable of scanning within a constant distance. It would be also interesting to study randomized protocols for some of the unsolvable cases, and also to study this problem in synchronous networks with unit-speed cooperating agents and unit-speed/infinite-speed malicious agents. Finally, it would be interesting to study the problem in $(m+1)$-connected graphs in the presence of m malicious agents, or in the solved cases presented in this paper in the presence of malicious agents that show a more severe behaviour.

References

1. Agmon, N., Peleg, D.: Fault-tolerant gathering algorithms for autonomous mobile robots. SIAM J. Comput. **36**(1), 56–82 (2006)
2. Alpern, S., Gal, S.: Searching for an agent who may or may not want to be found. Oper. Res. **50**(2), 311–323 (2002)
3. Bampas, E., Leonardos, N., Markou, E., Pagourtzis, A., Petrolia, M.: Improved periodic data retrieval in asynchronous rings with a faulty host. In: Halldórsson, M.M. (ed.) SIROCCO 2014. LNCS, vol. 8576, pp. 355–370. Springer, Heidelberg (2014)
4. Barriere, L., Flocchini, P., Fomin, F.V., Fraigniaud, P., Nisse, N., Santoro, N., Thilikos, D.: Connected graph searching. Inf. Comput. **219**, 1–16 (2012)
5. Bouzid, Z., Das, S., Tixeuil, S.: Gathering of mobile robots tolerating multiple crash faults. In: IEEE 33rd International Conference on Distributed Computing Systems, ICDCS 2013, 8–11 July 2013, Philadelphia, Pennsylvania, USA, pp. 337–346 (2013)

6. Chalopin, J., Das, S.: Rendezvous of mobile agents without agreement on local orientation. In: Abramsky, S., Gavoille, C., Kirchner, C., Meyer auf der Heide, F., Spirakis, P.G. (eds.) ICALP 2010. LNCS, vol. 6199, pp. 515–526. Springer, Heidelberg (2010)

7. Chalopin, J., Das, S., Santoro, N.: Rendezvous of mobile agents in unknown graphs with faulty links. In: Pelc, A. (ed.) DISC 2007. LNCS, vol. 4731, pp. 108–122. Springer, Heidelberg (2007)

8. Chalopin, J., Dieudonné, Y., Labourel, A., Pelc, A.: Fault-tolerant rendezvous in networks. In: Esparza, J., Fraigniaud, P., Husfeldt, T., Koutsoupias, E. (eds.) ICALP 2014, Part II. LNCS, vol. 8573, pp. 411–422. Springer, Heidelberg (2014)

9. Cohen, R., Peleg, D.: Convergence of autonomous mobile robots with inaccurate sensors and movements. SIAM J. Comput. 38, 276–302 (2008)

10. Czyzowicz, J., Labourel, A., Pelc, A.: How to meet asynchronously (almost) everywhere. In: Proceedings of 21st Annual ACM-SIAM Symposium on Discrete Algorithms (2010)

11. Das, S., Mihalák, M., Šrámek, R., Vicari, E., Widmayer, P.: Rendezvous of mobile agents when tokens fail anytime. In: Baker, T.P., Bui, A., Tixeuil, S. (eds.) OPODIS 2008. LNCS, vol. 5401, pp. 463–480. Springer, Heidelberg (2008)

12. Dieudonne, Y., Pelc, A., Peleg, D.: Gathering despite mischief. ACM Trans. Algorithms 11(1), 1 (2014)

13. Dobrev, S., Flocchini, P., Prencipe, G., Santoro, N.: Multiple agents rendezvous in a ring in spite of a black hole. In: Papatriantafilou, M., Hunel, P. (eds.) OPODIS 2003. LNCS, vol. 3144, pp. 34–46. Springer, Heidelberg (2004)

14. Dobrev, S., Flocchini, P., Prencipe, G., Santoro, N.: Mobile search for a black hole in an anonymous ring. Algorithmica 48(1), 67–90 (2007)

15. Flocchini, P., Huang, M.J., Luccio, F.L.: Decontamination of chordal rings and tori using mobile agents. Int. J. Found. Comput. Sci. 3(18), 547–564 (2007)

16. Flocchini, P., Huang, M.J., Luccio, F.L.: Decontamination of hypercubes by mobile agents. Networks 3(52), 167–178 (2008)

17. Flocchini, P., Ilcinkas, D., Santoro, N.: Ping pong in dangerous graphs: optimal black hole search with pebbles. Algorithmica 62(3–4), 1006–1033 (2012)

18. Flocchini, P., Santoro, N.: Distributed security algorithms for mobile agents. In: Cao, J., Das, S.K. (eds.) Mobile Agents in Networking and Distributed Computing, Chap. 3, pp. 41–70. Wiley, Hoboken (2012)

19. Klasing, R., Markou, E., Radzik, T., Sarracco, F.: Hardness and approximation results for black hole search in arbitrary graphs. TCS 384(2–3), 201–221 (2007)

20. Královič, R., Miklík, S.: Periodic data retrieval problem in rings containing a malicious host. In: Patt-Shamir, B., Ekim, T. (eds.) SIROCCO 2010. LNCS, vol. 6058, pp. 157–167. Springer, Heidelberg (2010)

21. Kranakis, E., Krizanc, D., Markou, E.: The Mobile Agent Rendezvous Problem in the Ring. Synthesis Lectures on Distributed Computing Theory. Morgan and Claypool Publishers, San Rafael (2010)

22. Luccio, F.L.: Contiguous search problem in sierpinski graphs. Theory Comput. Syst. 44, 186–204 (2009)

23. Yamauchi, Y., Izumi, T., Kamei, S.: Mobile agent rendezvous on a probabilistic edge evolving ring. In: ICNC, pp. 103–112 (2012)

24. Yu, X., Yung, M.: Agent rendezvous: a dynamic symmetry-breaking problem. In: Meyer auf der Heide, F., Monien, B. (eds.) ICALP 1996. LNCS, vol. 1099, pp. 610–621. Springer, Heidelberg (1996)

25. Zeng, Y., Hu, X., Shin, K.: Detection of botnets using combined host- and network-level information. In: IEEE/IFIP DSN 2010, pp. 291–300, June 2010

Author Index

Printed in the United States
By Bookmasters